Human Factors in Certification

HUMAN FACTORS IN TRANSPORTATION
A Series of Volumes Edited by
Barry H. Kantowitz

Human Factors in Certification

Edited by

John A. Wise
V. David Hopkin
Embry-Riddle Aeronautical University

LEA LAWRENCE ERLBAUM ASSOCIATES, PUBLISHERS
2000   Mahwah, New Jersey   London

The final camera copy for this work was prepared by the author,
and therefore the publisher takes no responsibility for consistency
or correctness of typographical style. However, this arrangement helps
to make publication of this kind of scholarship possible.

Lawrence Erlbaum Associates, Inc., Publishers
10 Industrial Avenue
Mahwah, New Jersey 07430-2262

Cover design by Kathryn Houghtaling Lacey

Library of Congress Cataloging-in-Publication Data

Human factors in certification / edited by John A. Wise and V. David Hopkin.
p. cm. — (Human factors in transportation)
Includes bibliographical references and index.
ISBN 0-8058-3113-4 (cl : alk. paper)
1. Aeronautics—Human factors. 2. Human engineering. 3. Aircraft
industry—Certification. 4. Aeronautical engineers—Certification. 5.
Nuclear industry—Certification. I. Wise, John A., 1944–. II. Hopkin, V.
David. III. Series.

TL553.6 .H8624 1999
629.13—dc21   99-049304

Books published by Lawrence Erlbaum Associates are printed on acid-free
paper, and their bindings are chosen for strength and durability.

Printed in the United States of America
10 9 8 7 6 5 4 3 2 1

DEDICATION

To Nancy, Mark, and Suzanne
—*John A. Wise*

To Betty
—*V. David Hopkin*

*　　*　　*

To all of those who attended
the original Bonas meeting

CONTENTS

SELECTION AND TRAINING

PARALLEL VIEWS AND TOPICS

CONCLUSION

SERIES FOREWORD

Barry H. Kantowitz
Battelle Human Factors Transportation Center

The domain of transportation is important for both practical and theoretical reasons. All of us are users of transportation systems as operators, passengers, and consumers. From a scientific viewpoint, the transportation domain offers an opportunity to create and test sophisticated models of human behavior and cognition. This series covers both practical and theoretical aspects of human factors in transportation, with an emphasis on their interaction.

The series is intended as a forum for researchers and engineers interested in how people function within transportation systems. All modes of transportation are relevant, and all human factors and ergonomic efforts that have explicit implications for transportation systems fall within the series purview. Analytic efforts are important to link theory and data. The level of analysis can be as small as one person, or international in scope. Empirical data can be from a broad range of methodologies, including laboratory research, simulator studies, test tracks, operational tests, fieldwork, design reviews, or surveys. This broad scope is intended to maximize the utility of the series for readers with diverse backgrounds.

I expect the series to be useful for professionals in the disciplines of human factors, ergonomics, transportation engineering, experimental psychology, cognitive science, sociology, and safety engineering. It is intended to appeal to the transportation specialist in industry, government, or academic, as well as the researcher in need of a test bed for new ideas about the interface between people and complex systems.

The present book is devoted to the role of human factors in certification. Its emphasis on the domain of aviation justifies inclusion in this book series. The book nicely fulfills a major goal of this series: to explore how people function in complex systems. The first section on Philosophies of Human Factors Certification discusses broad principles as applied to aviation technologies, air traffic control systems, and nuclear power plants. This is followed by more specific treatises on practical approaches, quality assurance, training, workload assessment, and several chapters related to aviation certification. The editors have done an outstanding job of assembling a group of chapters that together offer a systems approach emphasizing the role of human factors in large systems where errors have serious consequences. This book demonstrates how human factors can integrate the varies and diverse aspects of complex transportation systems.

PREFACE

In several nations during the past decade there have been authoritative and independent statements advocating the application of human factors principles and evidence to certification. Although most of these statements have originated in aviation contexts and this text is biased toward aviation examples, the statements themselves have generally envisaged that human factors should be applied wherever certification is currently conducted. Most of the human factors products of such broad applications would not be specific to any particular kind of system but could be used in many certification contexts, perhaps with some modification to cover any unusual system characteristics.

The problems that the application of human factors to certification should help to resolve appear to be most serious in large human–machine systems with certain definable attributes. Such systems have a combination of complexity and integrality, are safety critical in that the penalties for any fallibility in certification are high, and would benefit from some extension and unification of any existing validation and certification procedures. Different aspects of such systems, such as their structure, hardware, software, and personnel, have usually evolved different validation and certification procedures independently for each aspect, and none of them can deal in detail with the whole system as a functioning entity, including its human components.

An initial attempt to launch the application of human factors to certification was made in July 1993 at the Chateau de Bonas near Toulouse, France, where specialists with relevant knowledge and experience met for a few days and were encouraged to present their ideas on the main theme. They reviewed relevant subject matter, defined and evaluated feasible approaches, considered applications, specified topics and issues, and recommended productive courses of action. The intention was to provide a forum for informed discussion and the evaluation of ideas rather than to prejudge their value or durability. An objective was to start to define what the subject matter, approaches, and techniques should encompass, implying consideration of what should be ruled out as well as what should be included.

Much has happened to certification and to human factors during the past few years. For this volume, the editors invited some of those who attended the original meeting and who have remained involved with the subject to revisit the topic of human factors applied to certification and to prepare new and updated material on it. The outcome is this volume, which represents the current state of knowledge and advocacy of a topic that becomes ever more urgent as system complexity and automation continue to increase and the associated human factors problems show comparable increases in complexity and profundity. Further postponement of the routine application of human factors to certification will increase the probability that systems could evolve to an operational state with major human factors problems still within them, and with many of the most effective solutions of such unidentified problems rendered inapplicable because the system is no longer flexible enough to accommodate them.

An introductory chapter and a group of chapters presenting propositions and philosophies about human factors contribute to a framework for human factors certification. Hopkin (Chap. 1) is concerned that human factors contributions to certification should not only benefit certification but also reflect credit on human factors through their high quality. Wise and Wise (Chap. 2) compare a bottom-up approach with the more top-down approach that they equate with a systems approach and prefer. Wilson (Chap. 3) explores what is currently valuable in certification processes as a step toward realizing further benefits such as better designs and more professionalism within human factors. Hancock (Chap. 4) draws parallels between certification and legislation as processes that constrain diverse and unpredictable entities by imposing frameworks on them. Hanes (Chap. 5) approaches human factors certification by coupling the need to increase human factors participation in design processes with the practical means by which this could be achieved. Stein (Chap. 6), in considering certification in the context of system validation, also emphasise the legal status and functions of certification procedures and processes. Koelman (Chap. 7) discusses the operational concept of future air traffic management systems in Europe in terms of their combination of tactics and strategies. He notes the intention to validate but not to certify the operational concept.

A series of four chapters all adopt a more direct approach to certification activities. Stager (Chap. 8) believes that human factors issues in the design and development cycle of the system are crucial for its successful human factors certification because the primary objectives of that certification must be accomplished within those cycles. Taylor and MacLeod (Chap. 9) draw on their experience of compliance with human engineering standards in the procurement of advanced aviation systems. .They emphasize that certification should require proof of process as well as proof of content and performance. Gilson and Abbott (Chap. 10), in proposing that flight crews be certified for attaining mastery of sophisticated control systems, also propose criteria through which flight crews would be able to demonstrate that they had attained the requisite deeper levels of understanding. Small and Bass (Chap. 11) consider system evaluation as a precursor to certification and draw the distinction that whereas evaluation can show that the system behaves correctly, certification must show that it can only behave correctly.

Three chapters then deal with aspects of human–machine integration. Gibson (Chap. 12) notes that all the models for the certification of flight training have to rely ultimately either on expert opinion or on the actual outcome of training. Haglund (Chap. 13) describes a project in Sweden that sought to improve the selection of air traffic controllers and reduce training costs. Instead it revealed the limited value of the selection procedure and a need to revise it. MacLeod and Taylor (Chap. 14) argue that because the human's role in human–machine systems has evolved into being primarily cognitive, human factors certification must include a substantial understanding of and provision for human cognition to be effective.

Three chapters address topics that should be featured in any established human factors certification of advanced aviation systems. Tattersall (Chap. 15) discusses the effects of workload, in terms of its methods of measurement and of alternative patterns of adjustment to variations in workload demands. Westrum (Chap. 16), drawing on the role of the test pilot, suggests that a test controller for air traffic control systems could bring comparable advantages, particularly with respect to user involvement in system evolution. Bukasa (Chap. 17) views the application of human factors to certification as a welcome sign of increased formalization and institutionalization of human factors contributions in regard to complex systems.

The next six chapters all use ideas that already exist in aviation as a basis for discussing certification issues. Baldwin (Chap. 18) addresses the question of how human factors certification should be managed and organized. He notes that the reluctance of human factors specialists to commit themselves must be overcome. Harwood and Sanford (Chap. 19) point toward possible fresh approaches to certification by describing the field exposure of a forthcoming air traffic control automation system early in its development cycle instead of immediately prior to its implementation. Maurino and Galotti (Chap. 20)suggest how human factors certification requirements might be integrated into current certification processes, on the basis of existing International Civil Aviation Organization. regulatory requirements and guidance material. Gaillard and Leroux (Chap. 21) contrasts cognitive engineering with traditional methods of human–machine system evaluation and validation, as approaches to designing new tools. McClumpha and Rudisill (Chap. 22) consider human factors aspects of civil flight deck certification, with particular emphasis on the human–computer interfaces in automated cockpits. Pariès reviews some aspects of current airworthiness regulations and certification processes related to human factors and cockpit design, from the perspective of the occurrence of accidents with aircraft that have been properly certificated.

Two chapters consider issues that arise in the certification of complex future systems. Amalberti and Wibaux (Chap. 24), on the basis of French experience with human factors in the certification of advanced automated cockpits, identify the three main difficulties as the relation between human error and accident risk, the evolutionary nature of pilot expertise in contrast to the nonevolutionary certification requirements, and the status of human factors findings in relation to certification goals. Javaux, Masson, and De Keyser (Chap. 25) point out that the combination of complexity and lack of transparency that characterizes much current automation also limits the user's ability to cooperate with such systems.

In a concluding chapter, Hopkin (Chap. 26) describes some current characteristics of human factors as a discipline that would influence its application to certification, and notes some further related topics not otherwise mentioned.

ACKNOWLEDGEMENTS

We would like to acknowledge the contributions of Eurocontrol, the U.S. Federal Aviation Administration, Direction Générale de l'Aviation Civile, the National Aeronautics and Space Administration, Embry-Riddle Aeronautical University, and the Research Institute for Information, Science, & Engineering, all of which provided support for the original 1993 meeting on Human Factors Certification, and hence helped to launch the subject.

The editors owe a significant debt to Ms. Suzanne Wise for her hours and hours of effort getting the book in camera ready format. In particular, for the time spent recovering from the "improvements" that were inflicted on the documents by the inventive "features" of Word '98.

I

Introduction

Optimizing Human Factors Contributions

V. David Hopkin
Embry-Riddle Aeronautical University
Human Factors Consultant, UK

Human factors as a discipline deals primarily, although not exclusively, with people at work. Its objectives are to promote safety, efficiency, well-being, and productivity by achieving a good match between the human and the job. Such a match makes the tasks, equipment, demands, and conditions of the job as compatible as possible with human strengths and weaknesses. Human factors, being based on a knowledge and understanding of human beings, can be applied to every work environment and to every human activity that constitutes work. Certification is human work. Accordingly, human factors can be applied to certification. The application of human factors to certification should result in improvements and in an understanding of why the improvements have occurred and of how further improvements might be made.

About ten years ago, several independent policy statements advocated that human factors should be applied to certification. Partly because of these, certification was included in the United States National Plan for Aviation Human Factors, which was published in draft form in 1990 and in final form in 1995 (Federal Aviation Administration, 1990; 1995). The inclusion in it of certification did not consist of a mere mention, but of quite a detailed program of proposed work, that identified some relevant human factors topics and provided a framework of studies that could be done. This part of the National Plan was duly noted, but it did not lead to a blossoming of human factors work on certification.

The editors of this volume helped to convene a meeting to try to define and promote the application of human factors to certification, and the proceedings were published with a limited circulation and are now out of print (Wise, Hopkin, & Garland, 1994). They have tried to maintain momentum through other papers in the meantime (e.g., Wise & Hopkin, 1997), but have themselves been remiss in failing to seize every opportunity to emphasize the importance of human factors in certification in their own work. For example, a text on air traffic control (Hopkin, 1995) did not deal with certification as a topic in its own right, a text on automation and human performance includes many applications of human factors but certification is not among them (Parasuraman & Mouloua, 1996), a recent handbook on aviation human factors (Garland, Wise, & Hopkin, 1998) does discuss certification briefly but mainly in relation to evaluations, and two recent human factors publications on air traffic control by the National Research Council have only short sections on certification, referring mostly to licensing (National Research Council, 1997) or maintenance (National Research Council, 1998).

Meanwhile there have been developments in certification itself. Many certification processes have become more remote from the certifier, mainly through an increasing trend to include built-in diagnostics in system specifications and designs. The human certifier has less direct and often more limited access to these processes. Improved system reliability and better data about the nature and frequency of component failures, coupled with more automated trend analysis, have tended to increase the intervals between certifications. Automation has been more widely applied to certification, through both the automated logging of data for certification purposes and the progressive development of automated means of support for the judgment strategies of the certifier.

In principle, human factors can apparently be related to certification in two different ways. One starts from current certification procedures and methods and suggests how human factors could be applied to them so that they would become more effective because they utilize human capabilities better. This approach does not seek drastic changes in the procedures and methods themselves, but is intended to improve them through the application of human factors knowledge to them. It can be construed as a form of validation of existing certification procedures and methods (Wise, Hopkin, & Stager, 1993).

The alternative application of human factors to certification emphasizes its final objectives and seeks to ensure that every step taken toward the achievement of those objectives is an example of good human factors practice. This approach can bring changes in the procedures and methods of certification to circumvent human limitations and capitalize on human strengths. The final objectives of certification remain unchanged, but at least some of the processes in attaining them can be expected to change. In this alternative approach, the human factors contributions require much more explanation and justification. This chapter looks primarily at this second alternative.

EXAMPLES OF ALTERNATIVE APPROACHES

All applications of human factors as a discipline should meet professional standards. They should constitute "good human factors" insofar as human factors is an independent discipline with its own criteria of what is and what is not safe, efficient, satisfactory, acceptable, optimal, and compatible with good professional practice. They should pass muster if subjected to peer review by human factors professionals. There will occasionally be circumstances when a human factors requirement seems, initially at least, to be incompatible with a requirement of another discipline. A way must then be found to reconcile these requirements to achieve practical agreement on what to do without compromising basic human factors principles or the basic principles of the other discipline. Although it is possible in the first approach to apply human factors evidence to existing certification processes, it does not follow that there is a solution that is acceptable in human factors terms, meets human factors requirements, or would pass a human factors audit. The second approach, with its greater flexibility, offers better prospects that solutions of human factors problems will be found.

A practical example can illustrate the difference between the approaches. It is possible to certify a three pointer altimeter by applying human factors knowledge to it, and by making it as good in human factors terms as any three-pointer altimeter can be. But as an altimeter it is still potentially dangerous under certain circumstances, with a history of known sources of human error that have led to serious misreadings because it remains too easy to misread by 10,000 feet (Hawkins, 1993). What is wrong with the three pointer altimeter, in human factors terms, is that it has three pointers. A human factors approach that concentrates on good human factors would emphasize that no three pointer altimeter can ever meet basic human factors requirements for the depiction of height in aircraft cockpits satisfactorily. Its well-documented deficiencies are intrinsic to it, and any changes to it that retain three pointers are at best palliatives that cannot achieve the optimum human factors objective of the safest possible depiction of height in cockpits. In this instance, it is necessary for human factors to condemn the three-pointer altimeter, but this is not sufficient. Good human factors practice must include the constructive contribution of recommending an alternative that does meet human factors requirements and can be engineered, the counter pointer altimeter. This is not subject to the misreadings of the three-pointer altimeter, and its safety has been proved through a series of human factors studies that do constitute good human factors practice.

Human factors also specifies the criteria to be satisfied in providing height information in the cockpit that meets pilots' needs for their tasks. It is necessary not only to provide the information that pointers can give quite well, such as rate of movement and rate of change of movement, but also to incorporate a digital read-out of altitude. The way in which this digital read out is portrayed must itself meet stringent human factors requirements in terms of character design, brightness contrast ratio, and size and legibility of numerals, taking account of the diversity of ambient lighting and pilots' minimum eyesight standards. Any residual error rates in readings must be demonstrably low, with the remaining sources of error either being readily noticed and correctable or tending toward safety rather than toward danger because, for example, an aircraft may occasionally be too high but will never be too low and fly into the ground because of an altimeter misreading. This example can encapsulate the difference between the two possible human factors approaches to certification.

A further example, now of historical interest, concerns the attempts to introduce various peripheral vision directors into cockpits. The theoretical premise behind these was that they could visually replicate key aspects of the expanding visual field on final approach to an airfield, and providing the pilot with intuitive and nondistracting information very much as a streaming peripheral world does (Hopkin, 1962). Unfortunately, this did not work out in practice. The principles of a peripheral vision display depended on human sensitivity to movement in the visual periphery which indeed is good, but they also required the abilities to sense direction of movement, rate of movement, and rate of change of movement, which are poor and fallible in the periphery. Thus the principles on which the instrument was based constituted an oversimplification and it never did provide additional information as a bonus without constituting a distraction or disrupting attention, a human limitation that it was intended to circumvent. Certification to meet objectives that tried to optimize in human factors terms the methods of portrayal could be done, but good human factors practice would suggest that such a device cannot attain its objectives fully because of known human limitations in perception, attention, and the processing of different information in parallel.

PREVIOUS NEW APPLICATIONS OF HUMAN FACTORS

Another argument for concentrating on good human factors in certification rather than the continuing achievement of certification objectives through the application of human factors to current practices is that the former has been the more customary practice when human factors has been introduced in the past to new applications for the first time. The human factors approach to the study of aviation maps exemplifies this (Hopkin & Taylor, 1979).

A problem arose when it was found that the legibility of maps in use could not survive the photographic processing required to present them in a projected moving map display. The categories of information in the map legend that were degraded too much by the photographic processing and projection were identified, and they were redesigned so that they would survive this treatment.

Checks were made to ensure that no map information would be lost through changed visual interactions between map information categories resulting from the redesign. The next step was to commission an experimental map of a geographical region. It was a real map, drawn by cartographic draftsmen, printed on cartographic presses, using standard cartographic paper and inks. This paper map included exactly the same cartographic information as the original map, and it was then subjected to the same photographic processing and projection. The specification of this experimental map was compiled on the basis of existing human factors evidence about information presentation. It was not treated primarily as a map, but as an information display. Relevant evidence in human factors handbooks, standards, and guidelines was applied to compile the map specification according to these perceptual principles for the portrayal of information, coupled with a task analysis of how the map was used and what it was used for. A full description was derived of how and to what extent the photographic processing and projection affected the cartographic information categories on the map.

The resulting product was received with considerable skepticism in the cartographic world, partly because it violated several traditional cartographic principles. However, it proved better than expected, and it provided an excellent experimental tool to ascertain which of the existing human factors visual principles could be extrapolated to maps and which could not. Maps were far more complex visually than the subjects of other previous applications of those human factors perceptual principles. The experimental map provided a kind of short cut (Taylor, 1976). It represented good human factors within the limitations of the human factors knowledge then current, and it revealed where there were incipient incompatibilities or discrepancies between existing cartographic practices, objectives, and forms of validation and those that would receive a human factors stamp of approval. In such circumstances it is vital not to presuppose that either discipline, whether cartography or human factors, has a monopoly of wisdom. Usually each discipline introduces further factors which the other has not hitherto considered to be relevant. A frequent source of confusion can be the absence of a common terminology for both disciplines. Many basic cartographic visual concepts have no direct equivalent in human factors parlance, and vice versa. Once recognized, such factors can explain disagreements and point the way toward optimum compromises.

CATEGORIZATIONS

If human factors as a discipline is applied solely to help the certification objectives to be met through present certification practices, then the existing categorization of certification practices would be acceptable and could be retained. The common distinction among the certification of systems, of subsystems, and of equipment would remain, and so would the custom of certifying equipment when it is first accepted for use, prior to being restored to use following an interruption or a period of maintenance, and regularly according to a schedule. The distinction between the certification of equipment, of procedures, and of personnel would also remain, although this distinction seems more dubious in human factors terms, partly because of the apparent extent of their interdependence.

However, an approach that tries to optimize human factors contributions would examine possible alternatives to existing classifications of certification processes for their compatibility with human factors objectives, with a view to identifying feasible alternatives and discussing any differences. The purpose here is not to call into question current certification categorizations or to seem obstructive. Such an approach would be totally counterproductive because the ultimate application of human factors to certification requires the collaboration of those concerned with certification, just as the introduction of human factors into cartography relied on the collaboration of the cartographic world. Rather, the purpose is to ensure that human factors as a discipline is not initially compromised, even before attempts are made to apply it. The known requirements that certification has to meet are also stated in their own right from the outset, without being compromised beforehand in the interests of reaching agreement. The best working compromises cannot be achieved if the requirements of either of the disciplines, whether human factors or certification, have been diluted before the attempts to reach practical solutions that satisfy both disciplines have even begun. Certification can be viewed as a legal attestation that whatever it is applied to is performing within predefined acceptable tolerances. It can also be viewed as a formal way of proving that performance is trustworthy.

HUMAN FACTORS PROFESSIONAL PRACTICES

The optimization of human factors in applications is also preferred because at some point the question arises of who has the authority to rule whether the human factors contributions to the certification process constitute good human factors practice. An authoritative view on this can scarcely come from outside the discipline of human factors itself. A procedure may be required that is somewhat analogous to submitting papers to journals for professional and peer review. Depth of knowledge of human factors and an understanding of certification are needed to pronounce authoritatively on the quality of any human factors work applied to it.

One issue in applying human factors to certification is whether human factors practices and recommendations from other applications can generally be transferred to it. Perhaps certification poses numerous human factors problems that are specific to it, for which new solutions have to be found and proved.

Maps again provide an example: initially their human factors problems tended to be treated as unique because maps had much more complex visual information coding than previous applications of human factors principles of visual information coding. The outcome was that some principles did transfer to maps and some did not.

Human factors data vary widely in terms of their validity and the strength of the evidence on which they depend. It is difficult to justify the transfer of existing human factors recommendations to a new application such as certification if they depend on limited data that have sufficed to meet objectives elsewhere but cannot be claimed to represent best human factors practice. If any recommendations are simply a means of optimizing in human factors terms solutions adopted by others without reference to their full human factors implications, then they are scarcely suitable for transferring out of context to another application.

Good human factors practice is essential to teachability. It is possible to devise certification procedures that may be fulfilled by the select few who devised them but are very difficult for others to implement because they cannot be taught, although they can be done. An aspect of good human factors is to insist that teachability is examined so that a new option is not rendered impractical because of insuperable training problems.

THE INTRODUCTION OF HUMAN FACTORS INTO CERTIFICATION

A central issue is how to get started in any process of applying human factors to certification. How could it be applied? What would it be necessary to do? How would it be possible to recognize what kinds of functions are suitable for human factors certification? What procedures must be followed to be able to successfully achieve a human factors certification of equipment, procedures, personnel, or all or some of them in combination? Commonly, one may start with a task analysis of the certification processes, but human factors also addresses those processes themselves and the extent to which the processes meet human factors requirements or could be improved or modified to do so. This is not to question the competence, skills, knowledge, and experience of those currently concerned with certification, but to bring out how knowledge of human capabilities and limitations might enhance these processes, and make them more efficient, more reliable, quicker, and more consistent. It follows if the emphasis is on good human factors that a human factors specialist must have the authority to decide whether the human factors is good (not whether the certification is good). It also follows that people with human factors specialist knowledge need to have a practical role in the certification processes. This does not mean, and is not intended to mean, that they should usurp any of the functions of those concerned directly with certification, or that they need to have the whole range of skill, knowledge, and experience of professional certification specialists. An aspect of that professionalism is that certifiers often have considerable freedom to choose how to implement the certification processes and what to apply them to. Comparable freedom may be accorded instructors who may observe and rate their students to judge when they are ready for certification.

It is common to find in various certification procedures that it is necessary to call on other professional expertise to judge particular aspects of what is being certified. Numerous disciplines may contribute in this way to check, for example, that various kinds of engineered items meet the requirements, that the software is reliable, that there are no medical problems, and so on. A human factors contribution would be to certify that there are no serious human factors problems either, and the role would be analogous to that of other professions whose knowledge may be needed to complete the certification process in certain circumstances.

Preferably, the human factors contributions to certification should consist of a formalized and structured process. Usually the human factors specialist is not only the most appropriate person to say whether human factors problems have been satisfactorily resolved, but also to identify the human factors issues present. People in other disciplines do not compromise the fundamentals of their disciplines to sanction a certification that may not be adequate. If the software is unsatisfactory, the software specialist must say so. Similarly, if the human factors aspects are unsatisfactory, the human factors specialist must say so.

This raises the question of the quality of existing human factors data, and whether they are suitable for the purposes of certification, which will not usually be among the original purposes of compiling the data (Boff & Lincoln, 1988). Obviously much of the evidence, for example, about portrayal of information, characteristics of input devices, and the criteria the communications channels must meet, is thoroughly reputable and well validated in most circumstances, but some of the cognitive recommendations about the roles of human memory and understanding, for example, may rely on fewer or less well-established data. Every discipline contributing to the certification process can only use the best data available, and those data become more suitable for certification as the discipline matures. Human factors is not an exception to this. However, one of the reasons for needing a human factors specialist, and one of the constituents of good human factors, is a knowledge of the evidence on which human factors guidelines and recommendations are based because the strength and generality of that evidence indicate how far that evidence may be compromised or modified to meet certification requirements and where it must not be.

RÉSUMÉ

The application of human factors to certification should be regarded primarily as a process rather than a product, and the human factors applications should exemplify good human factors practice even though they may prescribe changes in existing certification processes. Some human factors problems in certification can have no optimum solution unless the human limitations in which they originate can be traced and overcome. Certification will reveal relevant issues that are familiar to one discipline and unknown to another. These are a potential source of disagreement but also point toward compromises. Human factors specialists should make contributions to certification processes that guarantee recognition of human factors implications. The objectives are not to oust those whose profession is certification nor to undermine their authority, but to exercise human factors influences throughout certification and to demonstrate the acceptability of certification processes in human factors terms. The relation between human factors and certification should be collaborative and mutually beneficial, since both share the same ultimate system objectives.

REFERENCES

Boff, K. R., & Lincoln, J. (Eds.). (1988). *Engineering data compendium: human perception and performance.* Wright Patterson Airforce Base, OH:: Harry G. Armstrong Aerospace Medical Research Laboratory.

Federal Aviation Administration (1990). *The national plan for aviation human factors.* Washington, DC: U. S. Department of Transportation.

Federal Aviation Administration (1995). *National plan for civil aviation human factors: An initiative for research and application.* Washington, DC: U.S. Department of Transportation.

Garland, D. J., Wise, J. A., & Hopkin, V. D. (Eds.). (1998). *A handbook of aviation human factors.* Mahwah, NJ: Lawrence Erlbaum Associates.

Hawkins, F. H. (1993). *Human factors in flight.* Aldershot, UK: Ashgate.

Hopkin, V. D. (1962). Peripheral vision and flight information. In A. B. Barbour & H. E. Whittingham (Eds.), *Human problems of supersonic and hypersonic flight.* (p 250-255). Oxford, UK: Pergamon.

Hopkin, V. D. (1995). *Human factors in air traffic control.* London: Taylor & Francis.

Hopkin, V. D., & Taylor, R. M. (1979). *Human factors in the design and evaluation of aviation maps.*(AGARDograph No. 225). Paris: NATO.

National Research Council. (1997). *Flight to the future: Human factors in air traffic control.* Washington, DC: National Academy Press.

National Research Council. (1998). *The future of air traffic control: Human operators and automation.* Washington, DC: National Academy Press.

Parasuraman, R., & Mouloua, M. (Eds.). (1996). *Automation and human performance: Theory and applications.* Mahwah, NJ: Lawrence Erlbaum Associates.

Taylor, R. M. (1976). *Human factors in the design and evaluation of an experimental 1:250,000 scale topographical map.* (Report No. 545) Farnborough, England: Royal Air Force Institute of Aviation Medicine .

Wise, J. A., & Hopkin, V. D. (1997). Integrating human factors into the certification of systems. In M. Mouloua & J. M. Koonce (Eds.), *Human–automation interaction: Research and practice* (pp. 181–185). Mahwah, NJ: Lawrence Erlbaum Associates.

Wise, J. A., Hopkin, V. D., & Garland, D. J. (Eds.). (1994). *Human factors certification of advanced aviation technologies.* Daytona Beach, FL: Embry-Riddle Aeronautical University Press.

Wise, J. A., Hopkin, V. D., & Stager, P. (Eds.). (1993). *Verification and validation of complex systems: Human factors issues.* (NATO ASI Series F, Vol. 110) Berlin: Springer Verlag. .

II

Philosophies of Human Factors Certification

2

The Use of the Systems Approach to Certify Advanced Aviation Technologies

Mark A. Wise
University of Central Florida, USA

John A. Wise
Embry-Riddle Aeronautical University, USA

The field of human factors is as varied and diverse as the human subject itself. One of its most important applications is the facilitation of safety and efficiency in a particular working environment, through the implementation of paradigms known about humans and their working relation with machines and systems. During the period since World War II (which is often viewed as the birth of human factors) no area has been the subject of more human factors research than aviation. At no time during that epoch is the influence of human factors more important or more imperative than it is today.

As technology-driven designs have been finding their way into the National Airspace System (NAS), there has been growing concern within the aviation industry itself, the Federal Aviation Administration, and the general public for a means by which to certify complex systems and the advanced aviation technologies that will be responsible for transporting, directing, and maintaining our airborne travel. It is widely agreed that human factors certification is desirable, but the philosophy that will underlie the approach is debatable.

There are, in general, two different approaches to certification: (a) the top-down or systems approach, and (b) the bottom-up or monadical approach. The top-down approach is characterized by the underlying assumption that certification can be best achieved by looking at the system as a whole, understanding its objectives and operating environment, and then examining the constituent parts. In an aircraft cockpit, this would be accomplished by first examining what the aircraft is supposed to do (e.g., fighter, general aviation, passenger), identifying its operating environment (weather, combat, etc.) and looking at the entire working system, including the hardware, software, liveware and their interactions; then, evaluative measures can be applied to the subsystems (e.g., individual instruments, electronic displays, controls).

The bottom-up approach is founded on the philosophy that the whole can be best served by first examining its constituent elements. This approach would perform the certification completely antithetically, by looking at the individual parts and certifying good human factors applications to those parts, under the basic assumption that the whole is equal to the sum of its parts.

In this chapter we will attempt to form an argument for the top-down (systems) approach, while addressing arguments against it, and pointing out the shortcomings and erroneous assumptions inherent within the bottom-up approach.

CERTIFICATION

To develop a cogent argument outlining the advantages of the top-down approach to certification, it must first be established what the goals of the certification process are in general, and the certification problems that human factors will attempt to overcome. Certification, in a generic sense, is the process by which a product is declared appropriate for a particular task in that it matches or exceeds a previously defined set of "design to" criteria. In being certified, it is implicitly understood that the product will perform the task for which it was designed safely and effectively.

Of Aviation Technologies

Certification of advanced aviation technologies involves, many times, the evaluation of products that are technologically new and previously unused or untested. This in itself poses an interesting problem, because with uncharted equipment the standards by which previous, like products had been evaluated now become obsolete and inapplicable.

Certification of aviation technologies is also unique from some other certification problems because of the extensive and unavoidable interplay between many systems, so that the systems themselves can be looked on as subsystems of a larger system. For example, air traffic control, the fleet of aircraft, and maintenance could be seen as systems unto themselves, but on a larger, more universal scale, their boundaries are not so narrowly defined. All of the aforementioned systems are merely players in the entire NAS. Therefore, the certification of each of these interrelated entities by itself falls, trapped, into the quagmire of "fuzzy" certification, which is neither desirable nor acceptable.

The challenge, then, is to overcome these obstacles. The means for doing so appears to be the implementation of the systems approach to certification.

Because the foundation of this theory is built on the premise that the system as a whole is more important than the parts of that system, it could be argued that, by its nature, it avoids the aforementioned problems.

THE SYSTEMS APPROACH

A *system* can be defined (in a broad sense) as the collaboration of functionally similar objects (humans, machines) working toward a common goal within its respective environment. The first and most important aspect in designing any system is to clearly define its goals and objectives (Christensen, 1987; Meister, 1987). In an automobile, the goal is safe, efficient, land travel; in the government, the goal (at least hypothetically) is to serve and protect the citizens; and in the airspace system, the goal is to provide safe, expedient, air transport.

Because the first step in developing a system is to identify the goals, so should the first step in certifying that system be to identify the goals, then certify that system based on those goals.

What Came First, The Product or the Idea?

Thomas Edison once said that a new invention consisted of "1 percent inspiration and 99 percent perspiration." But the inspiration, the invention's objective or goal, comes first. The first step in developing any system is to define what the goals will be; it is impossible, if not inconceivable, to begin to build a system without first deciding what it is going to do. The Wright brothers did not just begin to assemble pieces of wood and paper, only to find out, to their amazement, that the "thing" they built could fly.

Just as the goals of the system are a prerequisite for its development, so should they be a prerequisite for the system's certification. The definition of a system's goals dictates the means by which the certification of the subsequent subsystems should be handled. It seems illogical and erroneous to attempt certification of a product without first considering what the ultimate goal of the product is when it is placed in the context of the system.

Starting with the system's goal and working down provides a framework within which an evaluator can examine the parts as contributing factors toward the said goal. This eliminates excessive redundancy among components and does not leave room for certain vital components to be left out of the system.

Working Environment

The systems approach looks at the system and the elements of the system in their working environment, and therefore can evaluate the system's ergonomic layout. By this we are referring to the positioning of controls, instruments, displays, and so on ., within the system and their functional relations to each other. This would be similar to a task analysis, in which one wants to examine the physical relation of functionally similar and operationally dependent objects in the workplace. This certification is particularly applicable when introducing a new instrument or device into a workstation, where it ends up being placed wherever an opening is available.

Going from a bottom-up approach, a yoke, for example, could be certified to possess all of the characteristics of a sound human factors yoke, and it is certified on this criterion, but the evaluation ends there. The yoke, in all of its glory, is not optimally usable if it is placed behind the pilot's seat by the engineers. It is analogous to writing a book of poetry in Egyptian hieroglyphics in the 20th century: Although it may contain brilliant rhyme schemes and flowing poetic prose, no one can read it—it is merely wasted paper. Something is only as good as it is usable.

In contrast to the bottom-up approach, a top-down approach would look at a working simulation of the aforementioned cockpit, and ergonomic problems like the one mentioned earlier would be recognized. It is not at all likely that such a huge error would ever come about, the point is still valid, and the problem is still real.

Not only does a system's ergonomic environment need to be considered in certification, but its operational environment needs to be considered as well. By operational environment we are referring to lighting, weather, temperature and so on. The minimum light emittance needed for a display or instrument is directly related to the environment that it will be used in. Therefore it is necessary to evaluate the instrument in those conditions. This is not easily accomplished through a bottom-up technique, but is easily evaluated within the systems approach.

It could be argued by the bottom-up proponents that the product's environment could be replicated during the evaluation to take into consideration the lighting, for example. However, this only really takes care of half of the problem, because other considerations are the light being emitted from other displays, the glare due to the angle of the display in the system, and so on. All of these environmental factors would theoretically be observed in the systems approach through a simulation or mock-up.

Money, Money, Money

Major drivers in any system, whether it be in the developmental, evaluative, or production stages, are cost, and cost efficiency. The top-down approach is cost-effective in two ways: a) the certification personnel are not required to spend the same amount of time on every product in the system, because not every part of the system is forced to meet the same criteria of human factors engineering; and b) money is saved in the production of the products because of weighted criteria.

Toilet Seats and Tool Boxes. The U. S. government was under a great deal of scrutiny in the mid-1980s for purchasing miscellaneous items for its fleet of C-130s (a military transport plane), that appeared to have greatly inflated prices attached to them. Some examples were $15,000 toilet seats and $5,000 wrenches. The government justified the purchases by claiming that the equipment had to be "perfect" to be usable and safe in its operational environment. Although the prices paid for those products were probably justifiable (because of the research and development costs for a few production items), they give insight to a problem that could arise out of bottom-up certification: setting outrageously high standards for a product before its relative importance in the system and the system's environment is known.

If every product in the NAS had to be evaluated by the same standards, the prices for the products would be exorbitant. No one would argue that the toilet seats on a Boeing 757 should have to comply with the same human factors standards as the plane's navigation system, to take it to the logical extreme. But where is the line drawn? How can one judge which products need to pass strict human factors and ergonomic tests without first looking at their roles within the system? The point is that one cannot. As a consequence, every product, display, control, and widget would be subject to the same meticulous human factors standards. This would result in exorbitant prices for the products, which would be felt directly by manufacturers and indirectly by the paying airline passengers.

The systems approach would look at the system and the parts that make up that system and do something that the bottom-up approach cannot do: decide the relative importance of each part, and be able to make a well-informed decision as to what standards each needs to be evaluated by. Therefore, a rarely used, unimportant product does not have as much time and money spent on its certification as a relatively vital, often used product.

In this way, the systems approach would require a reevaluation and alteration of current certification standards. Products should be evaluated on their functional importance first and human factors standards second, where depending on the first criterion human factors certification may or may not be necessary. For example, if a job required an employee to shovel 3 pounds of coal from the coal pile to the furnace each day, there would be no need to certify that shovel to optimize ergonomic standards. Any shovel from the local hardware store would satisfy the requirements sufficiently, and to require anything else would be superfluous and cost inefficient. On the other hand, if the shovel were to be used 8 hours a day, 5 days a week, then it should be subject to more stringent encompassing human factors standards.

Workload

One of the main objectives in any human factors effort is to ensure a good fit between the humans and the machines they operate. This is done in several ways, including personnel selection, training, manning, and so on. Underlying all of these processes is an evaluation of the workload incurred by the human while operating the system, whether it be psychological or physiological.

Workload is a very important aspect of any system design, and it is something that must be examined in the certification process. The top-down and bottom-up approaches address the issue from different angles. The bottom-up approach would look at each part of the system and measure the workload involved with running that particular part. Then by summing all of the measurements, a gauge as to the total amount of workload that would be present in the entire system should be obtainable. The problem with this is that, once again, there should be significant differences between the parts by themselves and the human's management of those parts when they are incorporated into the system. This method does not take into account secondary tasks. There could be a significantly different amount of workload incurred by the human when he or she is operating the entire system than when the original estimate was made; and this number could be either high or low. Because, neither a high nor low workload is desirable, due to the associated performance deficiencies there needs to be a more accurate method: the systems approach.

The systems approach would have the advantage of observing the operator managing the entire system. Subjective and objective tests could be applied to determine the amount of workload, and appropriate measures could be taken to increase the crew, decrease the crew, or leave it the same.

BOTTOM-UP APPROACH

Several of the problems that are inherent within the bottom-up approach have already been described above. In addition to those problems there are others which not only show this evaluative philosophy to be incompatible with certification, but indeed prove it to be undesirable as well.

Inductive Conclusions

It has been argued by many well-respected modern philosophers, including Hume and Kant, that inductive arguments and assumptions can never be validated. The nature of an inductive assumption is that by observing past examples of a particular event, one concludes that in the future, a similar cause will produce the same effect. For example, if we drop a penny at Time T_1, and it falls toward earth, it will necessarily fall to earth at Time, T_2. According to Hume (1748/1977) there are no necessary causal connections, and any attempt to predict the future from the past is a fallacious one that is built on a circular argument. Kant (1781/1926), not as critical, said that there is causality, but we can still never validate a future event based on similar past events.

The process of certification requires us to employ inductive logic. We are, essentially saying that If X works now, at T_1, then it will work in the future at T_2. Induction is a necessary part of certification and cannot be overcome or circumvented, but the number of times that an inductive conclusion must be drawn can be minimized. As with all necessary evils, the fewer the better.

Top-down certification must only use inductive logic once; that is, in certifying that because the system works well during the evaluation, then it will work well in the future (in production use).

Bottom-up certification employs inductive reasoning early as well as later in the certification process, thereby making the probability of error more than twice as great. The human factors certification personnel must not only certify that Part X will work when placed in the system—the first case of induction—but they must also assume that the system will work when fully implemented—the second occurrence of induction.

Whole Sum of the Parts

> ...for owning to the fact that the destruction of the foundations of
> necessity brings with it the downfall of the rest of the edifice...
> Descartes (1641/1952 p. 75)

It does not take an advanced degree in engineering physics to deduce that a house that is made out of bricks could not be built on a foundation made out of straw without collapsing under its own weight. Basic physics (and common sense) tells us that any physical structure is only as sturdy as the foundation on which it is built. Similarly, in logic, an argument is only valid if the premises on which it is "built" are true. If the foundation is weak or the premises are shown to be untrue, the argument crumbles under the weight of invalidity. Bottom-up certification is guilty of being built on a straw foundation.

Bottom-up certification uses as its foundation the premise that the whole is equal to the sum of its parts. Although this statement may be true with a jigsaw puzzle, it is certainly not true in certification or any scientific endeavor. As far back as Aristotle—one of the first enquiring scientific minds and logicians—it has been recognized that:

> we often fall into error because our conclusion is not in fact primary and
> commensurate universally in the sense which we think we prove it so.
> We make this mistake when... the subject (element) which the
> demonstrator takes as a whole is really only part of a larger whole.
> (Aristotle, trans. 1952, p. 101)

In certification terms, Aristotle would be saying that we often err when certifying a part as an entity in and of itself, when it is truly only a part of a larger whole—the system.

Later in *Posterior Analytics*, Aristotle's argument further repudiates the use of the bottom-up approach in certification. It says that although a part can be certified by itself, the truth of that certification is only applicable to the part individually. It would not be true universally, because the part is in fact different when it is placed in the system.

With this weakness exposed, the foundation on which bottom-up certification rests is undermined—the theory is invalid.

A Bad Product With Good Certification Criteria

Another less philosophical, more practical, problem has to do with certification criteria and the role of the certification personnel in the process. It is conceivable for a product to be valid by human factors standards without being desirable by them. The three-pointer altimeter provides an excellent example (Hopkin, 1994). The altimeter could be certified on the grounds that it provides excellent contrast, brightness, and font size. It is judged that it could be visible from every part of the cockpit, and from a performance standpoint it is accurate to +/- 0.5 feet. The problem with this instrument obviously does not lie within its design, but within the instrument itself.

Incident reports and experimental analysis of the three-pointer altimeter have shown that it is responsible for pilot-induced errors a dangerous amount of times. Misreading by 10,000 feet is not uncommon. Therefore, an instrument that can meet many human factors requirements may nevertheless not be a good instrument. This once again ties back to the problem of certifying a product without looking at it in its working environment. In the bottom-up approach, this instrument could be certified; not to say that it would not be in the systems approach, but it is much less likely.

A Portrait of the Artist as a Certifier . To illustrate (quite literally) an error that can occur by using a bottom-up certification process we use an analogy: the analogy of an artist as certifier.

Imagine that you were hired out as a professional art certification consultant. This job required that you look at different pieces of art, then certify whether or not they represent what they are supposed to (e.g. an eye looks like an eye; a cow looks like a cow). One of your clients, a not too bright artist, comes to you and asks you to certify an eye for him that he has recently sketched (it looks like Fig. 2.1). The eye looks good—good proportions, proper relation between the pupil, iris, and so on—so you give it your stamp of approval: a good eye.

FIG. 2.1. A good eye

Over the course of the next 2 months the same dimwitted artist shows you another eye, then a nose, and then a mouth, all of which look like what they should represent. Again the obligatory stamp of approval is given for each feature. Finally, a couple of weeks later he shows you the whole thing, which looks like Fig. 2.2.

The picture is poorly drawn, not because any of the parts themselves are poor, or because the features are improperly aligned. The picture is poorly drawn because each feature was certified without knowing what the ultimate goal of the painting would be. Each feature by itself is a good drawing (open for debate) and accurately represents its respective object. But, when summed together, the whole is wrong.

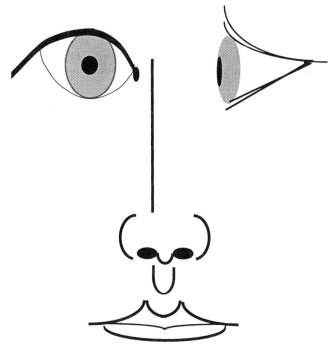

FIG. 2.2. A bad face.

CONCLUSION

Certification of advanced aviation technologies should not only pose a unique professional challenge for human factors experts as scientists, but also as consumers who want to have the safest air transportation for themselves and their families. To ensure this safety, the best possible method of certification should be employed: the systems approach. This is not to say that the systems approach is infallible, but it certainly is superior to the bottom-up approach.

The effectiveness of any certification is only as good as the individual(s) performing it. But, taking human error or misjudgments out of the picture, the systems approach is more sound fundamentally, practically, and philosophically.

It is because of its superiority and not its infallibility that the top-down approach is better suited to certification. Hume (1748/1977), a philosophical empiricist, argued that human judgments and scientific decisions are always made after one entertains two or more opposing arguments, examining the possibility of each by weighing their relative proofs, then believing the strongest case: "In all cases we must balance the opposite experiments, where they are opposite, and deduct the smaller number from the greater, in order to know the exact force of the superior evidence" (p. 74).

In this situation, the top-down approach provides the stronger case toward its cause. Although it brandishes some problems, the positives highly outweigh the negatives. The Humian approach to decision making should rightly be chosen over its counterpart.

REFERENCES

Aristotle (1952). Posterior analytics. In: R. M. Hutchins, *Great books of the Western world* (G. R. G. Mure, Trans.). Chicago: Encyclopedia Britannica 95–137.

Christensen, J. M. (1987). The human factors profession. In: G. Salvendy Ed. *Handbook of human factors*(pp. 3–16) New York: Wiley.

Descartes, R. (1979). Meditations on first philosophy. In: R. M. Hutchins, *Great books of the Western world* (G. R. G. Mure, Trans.). Chicago: Encyclopedia Britannica.

Hopkin, V. D. (1994). Optimizing human factors contributers In J. A. Wise, V. D. Hopkin, D. Garland. Human Factors Certification of Advanced Aviation Technologies. Daytona Beach, FL: Embry-Riddle Aeronautical University Press.

Hume, D. (1977). *An inquiry into human understanding.* Indianapolis, IN: Hackett. (Original work published 1748).

Kant, I. (1926). *Critique of pure reason* (N. K. Smith, Trans.). New York: St. Martin's. (Original work published 1787)

Meister, D. (1987). Systems design, development, and testing. In:G. Salvendy, *Handbook of human factors*(pp. 17–42) , New York: Wiley

3

The Gains From Certification Are in the Process

John R. Wilson
Institute for Occupational Ergonomics, UK

Most of my colleagues tend to think of certification as applying to standards and verification of ergonomics professional practice, rather than to the systems we help design. This may illustrate both the self-referential nature of professional groups and also genuine concerns about professional certification. These concerns are germane to the present argument because certification of systems may reduce in the end to certification of professionals in the process (an issue returned to later). Any misunderstandings over the notion of certification may be in part responsible for misgivings amongst some ergonomists and also their clients. If certification is seen only as a formalization and standardization of ergonomics activities, then there can be considerable opposition. When seen as a design review and approvals procedure, response seems more favorable. Certification, strictly, is merely the provision of a (written) official declaration of qualification or status. More usefully, it should be the "result of an applied examination process devised to formally test and affirm that the system being inspected satisfies certain accepted criteria" (Taylor & MacLeod, 1994, p. 164).

Green (1990) believed that the two main factors which safeguard flying from human error are both related to certification and regulation. First is the proceduralized nature of flying whereby as much activity as possible has been reduced to rule-based performance. Second is the emphasis placed on training and competency checking of aircrew in simulators and in the air. However, Green believed that other human factors are rarely addressed within the operating procedures and simulator training and can give rise to human reliability problems. These include: hardware factors, the compatibility of control–display relation and the way information is presented in relation to pilots' expectations; social factors and especially pilot–co-pilot relationships; and system factors including fatigue and cost–safety trade-offs. To these we might add potential problems with the integration of the "electronic crew member" following increased automation, moves from voice to data links, and pilot (rather than air traffic control) directed flight paths. Human reliability failures with artificial intelligence and automation—due to overreliance on the system fail-safe mechanisms, or to lack of operator confidence in the integrity or self-regulating capacity of the system, or to out-of-loop effects—are widely accepted as being due to deficiencies in plant design, planning, management and maintenance rather than to "operator error," Reason's (1990) latent error or organization pathogens argument. Reliability failures in complex systems are well enough documented to give cause for concern and at least promote a debate on the merits of examination and certification.

The purpose of this chapter is to explore what is valuable in certification, at least to show that the benefits can outweigh the disadvantages and at best to identify positive outcomes that may not be obtainable otherwise. On both sides of the debate on certification there is general agreement on the need for better human factors perspective and effort in complex systems design. What is at issue is how this is to be promoted and what role certification may play. There is also wide agreement on the very different approach and needs of human factors standards and certification as compared to those of engineering and the particular needs related to cognitive and collaborative work performance (e.g., Barnes,Orlady, & Orlady, 1996; McDaniel, 1996).

This exploratory review draws from many fields of application to try to provide a balanced position on certification. An assumption is that it is the total work system (people, hardware, software, environment, procedures, organization) that is being assessed, but that the methods and tools to do so, and indeed the assessors themselves, may also require certification.

PARALLELS

There used to be an unwritten law in work study (motion and time study in the United States) that noone was so enthusiastic about work measurement and standardization as those whose own jobs, they felt, precluded any possibility of such a process being applied to them. As a corollary, noone was quicker than management to oppose, utterly, any attempt to assess their own work when this was suggested, on the grounds that this was inappropriate, represented an unnecessary effort, and anyway it was impossible for any analyst to understand what they really do. (The parallel in society as a whole is NIMBYism [not in my backyard] when faced with proposals for road or building development, refuse dumps, or nuclear power plants.) One of my students was attached to work with a "high flier" in a major U.K. consultancy; they were charged with applying quality assurance procedures to the activities of the consultancy itself. This involved vetting all areas of their operation for compliance with BS5750 and ISO9000, the relevant service quality assurance standards. The student, and the formerly popular and high-achieving consultant, quickly became outcasts from the group because they were introducing some formality to the consultants' work and some prescription about how they should operate. The irony is that the core activity of this consultancy group was advising industry on the need, processes, and procedures for quality assurance!

This illustrates the difference that viewpoint and perspective make to opinions on formal systems, appraisals, standards, and review processes. The reviewers or appraisers see them as bringing about order and rationality, and as ensuring that "the best" is retained and "the worst" is identified and eliminated. Those being reviewed or appraised, on the other hand, may see formal systems as restrictive, petty, and unnecessary interferences with their activities, and as leading to "throwing the baby out with the bath water."

For many in the human factors community the most salient example is certification for human factors and ergonomics professionals themselves. This is taking place, for example, in the Committee for Registration of Ergonomists in Europe (CREE) scheme, and the Board of Certification in Professional Ergonomics in the United States. In a "Provocations" article in *Ergonomics in Design*, Senders and Harwood (1993) took contributions from both sides of the argument. They pointed out that one danger is of formality driving out reality, in that any degree or certificate may become more important as an end in itself than the competence it implies. Educators often are faced with this from their students, when attempts to discuss and explore are met by the students' desire to digest directive information tailored to an examination. In the same piece, Schumacher and Dorsted question certification in terms of need, process, and impact. If we summarize and generalize some of their objections, these are:

- Certification does not ensure quality (or integrity or ethics).
- What position will certification have in law?
- Do we want the homogeneity that certification might bring?
- Prescriptive criteria may stifle innovative designs.

A response from Hendrick in the same article made little attempt to answer directly some of the questions about whether professional competence certification is needed at all, concentrating instead on defending the processes involved. He did argue, however that certification can promote growth of a discipline and its image, although admitting that the process cannot guarantee "worthwhile job performance ... competency ... [or] ... conformance to ethical, moral, and professional standards" (Senders& Harwood, 1993)

What can possibly be the arguments for certification then, if it cannot even guarantee basic compliance? To find value in certification, perhaps we can look at other areas of ergonomics endeavor. In work organization and job design, and in particular the implementation of changed work systems, it is often argued that the content of any change is of minor importance for successful outcomes compared to the quality of the change process itself. If the best process possible is put in place then we can afford to "change the change," iterating the actual content as required. Translating this idea to certification in complex systems, we could look on certification as the means by which an improved development process is enabled, rather than as a limitation on and detailed specification of the content of the development.

From the fields of product liability (in the 1970s and 1980s) and health and safety at work (in the 1990s), we can see some of the more systemic benefits possibly accruing from certification. Out of the imposition of regimes of strict liability have come better processes and systems of design amongst producers and moves to a more rational standards regime (horizontal standards) amongst the lawmakers. Consequences as a result of the health and safety (ergonomics) legislation implemented in the 1990s across member states of the European Union (EU), as a result of EU Directives, are even more marked. Although it would not be appropriate to be too naive about beneficial outcomes, it certainly seems as if the need for employer conformance with ergonomics criteria and practices has stimulated the production of tools, techniques, instruments, and methodologies for investigation and diagnosis that will be of value across a range of concerns. Not all of the consequent developments are to be widely welcomed—I imagine we have all been shocked by some of the so-called ergonomics aids now on the market—but the overall effect has been one of dynamic growth in the discipline. The ergonomics community itself has had to produce new approaches and techniques, improve the validation of existing ones, and generally ensure greater justification for its guidelines and recommendations. Even ergonomists who are dubious about the value or validity of specific requirements in the relatively new regulations have been pleasantly surprised by the consequent pressures for quality in methods.

MANPRINT—Lessons Learned

Can we learn from human factors certification in nuclear or military systems? In the nuclear industry there has been increased use of human factors (and especially human reliability) guidelines, tools, methods, and (licensing) review processes, with some reported success (see Bongarra, 1997; Herman, 1993). The well-known MANPRINT developed for procurement in the U.S. Army has been adopted by both the British Army and, in modified form, the Royal Navy. It is useful to examine the reasons given by the British Ministry of Defence (1992) for its adoption :

- Perceived success of MANPRINT in the United States, including improved maintainability of equipment and use of analytical techniques to ensure wider and better usability.
- A desire to identify and achieve the best balance between people and equipment. It is recognized that "high-quality, multi-capable... better

equipped, motivated, and properly trained" personnel will only result if manpower issues are considered as a part of the equipment procurement process.

- Personnel costs now outweigh equipment costs and it is hoped to provide more control over these by being better able to anticipate, budget for, or reduce costs through design improvements. It is also anticipated that given a better specification generally and thus more cost-effective products, it will reduce overmanning, poor performance, and errors.
- Better working conditions and reduced training costs.
- Greater requirements for cognitive skills rather than physical, and the shrinking pool of skilled labor available, mean it is desirable to constrain designers to produce operable equipment for specified personnel. These reasons have been labeled *skills drift* and *demographic trough or slide* respectively (Goom, 1993).
- New health and safety legislation applies to military as well as civilian systems and "covers areas which have traditionally been regarded as usability rather than safety issues," again an argument for the broad approach of MANPRINT.
- MANPRINT covers manpower, personnel, training, human factors engineering, health hazard assessment, and system safety (and habitability and environmental ergonomics for the Navy). This ensures a single source of human factors issues, thus preventing or reducing suboptimality in systems design.

This support for MANPRINT, and presumably for similar certification systems, raises three questions: Are these claimed advantages real? Are they important? Are they generalizable, especially to civil systems? A skeptic might answer "no," "partly," and "no" to these questions; it is easy to be cynical about any claims on the part of the military establishment to be making efforts to reduce costs, for instance. Nonetheless, a more reasonable view might be to answer the three questions by "the claimed advantages seem reasonable," "yes, they are potentially very important" and "they might be generalized to other situations in other industries."

Bearing in mind the preceding argument, in fact the strongest support for the certification process might derive from the fact that the claimed advantages are as much or more concerned with process as they are with content. If the key gains reported for MANPRINT are summarized and generalized they appear to be:

- Better human–machine systems designs and improved usability.
- Improved cost-effectiveness and cost control.
- Widening of the user base.

- Compliance with ergonomics and health and safety legislation.
- More efficient design process.

The indirect benefits of certification, felt within systems design, might be as important as any direct gains in judging the value of the process.

STANDARDS FOR CERTIFICATION

If we are to have certification then this must be related to some norms, standards, or standardized procedures. A system might be certified if it can be shown to have attributes that meet certain recommended values or if its performance meets acceptable limits on certain defined criteria or if it is shown that defined analysis or test methods have been applied to the design. In this last case, of course, the methods themselves will have to be certified first. Of relevance to complex systems is a particular case of the last, "there is little doubt that a principal future use of simulators will be for licensing and certification" (Jones et al., 1985, p. 88). Mainly used now for pilot training and proficiency approval, there seems no reason why artificial intelligence in the cockpit, and particularly its interaction with crew, cannot also be assessed in simulators. The interesting issue then is certification of the simulation system itself.

Debate over type and coverage of standards has a long history in the field of product safety regulation. "The trade-off between voluntary and mandatory standards [concerns] acceptability, applicability and ease of formulation versus possible non-compliance ... standards enshrined in legislation are of little value unless there is strict enforcement" (Wilson, 1984, p .204). Much support has been given to the notion of performance standards rather than construction or dimensional standards. Problems with safety standards have been identified as: inadequacy in scope and permissible levels of risk; not addressing all foreseeable hazards or types of behavior; specifications that tend to be generalized, partial, and inadequate; and a general lack of a standardized format. Although more than a decade later these criticisms are still valid for product ergonomics standards, there has been a major change in direction away from vertical and product-oriented standards toward horizontal and hazard-oriented standards. Advantages of these are faster development, easier updating, greater applicability, more consistency, and better clarity and understanding about necessary safety levels (van Weperen, 1992).

Meister (1985) differentiated "attribute" standards that describe how the product should appear or should function and "performance" standards that describe how the designed product should perform. He saw the former as being general and applying mainly at the component or equipment level and the latter as relevant to the subsystem and system levels. He criticized the state of human factors standards in much the same way as consumer product standards have been criticized. As a consequence, Meister (1989) subsequently stated that "whether because the standard lacks substantive data support or because human factors is generally viewed ... as a constraint on ... freedom to design, MIL-STD 1472C [for instance] is honored as much in the breach as in the observance" (p. 139). The reader is referred to *Applied Ergonomics* (1995), Dul, de Vlaming, and Munnik (1996), Parsons (1995), and Stewart (1998) for reviews of ergonomics standards and of the process of generating and approving them.

STANAG 3994 AI (NATO, 1996) is the NATO Standardization Agreement on the Application of Human Engineering to Advanced Aircrew Systems. The intention of this is to "establish criteria for the application of human engineering considerations in the design, development and evaluation of advanced aircrew systems" The purpose also is to provide a basis for agreements between contractors and procuring agencies on human factors scope, context, techniques, and criteria. A general human engineering program model is defined, with core requirements including those to do with:

- System analysis—mission analysis, function analysis, potential operator capability analysis, potential equipment identification, function allocation.
- Analysis of operator/maintainer tasks—time line analysis, task analysis, critical task analysis, decision analysis, error analysis, loading analysis.
- Preliminary system and subsystem design—information requirements analysis, control requirements analysis, workspace requirements analysis, environmental analysis.

A number of other issues are defined as requiring agreement between agency and contractor, including programs of research involving experiments, testing, and dynamic simulations with human participants; detail on application of relevant human factors standards; development of software and hardware procedures in operation and maintenance; production of mock-ups and models for conformance testing; and development and planning for test and evaluation, including criteria justification and test interpretation details.

Perhaps most critical in terms of any sustainable argument for certification are the requirements to prepare a human engineering program plan. This must identify standards of relevance to the system and must identify what human engineering activities will be involved, timescales and criteria. It should indicate how formal interaction between human factors specialists and other relevant design specialists will be achieved. Finally, provision is made for a tailoring of details in the standard to meet any specific requirements of the system under development.

For someone not involved in military ergonomics, several things seem apparent about STANAG 3994. It appears at first sight to be complex and unwieldy, with potential for great overlap in particular among the many different analyses. Even coordinating these, making sure they are complementary but not duplicated (or even contradictory) will be a considerable project management task. On the other hand, emphasis placed on analytical activities and not on prescriptions of design detail is welcome. However, some of the analyses imply checklist comparison procedures, for instance "multi-function control modes, and display menu selections shall be analysed .... and the resulting .... structure shall be plotted and analysed for ease and effectiveness of use" and "workspace [requirements] shall be analysed in terms of their access, vision, reach, egress and emergency requirements, for the range of body sizes, clothing and protective equipment" (NATO, 1996, p. 5). This will presumably bring into play a plethora of other standards, guidelines, and recommendations, the effect of which may be to impose a degree of complexity and restriction on development that is not commensurate with finding innovative design solutions. Nonetheless, the document does allow for flexibility in its provisions according to circumstances. Thinking again about systemic gains, it may be that the most important benefit will be the integration and collaboration required between ergonomists and engineers.

   Undue complexity in certification, as suggested earlier, may have excessive cost implications. For instance, and even under present regulation systems:

> It is relatively easy for the profitable airline .... but the airline operating in a more competitive area of the aviation system, where economic margins are extremely constrained, may simply be unable to undertake all of the desirable training and standardization of equipment without going out of business. The regulatory authority may have considerable difficulties in compelling such airlines to undertake costly procedures as the airlines may accurately point out that by doing so they will be made less cost efficient vis à vis foreign operators (possibly operating in a less regulated environment) with whom they compete directly. ... The temptation for operator and regulator alike, when faced with an acknowledged but intractable problem, is to undertake some unconscious dissonance resolution by regarding the problem as less serious than they might if it were readily soluble. (Green, 1990, p. 510).

If this concern applies to aviation systems developers and suppliers as well as to the operating companies, then the impact on the workability of any certification process may be serious.

TO CERTIFY OR NOT TO CERTIFY?

Why is someone who typically dislikes regulation, systematization, and quality assurance generally writing here to support certification in complex systems development? The answer lies in what has been stressed already, namely that the systemic outcomes of having a certification program in place may be advantageous enough to outweigh any drawbacks of the regulatory regime and even any inappropriate content or application of standards.

If human factors certification is to work in any domain we need first to consider why it has not been in place there previously and address all potential reasons very seriously. I worked with a very large, reputable transnational company, where the design engineers pass their designs through every conceivable review and approval process—HAZOP, P & I approval, environmental impact assessment, engineering audit, and so on. There had not been any systematic human factors approval process and we had to ask why. Possible reasons included:

- Ergonomics is not seen as important historically by engineers or managers.
- Ergonomics is assumed to be included in all the other types of approvals and standards.
- Certification of human factors is not seen as cost-effective; there will be few gains, there will still be problems afterward (because people are seen as fallible), but much time and energy has been expended in the interim.
- Certification is genuinely seen as not required by engineers and ergonomists.
- Human factors certification may be resisted by ergonomists themselves, perhaps because of the requirements or restrictions it may put on them.
- Certification is seen as impossible to do or impossible to do well.

Before even beginning to introduce a process of ergonomics design review to this company and before even planning what might be included, we needed to examine these potential reasons, see which were relevant, and address the perceptions and organizational issues involved.

THE CASE AGAINST CERTIFICATION

A number of arguments might be made against certification. Certification may be seen as unnecessary. Presumably in this view there is seen to be little or no room for improvement in ergonomic design of systems (hardly a sustainable argument) with the alternative that the complex system domain (say aviation) is self-correcting. If it is, the argument goes, then human factors deficiencies will be rectified anyway during development and commissioning and where they are too large or deep seated then the system itself will not remain in operation. I find this argument unconvincing unless major failures in operation are deemed to be a part of this self-correcting process (it must be accepted that, currently, aviation does have remarkably reliable and safe systems of hardware, software, procedures, communications, and people, in part through built-in error recovery mechanisms).

An extension to the first objection is that certification is unnecessary for simple or relatively stable systems and is impossible to do adequately for complex systems. There may be some validity to this argument but the advantages to be gained from improved quality in the development process might counter it to some extent. It must be recognized that any system of approvals must allow for trade-offs between different human factors and between human and technical factors.

Further criticisms might be that a particular certification regime is restrictive and cumbersome. If we replace *is* by *can be* then I would agree. However, if we aim certification at performance and assessments, instead of design specifications, allow tailoring of standards to meet circumstances, and—most importantly—make ergonomics design review and approvals an intrinsic part of development rather than an extraneous add-on, this criticism can be addressed. In a similar way, we can meet objections that certification might stifle innovation and lead to homogeneity in design.

Certification might be criticized on the grounds of being misdirected, with formal standards aimed only at reducing the incidence or consequences of active errors. In this view, standards may be much less help with latent failures or resident pathogens in the system; these are the system problems that may have lain dormant in the system for a long time and are spawned by the activities of designers, managers, and, indeed, regulators themselves (Reason, 1990). One could take a positive view however, that in fact it is these violations, giving rise to latent errors, that are best attacked through a process of certification, due to the process itself being a good discipline on all involved in high-level planning and decision making.

As for a fourth set of criticisms, that certification is untestable, widely unacceptable, and thus unworkable or unenforceable, this will largely be a function of the particular regulatory regime. To repeat again an earlier point, if a system of certification can be constructed such that it is seen to improve and streamline the development process and time, as well as increase systems integrity, then acceptance will be more widespread. With a similar perspective, Longridge (1997) described the voluntary Advanced Qualification Program for pilot training and qualification.

THE CASE FOR CERTIFICATION

The case for certification can be made in terms of positives as well as by questioning the validity of criticisms. Three areas of benefit may be defined, all systemic in nature in that they emanate from the fact that human factors certification of complex systems will have effects beyond pure definition and assurance of compliance with human factors standards.

First we have the improvements in the design process that might be expected. Knowledge that a system must be certified in terms of human factors may not ensure that all correct detail design alternatives are chosen; there is no such thing as a perfect design, given all the trade-offs that must be made. However, it will mean a greater likelihood that all relevant issues are addressed and their consequences assessed much earlier in system development. Costly changes after prototyping or even during commissioning trials can be reduced both in number and in their consequences. A related consideration is that for any suppliers to be able to meet future certification requirements, technical and financial decision makers will have to coordinate much earlier and better with those responsible for human factors. Norris and Wilson (1997) for instance, made a strong case for the improvement in designs and development systems that will result from integrating ergonomics and safety evaluations within the product design process.

A second benefit is predicted based on experience in other domains in which ergonomics standards have been introduced or toughened. In the act of formulating, specifying, and testing the processes and procedures necessary to allow systems to be certified, the human factors community will have to respond to pressures for better methods, techniques, and criteria and will have to validate, justify, and communicate them better.

Finally, an improved professionalism in human factors, the perceived benefits for the design process, and—if experience in industrial health and safety and ergonomics is any guide—raised interest among engineers in human factors problem solving, will all act together to produce a more human-centered design approach. Thus, through both the very existence and also the process of certification, even more than through its content, the design of complex aviation systems will be improved.

CONCLUSIONS

We should not talk of certification only as a choice between two options—to certify or not to certify. If we draw an analogy in politics this is like saying that electorates have a choice only between the authoritarian (prescription, control, punitive consequences of noncompliance) and the libertarian (the individual has an absolute right to do as he or she pleases, and the market will ensure instability is kept in bounds). Such debates see anarchy as the only outcome if a choice between the two options is not made. There are other paths in government, however, whereby individuals have both rights and freedoms and also responsibilities toward society, and whereby society self-corrects in attempts to redress imbalances in power. Thus, a regime of certification can be implemented such that it provides a framework for complex systems design, a benchmark to aim for, and a bulwark against very poor design, still allowing room for innovation and creativity and not imposing too costly or cumbersome a design regime.

For a formal certification system to have a chance of success, early consideration must be given to the following

- Distinctions to be made and balance to be found between certification of attributes, performance, personnel, and process.
- Desirable degree of prescription or latitude for design.
- Identification, definition, agreement and validation of test methods, measures, criteria, and so on.
- Provision for flexibility and updating of requirements.
- Examination of trade-offs between value and the resources required for the certification process.
- Systems to certify the certifiers and certification systems.
- Communication of outcomes of the certification process in more useful terms than just pass–fail or yes–no judgements.
- Consideration of implications of non-conformance, and thus enforcement

It must be stressed once again that the process of certification can be of value even if we are at first unsure of, or unhappy about, the content. More than this, if we get the process right then content problems—in terms of appropriate requirements or missing tools or data for instance—can be rectified as part of the process being put into operation. We must remember though that the individuals who produce certification processes or who test and approve systems are themselves fallible, as also will be any intelligent systems built to help with certification. Perhaps this is the key issue for acceptance of certification—*Quis custodiet ipsos custodes?*

REFERENCES

Applied Ergonomics. (1995). Special issue on ergonomics standards.

Barnes, R. B., Orlady, H. W., &Orlady, L. M., (1996, April). *Multi-cultural training in human factors for transport aircraft certification.* Paper presented at the 3rd Global Flight Safety and Human Factors Symposium, International Civil Aviation Organization, Auckland, New Zealand.

Bongarra, J. P., (1997, June). *Certifying advanced plants: A US NRC human factors perspective.* Papers presented at theIEEE Sixth Annual Human Factors Meeting, Orlando, FL.

Dul, J., de Vlaming, P. M., & Munnik, M .J. (1996). A review of ISO and CEW standards on ergonomics. *International Journal of Industrial Ergonomics, 17,* 291–297.

Goom, M. K. (1993), An industrial view of MANPRINT. In E. J. Lovesey (Ed.), *Proceedings of the annual conference of the Ergonomics Society* (pp. 34–39). London: Taylor & Francis.

Green, R. (1990). Human error on the flight deck. *Philosophical Transactions of the Royal Society of London, Series B, 327,* 503–512.

Herman, L.,(1993). An international survey of human factors involvement in the nuclear industry. In E. J. Lovesey (Ed.), *Proceedings of the annual conference of the Ergonomics Society* (pp. 215-220). London: Taylor & Francis.

Longridge, T. M., (1997). Overview of the advanced qualification program. *In Proceedings of the Human Factors and Ergonomics Society 41st annual meeting,* (pp. 898-901).

McDaniel, J. W. (1996). The demise of military standards may affect ergonomics. *International Journal of Industrial Ergonomics, 18,* 339–348.

Jones, E. R., Hennessy, R> T. & Deutsh (Eds.) (1985). *Human factors aspects of simulation.* Washington D.C: National Academy Press.

Meister, D.,(1985). *Behavioral analysis and measurement methods.* New York: Wiley.

Meister, D. (1989). Conceptual aspects of human factors. Baltimore: John Hopkins University Press.

Ministry of Defence, (1992). *The MANPRINT handbook* (2nd ed.). London: Her Majesty's Stationary Office.

NATO. (1996). STANAG 3994 Al: The application of human engineering to advanced aircrew system

Norris, B.& Wilson, J.R., (1997). Designing safety into products. Product Safety and Testing Group, University of Nottingham. ISBN 0952257122.

Parsons, K., (1995). Ergonomics and international standards. *Applied Ergonomics, 26,* 239–247.

Reason, J. (1990). The contribution of latent human failures to the breakdown of complex systems. *Philiosophical Transactions of the Royal Society of London, Series B, 327,* 475–484.

Senders, J. W. & Harwood, K. (1993). Provocations: To certify or not to certify. *Ergonomics in Design, 1,* 8–11.

Stewart, T. (1998). Ergonomics standards: The good, the bad and the ugly. In (E. Lovesey & S. Robertson (Eds.) Proceedings of the Ergonomics Society annual conference. (pp. 3–7) London: Taylor & Francis.

Taylor, R.M.& MacLeod, I. S. (1994). Human problems with certification of man-machine systems. In S. A. Robertson (Eds.) Proceedings of the Ergonomics Society Annual Conference (pp. 161-166) London: Taylor & Francis.

Van Weperen, W. (1992). A hazard-oriented approach to product safety criteria. *Proceedings of ECOSA Workshop on Product Safety Research in Europe.*Amsterdam, July, 10-19.

Wilson, J. R. (1984) Standards for product safety design: A framework for their production. *Applied Ergonomics, 15,* 203-210.

4

Certifying Human–Machine Systems

Peter Hancock

University of Minnesota, USA

"You see, one thing is, I can live with doubt and uncertainty and not knowing. I think it's much more interesting to live not knowing than to have answers which might be wrong. I have approximate answers and possible beliefs and different degrees of certainty about different things, but I'm not absolutely sure of anything and there are many things I don't know anything about, such as whether it means anything to ask why we're here... I don't have to know an answer. I don't feel frightened by not knowing things, by being lost in a mysterious universe without any purpose, which is the way it really is as far as I can tell. It doesn't frighten me.

(Richard Feynman)

PREFACE

I want to preface this work with a brief account of its history. Through the very kind invitation of the organizers, I was able to attend the meeting on which this volume is based. Like others, I was asked to provide a contribution to be brought to the meeting. This originally appeared as 'certifying life' (Hancock, 1994a) and during the meeting I developed a second commentary on the problem of certification and legislation that was the topic of one of the individual workshops (Hancock, 1993b). Some time after the meeting, I revisited these works and combined them into a single chapter for a book I was writing entitled *Essays on the Future of Human-Machine Systems* (Hancock, 1997). I had divined, with the editors of this volume, that the original publication would unfortunately have limited circulation and as the questions of certification are so crucial, it might be beneficial to solicit a wider audience for the work. This offering is based on this synthesized effort with some brief requested revisions. I offer it for consideration in large part because it was written at such a traumatic time for me and raises issues that are, for me at least, still to be resolved.

PREAMBLE

Systems have three possible states. They are stable, transient, or failed. When a system is stable, no certification is necessary. When a system is in transition, no certification is possible. When a system has failed, no certification is needed. I argue, therefore, that certification is a palliative and an anodyne for societal concerns over the potential destruction that advanced systems can wreak. I further submit that the manifest need for certification is part of our occidental view that nature must be tamed, constrained, and controlled. As a consequence, certification is directly related to legislation and the simplistic worldview that such legislation promotes. It is unlikely that our cultural myopia will be excised easily. However, mutual coevolution and validation by nature itself will fulfill the argument for me; within the fullness of time.

INTRODUCTION

When my father died, I was some 40,000 feet above Iceland. I am today still unable to reconcile myself to the fact that I could not talk to him at length one final time. So when I saw him in Cheltenham Hospital's Chapel of Repose much of what I felt was anger and frustration diffusely directed. I realize now that part of that frustration had to do with life itself. As I stood in front of his body, I could not help but feel that he was only asleep. After all, he had not changed substantively since I had last seen him. However, physical appearance belied what I knew and what we all must eventually face ourselves: What had made my father my father had gone. As I started to write on the issue of certification, I realized that some doctor had been asked to certify that my father was dead. Indeed, many agencies required evidence (the death certificate) to remove him from the lists of the living. As you might imagine, and I hope you do not experience, the bureaucracy of dying is as obscene as the event is disturbing. As I sat in the Department of Health and Social Services, I pondered the comparison between my father's death and the demise of any system in society, biological or technical.

CERTIFYING FAILURE

My immediate thought was, why bother? Nothing in the process of certification was going to bring my father back and so the whole procedure seemed, to me, pointless. However, it became obvious that to society in general my father occupied a number of different roles and it was in respect of these roles that the process of certification inexorably rolled on. In system terms, my father had evolved from stable, through transient, to a failed state and the critical societal question was the establishment of the cause of that failure. As medical science seeks such cause through postmortems, we in human factors conduct accident investigations, essentially the same process.

The fundamental assumption is that knowledge of what went wrong last time will help us to avoid the same sequence of events leading to failure next time. With respect to my father, such reasoning is specious. There can be no next time. With respect to a theory of technical systems operation, such reasoning is also becoming more naive. Contemporary failures are multidimensional in terms of causation and therefore more idiographic in terms of character. Despite our continual attempts to extract or even impose patterns on failure, we are faced with the certainty that no two failures will ever be exactly the same. Hence, our search for patterns will devolve to ever higher metalevels of description until we provide the single parameter, unified field theory of failure; namely, "It broke!"

The alternative is dependent on the resolution of an empirical proposition concerning the demise or failure of natural ecological systems. Among others, Kaufman (1993) has suggested that there is a linear relation between the logarithm of the frequency of perturbing events and the logarithm of the severity of disturbing events in complex natural ecosystems. Reminiscent of Perrow's (1984) argument in relation to closely coupled systems, this proposition could imply that there is a lawful relation between the frequency of failure and the magnitude of failure in complex human–machine systems. If this law, founded in the application of nonlinear dynamics, does apply to technical systems as well as natural ecosystems, it would provide strong evidence that technical systems are natural and that they are subject to the same fundamental laws and constraints. In particular, it would imply that catastrophic failures of technical systems are the inevitable result of the magnification of minor forms of perturbation that occur all the time in complex structures and processes.

This vision has the dull cold specter of determinism hanging about it. Are failures inevitable? Are they progressively more idiosyncratic? Is the Markov chain of causation to grow like Scrooge's into a "ponderous heavy" construct? Are the subtleties and nuances of nonlinear dynamics to bedevil us as the weirdness of quantum mechanics has plagued physics? Is the end result of this the assertion that safety efforts are futile and certification a lost cause? Perhaps. However, science has done rather nicely in terms of prediction and we clearly still expect that our knowledge of regularities, sanely applied, can lead to safety gains and accident reduction. But always remember, at some level, we have to believe in regularity, because we humans at a fundamental level invent it. Therefore, we have to subscribe to the notion that the future is, at least partially, predictable from the past and therefore controllable. To subscribe to a radically differing version of this belief is to risk being labeled literally insane. Indeed, as

Schrodinger (1944) observed, "It is well-nigh *unthinkable* that the laws and regularities thus discovered should happen to apply immediately to the behavior of systems which do not exhibit the structure on which those laws and regularities are based".(emphasis mine).

At the heart of the schizophrenia of this position is our dissonance between a view of time as a linear dimension in which unique progression obviates exact repeatability versus time as cyclic phenomenon in which repetition and recurrence dominate (Toulmin & Goodfield, 1965). Our present *zeitgeist* is to believe that the future must be like the past in some way, but cannot be the past in a true sense. Laws that are structured from past experience should have a consistent influence in the future. This latter assumption is a belief,[11] not an empirically supportable statement, as is the predicate of regularity and consistency in a more subtle way.

I felt reasonably confident in asserting that the postmortem cause of my father's death was about as accurate as the most cursory of all accident investigations. In reality, doctors deal with death in much the same way they deal with illness. They do not have the time for exhaustive diagnosis of particular problems, hence they frequently treat symptoms by providing palliative agents of widespread capability that will cover the symptoms of the problem without ever necessarily identifying the specific cause. Cause of death is even less likely to receive detailed examination because the problem rarely proliferates. If it threatens others (e.g., contagious disease) or if there is some specific reason (e.g., homicide), they do what some analysts do anyway, passing the problem on to a specialist (e.g., a forensic pathologist). What is frequently not acknowledged is that because we do not fully understand the phenomenon of life, there are many cases in which we cannot always specify why life is extinguished. With respect to complex systems, as they grow less determinate in their actions (indeed as many such systems already are), their cause of death may become equally difficult to specify. Right now many professional medical personnel will acknowledge that some individuals die because they no longer wanted to go on living. Can we expect an analog of this in our machine systems? Could we, in all honesty, sign a certificate to that effect?

---

[11] *The foundation of these beliefs has been most eruditely articulated by Sheldon Glashow who in his New York Times article "We believe the world is knowable" (October 22, 1989) stated: "We believe the world is knowable, that there are simple rules governing the behavior of matter and the evolution of the universe. We affirm that there are eternal, objective, extrahistorical, socially neutral, external and universal truths and that the assemblage of these truths is what we call physical science. Natural laws can be discovered that are universal, inviolate, genderless, and verifiable. ... Any intelligent alien anywhere would have come upon the same logical system as we have to explain the structure of protons and the nature of supernovae. This statement I cannot prove, this statement I cannot justify. This is my faith."*

In summary, certifying death or failure is a losing cause at best and fundamentally irrelevant at worst. At a bureaucratic level there are many boxes to be checked and some superficial reasons why we need a paper that records demise. But as with the munchkin doctor in *The Wizard of Oz*, repeatedly asserting the absence of life is hardly an answer to the future of life. Certifying failure states in complex systems is similarly redundant. Postmortems identify a concatenation of circumstances that connote chains of interactive failures. A priori prediction of such failures has not been, and some would suggest, cannot be anticipated. The search for patterns in such failures will, inevitably, turn up some commonalties, because humans can turn up commonalties in the most diverse displays, including the fluffy clouds of the sky. However, prevention based on postmortem is inevitably a losing battle.

CERTIFYING STABILITY

If certifying failure is irrelevant, shouldn't we at least certify systems for their stable or everyday state of operation? Shouldn't we be able to assure ourselves that within the design parameters and operational envelope, the system reliably does all that we claim it does? In part this depends on what we mean by complex systems. Let us consider the nature of machines and consider indeterminacy in machines in the same manner we consider the potential for intelligence in machines. Some four decades ago, Scriven (1953) could be fairly unequivocal. He asserted that: "Machines are definite: anything which was indefinite or infinite we should not count as a machine". Today we cannot be as certain. As a result, Scriven's (1953) argument about the incompleteness of Godel's legendary incompleteness theorem is not without problem. However, the process of reasoning is instructive:

> Godel's theorem must apply to cybernetical machines, because it is of the essence of being a machine, that it should be a concrete instantiation of a formal system. It follows that given any machine which is consistent and capable of doing simple arithmetic, there is a formula unprovable-in-the-system - but which we can see to be true. It follows that no machine can be a complete or adequate model of the mind, that minds are essentially different from machines.

We understand by a cybernetical machine an apparatus which performs a set of operations according to a definite set of rules. Normally what it is to do in each eventuality; and we feed in the initial "information" on which the machine is to perform its calculations. When we consider the possibility that the mind might be a cybernetical mechanism we have such a model in view; we suppose that the brain is composed of complicated neural circuits, and that information fed in by senses is "processed" and acted upon or stored for future use if it is such a mechanism, then given the way in which it is programmed—the way in which it is "wired up"—and the information which has been fed into it, the response—the "output"—is determined, and could, granted sufficient time, be calculated. Our idea of a machine is just this, that its behavior is completely determined by the way it is made and the incoming "stimuli": there is no possibility of its acting on its own; given a certain form of construction and a certain input of information, then it must act in a certain specific way.

In arguing the mind cannot be like a machine, Scriven (1953) is limited in a number of ways. First, there is no rationale for suggesting that a mind can explore all possible states. That is, as we cannot know everything, it may well be the things we do not know that contain anomalies intrinsic to Godel's contention concerning unprovability. Second, the argument about seeing what is true but is unprovable in the system can rapidly become a tautology in which we ask how the seeing or realization is done. Thus the theoretical difference between mind and machine may be obviated by practical exigency. For the purpose of the argument here, we cannot then state all possible conditions within an operational envelope with certainty. What certification devolves to in this case is an assessment of probability. As a consequence, the heart of certification would seem to represent a customer warranty. For small individual objects, this interaction may be appropriate as the vendor and the customer in a capitalist society are frequently divorced in time and space. But suppose they were not. Suppose that you were a carpenter who had built perhaps a toy for your child or a working bench for yourself. Would you require certification from yourself? The answer is almost certainly not. In the same way, contemporary complex systems, although not the creation of one individual or bought by that individual, still possess the same character. That is, for systems such as the National Airspace System (NAS), society is at one and the same time both vendor and client. Even within this global perspective, it is frequently the same agency that operates a system that regulates and certifies it. Thus if certification is like insurance, society is its own carrier. However, unlike an individual insurance company, there is no higher level of society to appeal to, either in anger or in redress.

We would like to think that if all individual parts of a system were certified, then the overall system would itself be safe. This is bottom-up, wishful thinking. It is the *sine qua non* of design that objects and systems are created for stability of action and hence should be certifiable within the design space. Yet, here it is the combinatorial explosion of potential interactions, as much as nature's own test and evaluation of those interactions, that defeats the hoped-for assertion. I should note here that combinatorial explosion of interactions alone does not connote instability as represented by the transient states of operation. This is examined later.

More critically, what are we designing such systems for? It is the frequent observation of the more experienced members of the design community that you never get the opportunity to create complex systems from the groundup. Almost always they are evolutionary in nature in that new elements are added to older systems until the working environment is palimpsestreal. If this is the case, we will never be able to completely specify the parameters of a system that is itself chronically under-specified. More to the point, as we build systems that are beginning to cost trillions of dollars (e.g., the NAS, intelligent transportation systems), we will want them to deal not only with existing conditions but also with future anticipated and even unanticipated demands.

Hence, future complex systems must be generative and creative in exploring potential operational spaces to be cost-effective. In consequence, such systems must be underspecified, for not to do so would be to defeat their evolutionary capability and purpose. Systems that are intentionally under-specified cannot be certified for all phases of operation. Thus we arrive at an impasse. That is, the very systems that we seek to certify should, by design, defy certification! However if one seeks to justify certification for stable states of system operation, one will devolve to this paradox. *The paradox is that certification is a guarantee of future operation and it implies a predictive determinism about that future state.* If such deterministic foreknowledge could be achieved, the operation involved would be completely automatic and by definition not a complex system of the sort relevant here. However, as the future is conceived of as either partly deterministic or totally indeterminate, we want systems to adapt to unforeseen conditions and to explore strange new worlds, to justify their cost. Under neither circumstance is certification necessary or indeed feasible.

In his argument concerning the potentiality of machine intelligence, Turing (1950) examined the same issue from an inverted position and countered the argument that machines could not be intelligent because of the informality (or indeterminacy) of behavior. He indicated that:

> It is not possible to produce a set of rules purporting to describe what a man should do in every set of circumstances. One might for instance have a rule that one is to stop when one sees a red traffic light, and to go if one sees a green one, but what if by some fault both appear together? One may perhaps decide that it is safest to stop. But some further difficulty may well arise from this decision later. To attempt to provide rules of conduct to cover every eventuality, even those arising from traffic lights, appears to be impossible.

Given both the paradox of certification and the improbability of comprehensive future prediction, certification around stability appears a vacuous endeavor indeed.

CERTIFYING TRANSITION

If we do not need to or cannot certify failure and are excused from certifying stability, surely we have to explore certification in the intervening realm where systems fluctuate between stability and failure—the regions of transition. This appears most relevant, because it is the process of incipient failure and recovery from failure that represents the most critical active phase of operation. The problem again is one of predictability. That is, if a sequence of conditions prevails and a sequence of processes is in operation, a series of outcomes are guaranteed. However, when we step into transitional states, we enter regions that by definition provide increasing uncertainty.

I noted previously that the societal investment in large-scale complex technical systems implies that they should be generative and explorative. I extend this description to imply that such systems should also be skillful. I use *skillful* in a specific sense here. The context is one that has been used in examining adaptive systems (Holland, 1996). It has been posited that adaptive systems are structured in response to their initially experienced environmental contingencies. This means that adaptive systems, of which life is the pre-eminent example, grew at the edge of chaos. The latter condition is one in which the phase plane of operation devolves from a stable condition toward a chaotic one. It is at the edge of chaos that adaptation develops. Skill in this context is the ability to explore the edge of chaos and the advantages intrinsic to that region, without fallback to immalleability or trespassing into chaos itself.

By definition, systems in transition reside in the region between stability and chaos. Hence, certification of skillful systems in transition is to suggest that we can 'predict' the response of an adaptive system that has as its primary function coping with unanticipated conditions. The imperilment of such a procedure should now be clear. We cannot certify a system in such conditions. To do so would be to constrain the very ranges of response of a system that we want to be open and unconstrained to recover to a state of operational stability.

WHY CERTIFICATION?

If we are unable to certify systems in stable, transient, or failed states, why do we seek certification at all? It is my contention that certification grows from a societal need expressed in legislation and more particularly our occidental view of fault and derived litigation. In this last section, I want to compare the process of certification with how we develop and apply legislation, especially as it relates to behavior. Embedded in this parallel are analogical, metaphorical, and literal relations. Certification must, in principle, intersect with legislation at a national, international, and global level. Behind legislation, particularly that which applies to individual action, lie assumptions about moral and social responsibility and these are of course under continuous review and critique.

However, we accept that interpolated with such philosophical concerns and superimposed upon them, are human created laws. Law in this sense is not synonymous with law in a physical sense. There is continuous dispute about interpretation in legal circles and indeed judges, solicitors, lawyers, attorneys, barristers, and juries would be redundant if legal concepts were completely determined (this does not of course imply that scientific laws are themselves not without challenge). At a certain level, laws are empowered to control society. In a similar manner certification is viewed by some as an effort toward control over events and processes. If, in the face of an undetermined environment we cannot guarantee control, at least certification represents an expression of a desire for control. Whereas systems are faced with answering the untrammeled vagaries of an uncertain environment, humans can be somewhat more arbitrarily constrained in their behavior because legislation is arbitrary anyway. However, we can imagine situations where individuals engage in unintended actions that result in transgressions of the law (e.g., vehicle fatalities). We can also imagine circumstances in which we ask whether it is justifiable to transgress the moral basis of law (e.g., killing dictators). Therefore, the analogy between certification and legislation is partial because we seek certainty as an adjunct of control, but recognize that we cannot ubiquitously or even frequently achieve such an aim.

Control and certainty apply forward in time. That is, control is for a purpose and certainty statements always look to the future (as we believe the past to be determined anyway). However, the function of compliance is one predominantly of enforcement. In legal terms, compliance of behavior is achieved by the force of power and punitive action. The majority of legal compliance comes from self-reference and appropriate laws that recognize enlightened self-interest (e.g., traffic laws). Compliance in certification terms is similarly framed. There is much hope that self-compliance by professionals throughout the process means that self-interest will achieve the best possible result. However, certification can be used as a tool of penalization. In essence, we hope professionals in system design and certification behave like law-abiding citizens, not needing enforcement officials running around after them to ensure they behave properly. We fear that certification will be used as a bludgeon to cower individuals into compliance. The analogy is not complete because designers and system developers have to explore boundaries of what is known; however most individuals do not press the envelope of the law. The fallacious criticism of George Bernard Shaw is appropriate here: "Only fools and idiots seek to change society. Therefore all societal change is effected by fools and idiots".

At the heart of this fallacy resides the enlightened designer who seeks constructive ways to facilitate change. If compliance is an insurance function that oscillates between the past and the future, accountability is a historical process that seeks to attach individuals to the decisions and actions that they take. In law we can happily talk of diminished responsibility when individuals are unable to recognize the consequences of their actions. Similarly for certification we cannot indemnify completely a design as we cannot have complete knowledge of all potential future conditions. Therefore, certification is engaged at the end of the fabrication process and the beginning of the large-scale manufacturing process when the system is arbitrarily frozen for a moment in time, to imply that if all things remained without change, the system would occupy the best of all possible places in the best of all possible worlds. Accountability then can only be used when in the process of design test and evaluation, something was neglected that *could have been known at the time or even actually was known by the developers at the time of certification*. One of the most important facets of certification in complex systems is the problem of technical evolution and nonstationariness. That is, we are trying to hit a moving target with something, that is inherently static, and therefore time-locked. Systems evolve and change quickly and it is dauntingly difficult to keep certification going at the same pace. There is a parallel in law whereby for technologies like DNA and computer systems, legislation, literally, cannot keep up. Hence, the old question emerges: How to provide stability against the background of instability? For some, the greater the instability the greater the need for stability. For others, instability represents opportunity and the clear need for flexibility. These need not be mutually exclusive aims in adaptive systems.

I hope I have demonstrated strong parallels between certification and legislation. They each represent human attempts to place arbitrary frames of reference on diverse and unpredictable entities. We should recognize that such frameworks by constraint are arbitrary and therefore cannot provide a perfect fit. However, for most people, both certification and legislation appear preferable to their absence. In a real sense, science is also a member of this movement, which is part of the human appeal for comprehensibility in the face of the incomprehensible. In that way, certification seeks nothing new but is part of a time-honored tradition that begins when children must first make some sense of the blooming, buzzing confusion with which they are presented.

In drawing parallels between certification and legislation I do not desire to advocate either. I am still naive enough to be an optimist. In the end we can come down to one's opinion of people. Do we have to generate minimum standards of behavior and conduct against which to hold the lowest common denominator? Or, can we aspire to self-generated standards that inspire the application of a highest common factor? The aged, the jaundiced, and the worldly wise will sadly shake their heads and quietly admit to the former pessimism. Especially in light of the apparently immovable behemoth of social institutions and procedures, such as meeting the costs and the time lines of a technical project. However, we have seen massive social change in the past decade and if survival is a requirement, pluralism is essential. It is all a matter of social self-enlightenment. Sadly, to many we seem to be headed in the opposite direction. Although life has become more comfortable, at least to those of us who rely on and benefit from the higher realms of technology (although still a minority of the planet's population), it does not seem that life has become more fulfilling. Indeed, many of the turn-on, tune-in, drop-out generation still hanker after the elysian escape. However, as I started with my father, I end with a phrase that he was most fond of: "It only needs a small candle to change a large darkness." I think the nature of how we deal with our technology and the promises and indemnities with which we vest it directly relate to our view of the future. I like Kennedy's observation: "Some see the world the way it is and ask why? I see the world the way it can be and ask, why not?" Let us hope that seeing the world the way it is does not blind us to the world that can be, before it is too late.

HOPE FOR THE FUTURE

I have presented a discussion that has used an analogy between certification and life. Life is a successful, adaptive, complex system that is found within an environment but can itself make changes in that environment. I have suggested a parallel between the failure of a system and death. By extension, the parallel holds for health (stable states) and disease and trauma (transition states). I have not articulated these latter conditions in as much detail. I have, however, suggested that certification of stable and transient system states is a relatively futile exercise, as I posit that the very systems we are focusing on are ones that imply the need for open, explorative, and nondeterministic functions. Certification of failure is a time-honored societal endeavor to provide information on how to obviate failure in future systems. In deterministic systems with high frequency of occurrence in the same fundamental state (e.g., DC-10s), this can be a useful function. For one-off, large-scale systems of progressive indeterminacy such certification becomes impossible. Consequently, I suggest that certification serves a social role in activities such as apportioning blame or accountability. The latter function is a societal palliative for the fears that such indeterminacy brings. I further submit that this is an occidental preoccupation that stems from the notion of controlling and taming nature. I take all other aspects of certification to be lowest common denominator insurance.

In reviewing the preceding, it might appear to be a rationale for doing nothing with respect to the design, test, and evaluation of systems and to fatalistically accept the uncertain outcome that nature chooses to provide. I reject this fatalism wholeheartedly. What is objected to is an attitude of mind that proposes that we can know all the states of complex systems we have already created and are creating by the moment. Therefore:

- I advocate a greater exercise of humility, especially with respect to an understanding of the influence and effect of the technology we create.

- I advocate a change in attitude from the legalistic blame we seemed destined to fix, to a recognition of societal responsibility for the things we collectively desire to have and subsequently construct.
- I advocate a recognition of the explorative and adaptive nature of ourselves and by extension the technology we create to extend ourselves and our range of capability.
- I advocate the need for the immediate integration of those whose innovative work is enlightening complex adaptive system operation with those who design, test, and evaluate such technical assemblies.
- Finally, I advocate a strong thrust of research in the area of 'skillful' systems which possess an acknowledged degree of skill in recovering to stability.

In sum, I advocate the replacement of the procedures of certification with the exploration of training skill in complex human–machine systems. I am not foolish enough to believe such recommendations are likely to result in immediate change. However, time and nature are powerful things, and indeed, perhaps the same thing, and in the end they exert their influence.

REFERENCES

Hancock, P. A. (1993a). Certifying life. In:J. A. Wise, V. D. Hopkin, & D. Garland. (Eds.); *Human factors certification of advanced aviation technologies* (p. 207–214). Daytona Beach, FL: Embry-Riddle Aeronautical University Press.

Hancock, P. A. (1993b). Certification and legislation. In J. A. Wise, V. D. Hopkin, & D. Garland. (Eds.);. *Human factors certification of advanced aviation technologies.* (p 35–38). Daytona Beach, FL: Embry-Riddle Aeronautical University Press..

Hancock, P. A. (1997). *Essays on the future of human machine systems.* Eden Prarie, MN: Hancock/Banta.

Holland, J. H. (1996). *Hidden order: How adaptation builds complexity.*New York: Addison-Wesley.

Kaufman, S. A. (1993). *The origin order.* Oxford University Press: New York.

Perrow, C. (1984). Norma*l accidents: Living with high-risk technologies.* Basic Books: New York.

Schrodinger, E. (1944). *What is life?* Cambridge, UK: Cambridge University Press.

Scriven, M. (1953). The mechanical concept of mind, *Mind, 62, 230–240.*

Toulmin, S., & Goodfield,. J. (1965). *The discovery of time.* New York: Harper & Row.

Turing, A. M. (1950). Computing machinery and intelligence. *Mind, 59,* 433–460.

5

Human Factors Requirements in Commercial Nuclear Power Plant Control Rooms

Lewis F. Hanes
Consultant, USA

Establishing a human factors engineering (HFE) certification program for advanced aviation technologies and for complex systems in other industries presents many challenging questions. In developing answers to these questions, it is useful to learn from the experiences of organizations that have a human factors certification program already in place. The U.S. Nuclear Regulatory Commission (NRC) has such a program. The purpose of this chapter is to present a brief description of this activity.

INITIAL HUMAN FACTORS REQUIREMENTS

NRC interest in human factors issues associated with nuclear power plant (NPP) control rooms came into existence following the Three Mile Island Unit 2 accident in 1979. The NRC Action Plan (NRC, 1980) developed in response to the incident required NPP licensees and license applicants to perform detailed control room design reviews (DCRDRs) to identify and correct HFE design deficiencies. These reviews included assessment of control room layout, adequacy of the information provided, arrangement and identification of the important controls and displays, usefulness of the alarm system, information recording and recall capability, lighting, and other human factors considerations that have an impact on operator effectiveness and plant safety (Ramey-Smith, 1985).

The NRC (1981) issued NUREG–0700 (NRC, 1981) for use by the NRC staff in reviewing the DCRDRs conducted by the utility licensees.

NUREG–0700 consists of human factors guidelines adapted to NPP control rooms and additional guidelines as required (Ramey-Smith, 1985). The formal NRC technical and documentation requirements for the DCRDR are contained in NUREG–0737, Supplement 1 (NRC, 1982). The NRC required that licensees and applicants perform a DCRDR on their designs, consisting of the following (Ramey-Smith, 1985):

- Establishment of a qualified multidisciplinary review team.
- Function and task analysis to identify control room operator tasks and information and control requirements during emergency operations.
- Comparison of display and control requirements with control room inventory.
- A control room survey to identify deviations from accepted human factors principles.
- Assessment of human engineering discrepancies (HEDs) to determine which HEDs are significant and should be corrected.
- Selection of design improvements.
- Verification that the design improvements will provide the necessary correction and will not introduce new HEDs.
- Coordination of control room improvements with changes from other programs, such as operator training and upgraded emergency operating procedures.

The NRC pointed out that NUREG–0700 and similar NRC documents did not have to be used by utilities. It is believed, however, that every utility applied these NRC reports in performing the DCRDRs. Each utility prepared a DCRDR report that included proposed control room changes and the schedule for change implementation. The NRC human factors staff reviewed each DCRDR submitted and conducted on-site audits of control rooms. The DCRDRs are all completed with the final NRC safety evaluation report issued in January 1996 (Eckenrode & West, 1997). Many changes to control rooms were implemented based on the results of applying the DCRDR process. Several utility personnel and consultants involved in these reviews (Cooley, 1997; Herrin, 1997; Feher, 1997) provided observations regarding the value of the reviews. Some of the general observations are as follows:

- Reviews and resulting control room changes were expensive.
- Almost no objective evaluations were performed to determine if safety was improved.
- Most human factors professionals involved are of the opinion that plant safety is improved (probably based in part on the nature of the HEDs observed and the corrective actions taken).

REQUIREMENTS FOR NEW DESIGN

Utilities have begun to introduce improvements (upgrades) in control room human–system interfaces (HSIs) requiring NRC review. In addition, the utility industry has underway a program to develop standardized advanced commercial NPP designs. The NRC staff has three main documents to support the HFE review process and to provide criteria for evaluation: NUREGs 0800, 0711, and 0700, Rev. 1.

The Standard Review Plan, NUREG–0800, Draft Report for Comment (NRC, 1996b), is applicable to both review of existing control room upgrades and new control room designs. The review addresses the design process and the product of applying the process. NUREG–0800 provides acceptance criteria for 10 review areas: HFE program management, operating experience review, functional requirements analysis and allocation analysis, task analysis, staffing, human reliability analysis, HSI design, procedure development, training program development, and verification and validation. NUREG–0800 identifies NUREGs 0711 and 0700, Rev. 1 as the sources of specific details.

NUREG–0711 (NRC, 1994), the HFE Program Review Model, describes the HFE program elements required to develop an acceptable detailed design specification. The purpose of NUREG–0711, as reported by Bongarra (1997), is to ensure that: the HFE has been integrated into the plant development and design; (2) the HSI's, procedures, and training reflect state-of-the-art human factors principles and satisfy all other appropriate regulatory requirements; and (3) the HSI's, procedures, and training promote safe, efficient, and reliable performance of operation, maintenance, test, inspection, and surveillance tasks. (p. 6-20)

NUREG–0700, Rev. 1 (NRC, 1996a), the Human–System Interface Design Review Guideline, provides guidance for the review of HSI designs and for performance of reviews undertaken as part of an inspection or other type of regulatory review involving HSI design or incidents involving human performance. The guidance consists of a review process and HFE guidelines. Three companies have developed advanced NPP designs and have submitted applications to the NRC for design certification (Bongarra, 1997). Title 10, Code of Federal Regulations (CFR), Part 52 was issued to streamline the plant licensing process for these new designs. The licensing process of Part 52 consists of a final design approval by the NRC followed by a standard design certification that is issued as a NRC rule. Utilities have the opportunity to purchase the standard design and utilize it as already approved by the NRC. To ensure that an as-built plant conforms to the standard design certification, inspections, tests, analyses, and acceptance criteria (ITAAC) must be specified as part of the standard design certification. After certification, the NRC will ensure that the design has met the ITAAC. To obtain a standard design certification under Part 52, a plant designer must submit a Standard Safety Analysis Report (SSAR) to the NRC for review. The NCR's review of the SSAR is issued as a final safety evaluation report, which will form the basis for the final design approval.

It is stated in Title 10, CFR, Part 52 that an application for design certification must comply with the requirements contained in 10 CFR 50.34 (f). It is this rule that establishes that applicants for design certification must address HFE. The NRC staff evaluates the HFE material submitted as part of the certification process for new plant designs. The review process is very different from the DCRDRs evaluated in the past. A major reason is that detailed control room and instrument design information on which to make a safety determination is not available. Due to changing technology, much of the detailed design will not be completed prior to the issuance of design certification. Therefore, the NRC is performing the design certification evaluation using as guidance an implementation process described in NUREG–0711. The NRC review criteria also require definition of the minimum controls, displays, and alarms needed to carry out operator actions associated with performing emergency operating procedures and other risk important operator actions identified from Probabilistic Risk Assessment/Human Reliability Assessment analyses (Bongarra, 1997). In addition, the applicant must submit ITAAC/Design Acceptance Criteria that will ensure that the HFE design process is properly executed.

CONCLUDING REMARKS

The NRC has developed processes to review HFE associated with NPP control room designs and instrumentation. One process was developed and used to evaluate existing detailed designs following the 1979 accident at the Three Mile Island Unit 2 plant. A more recent process was developed in response to control room upgrades being submitted for NRC approval, and the NRC–nuclear industry effort to obtain certification for standardized advanced commercial NPP designs. Because of changing technology with the advanced designs, the detailed control room and instrumentation designs are not available for evaluation. Therefore, the NRC has developed a method to assess the HFE process used to develop the detailed designs.

NRC staff, consisting of human factors professionals sometimes supported by consultants, reviews and evaluates submissions by design groups. It approves, approves with comments, or disapproves the submissions. The NRC, in deciding whether to certify the overall plant design, considers results of this effort.

The NRC has many years of experience in certifying NPP designs. The NRC has also learned a great deal about how to implement a comprehensive design certification program. Aviation and other industries should study in detail the NRC lessons learned as they evaluate the need for and desirability of HFE certification.

REFERENCES

Bongarra, J. P., Jr. (1997). Certifying advanced plants: A US NRC human factors perspective. In D. Gertman, D. Schuman, & H. Blackman (Eds.), *Proceedings of the 1997 IEEE Sixth Conference on Human Factors and Power Plants.* (pp. 6-18–6-24.). New York: IEEE.

Cooley, S. H. (1997). Been there, done that. In D. Gertman, D. Schuman, & H. Blackman (Eds.), *Proceedings of the 1997 IEEE Sixth Conference on Human Factors and Power Plants.* (pp. 3-29 – 3-32). New York: IEEE.

Eckenrode, R. ,& West, G., Jr. (1997). Chronology and key NRC activities related to the detailed control room design reviews. *In* D. Gertman, D. Schuman, & H. Blackman (Eds.),*Proceedings of the 1997 IEEE Sixth Conference on Human Factors and Power Plants.* (pp. 3-19 – 3-21) New York: IEEE.

Feher, M. P. (1997). Detailed control room design reviews—were they worth the effort? In D. Gertman, D. Schuman, & H. Blackman (Eds.), *Proceedings of the 1997 IEEE Sixth Conference on Human Factors and Power* Plants. (pp. 3-19 – 3-21.). New York: IEEE.

Herrin, J. L. (1997). Duke Power Company's detailed control room design review. In D. Gertman, D. Schuman, & H. Blackman (Eds.), *Proceedings of the 1997 IEEE Sixth Conference on Human Factors and Power Plants.* (pp. 3-27 – 3-28). New York: IEEE.

Ramey-Smith, A. (1985). Nuclear power plant control room design reviews: A look at progress. *In Proceedings of the Third Conference on Human Factors and Power Plants* (pp. 117-118).

U. S. Nuclear Regulatory Commission. (1980*). NRC action plan developed as a result of the TMI-2 accident* (NUREG–0660). Washington, DC: Author.

U. S. Nuclear Regulatory Comission. (1981). *Guidelines for control room design reviews* (NUREG–0700). Washington, DC: Author.

U. S. Nuclear Regulatory Comission. (1982). *Clarification of TMI action plan requirements - requirements for emergency response capability* (NUREG–0737, Supplement 1). Washington, DC: Author.

U. S. Nuclear Regulatory Comission. (1994). *Human factors engineering program review model* (NUREG–0711). Washington DC: Author.

U. S. Nuclear Regulatory Comission. (1996a). *Human-system interface design review guideline* (NUREG–0700, Rev. 1). Washington D.:Author.

U. S. Nuclear Regulatory Comission. (1996b). *Standard Review Plan for the review of Safety Analysis Reports for nuclear power plants* (NUREG–0800, Draft Report for Comment). Washington, DC: Author.

6

A Critical Component for Air Traffic Control Systems

Earl S. Stein
Federal Aviation Administration , USA

Although systems break down due to mechanical failure, human error is often present to one degree or another (Nagel, 1988). The operator carries the responsibility and blame even though, from a human factors perspective, designers may have built error potential into the system. Human factors are not a new concept in aviation. The name may be relatively new, but the issues related to operators in the loop date to the industrial revolution and, later, to the aviation buildup for World War I. Poor selection and training led to drastically increased personnel losses. Hardware design became an issue later, but researchers focused early efforts on increased care in pilot selection and training. Personnel selection and training became very sophisticated, producing better person–machine systems. Also, aircraft designers began taking display and control human factors seriously. Despite this, however, they fielded systems such as the multipointer altimeter, which was difficult to read even when the crew were not under load and almost impossible when they were (Fitts & Jones, 1947-1960).

The National Airspace System (NAS) has become more complex since World War II. There is more demand on limited airspace, and designers are creating systems that can be difficult to use and maintain. Events leading to errors and accidents often involve both ground and airborne sides of the system. In addition to cockpit crew operations, the modern civil and military airspace system includes airport ground operations, air traffic control, and maintenance of both air and ground hardware and software. Airway facilities personnel in the Federal Aviation Administration (FAA) handle the ground maintenance of navigation aids. Airlines and aircraft owners must maintain their aircraft to standards subject to review by FAA inspectors. These activities focus on hardware and seldom question how human beings manage to accomplish them.

HUMAN FACTORS IN COMPLEX SYSTEMS

Developers have often created systems in which the operator was the last to know what was going on. When the designers realized they had a problem, they reached out to human factors professionals. It is not unusual for the designers to ask human factors specialists for a better way to select and train because the hardware and software designs are already frozen.

There are lessons from history that apply to systems development. Thomas (1985) described the development of air traffic control over a 7-year period in the 1950s and 1960s. He noted the difficulty in transitioning from the older broadband radars to the more modern narrowband digital systems. Design and development did not involve the users. Building new systems without taking users' needs and abilities into account is inherently risky.

CERTIFICATION

Before a new piece of equipment is installed and used or a piece of equipment that was taken out of service for maintenance is placed in NAS service, it must be certified. That is, a piece of equipment such as radar must be certified to be performing its intended function acceptably and within specified tolerances before it is used. Of course, this does not mean the equipment is well designed, user friendly, or even maintainable. It simply means that the required function is being accomplished.

Certification is a legalistic term. It implies that the certifying agency has the power to determine if a system can be used and under what conditions. Certification implies sound methods that have met the tests of time, reliability, and validity. It suggests protection of the public from hazards generated by systems that have been poorly designed and from operators who are unqualified or unable to perform. FAA Order 6000.39 (FAA, 1991a) defines certification to mean that the system is capable of providing functions needed by the user. It emphasizes that the system must be capable of providing an advertised service. The order does not describe any requirement for human factors involvement, which is either assumed or ignored by the authors without specification.

This process raises human factors issues. If a piece of equipment, a facility, or a system is accomplishing its intended function, is there a place for human factors in the evaluation leading eventually to certification and is it going to enhance this process? What are the criteria to be applied that say, "This piece of equipment or system is working better now (i.e., safer) than it would have if human factors had not been involved?" How might such a process include not only newer systems in which there is the possibility of human factors involvement but also older systems in which chances are high that human factors did not play a significant role in development?

As suggested earlier, human factors is not directly involved in the legal aspects of certification. However, if a system is to be all it can be, human factors should be in the overall system development cycle from the beginning of the process. This will increase the probability of user-friendly, productive systems that are ready to go though the final step of legal certification.

In the United States, the FAA has the responsibility for certifying aircraft and personnel in aviation. It also certifies its own equipment and the staff who control and maintain the NAS. Subject matter experts who are flight examiners, controllers, and hardware systems specialists handle most of this legal requirement. In the past, human factors personnel existed in the background, advising those who would listen when they actually certified an aircraft or other system. Generally, certifiers do not ask for human factors input.

There have been some exceptions to this. In 1984, developers of the Automated En Route Air Traffic Control System commissioned an empirical validation of a construct called *workload probe*. The probe theoretically predicted controller workload up to 20 minutes in advance of weather and anticipated traffic. Human factors specialists tested it using simulation at the FAA William J. Hughes Technical Center (Stein, 1985). They collected measures of real-time controller workload using a scale called the air traffic workload input technique (ATWIT). The results of the workload predictions from the probe significantly correlated with participants' self-ratings using ATWIT and with ratings by over-the-shoulder observers, demonstrating the potential usefulness of workload probe. This empirical testing is a potential model for future systems evaluation.

Human factors research like this has the support of Congress. The Aviation Safety Research Act of 1988 increased awareness of the possibilities of human factors in the NAS. It required the FAA to expend a finite portion of its annual budget on human factors related to systems under development. The *National Plan for Aviation Human Factors* (FAA, 1991b) was a product of this law. This was a very comprehensive document that theoretically defined the human factors research required for the present time and the foreseeable future. The plan outlined FAA standards of system maintainability, but it was notable that it did not address any issues related to certification of those standards or, for that matter, to the test equipment that personnel will use to accomplish maintenance tasks. It did imply, however, that additional work would be necessary to evaluate such factors as maintenance documentation and the approaches to diagnostic support requirements. Proposed solutions included the development of handbooks, checklists, and special courses. Since human factors personnel wrote the plan, the FAA has initiated some of the proposed improvements. For example, under the sponsorship of the Office of the Chief Scientist for Human

Factors, the FAA William J. Hughes Research Development & Human Factors Laboratory did produce the *Human Factors Design Guide* (Wagner, Birt, Snyder, & Duncanson, 1996) for the Airway Facilities organization. The laboratory has distributed the guide widely in both hard-copy and CD-ROM versions.

The FAA (1995), in cooperation with other agencies, developed a revised and consolidated plan. It stated the following:

> Human-centered automation research focuses on the role of the operator (active or passive) and the cognitive and behavioral effects of using automation to assist humans in accomplishing their assigned tasks for increased safety and efficiency. The research in this area addresses the identification and application of knowledge concerning the relative strengths and limitations of humans in an automated environment. It investigates the implications of computer-based technology to the design, evaluation, and certification of controls, displays, and advanced systems (p.12).

Air traffic control is constantly evolving and will require evaluation and certification of new systems.

Air traffic control specialists and operational test and evaluation personnel traditionally accomplish certification of air traffic control systems. In the past, human factors personnel did research and development of new systems and, to some extent, operational test and evaluation. However, human factors is not considered a major element of the testing process, but rather a necessary adjunct, or, from a program manager's perspective, a less-than-necessary component of the process. Program managers are often told to build and field new systems at cost and within very tight time schedules. Human factors is postponed or dropped from the schedule as part of this pressured environment. Systems developers justified this view because, when they built systems in the past without adequate human factors support, analysis of prototypes often uncovered unforeseen person–machine issues even if it was late in the development cycle, and that was enough for them.

Introducing human factors either in parallel or as a precursor to certification requires a cultural change of the magnitude suggested by the Department of the Army MANPRINT Program (Booher, 1990). In the introduction of his book, Booher writes:

> People are both the cause and the solution. People are both the benefactors and the victims. Through human error in design, operation, or repair of machines, others are hurt, killed, or made unhappy or, at the least, inconvenienced. On the other hand, it is through human intelligence and unique human skills that equipment, organizations, and knowledge-enhancing products are designed and operated effectively, efficiently, and safely. (p. 2)

Booher saw the solution to these issues as a reorientation from hardware to people.

The MANPRINT philosophy suggests that there is a long-term payback for good human factors in the initial development of a system. This is a life-cycle approach to new technology. One of the problems identified by Booher (1990) is that the benefits of early investment in good design for long-term system reliability do not usually accrue to the program managers personally because they move on to other developmental efforts. The key to success may be the education of all personnel involved in the process. This sounds reasonable in principle but is not easy to implement in practice.

HUMAN FACTORS ROLES

Warm and Dember (1986) described their concerns over systems that are designed in such a way that there may be attention breakdowns to the point that operators are no longer "awake at the switch." In such a situation, whether in aviation or not, the system is operating without any supervisory control. The fact that this can and does happen became clear several years ago when the Nuclear Regulatory Commission closed the Peach Bottom Power Plant in Pennsylvania. Inspectors found operators literally asleep at their stations.

System performance is the key. The human is a critical component of the system that cannot and should not be equated with a piece of hardware. There are issues in which machines do not become involved such as motivation, esprit de corps, fatigue, and human information processing. All of these factors can have an impact on both human and system performance. Integrated goals other than the global desire to improve safety and performance while reducing operator workload are not always apparent in system design. These are admirable yet nonspecific goals that lend to the complexity of defining both system and individual performance.

Can we define performance and use it as a criterion for evaluation during development and operational testing? Past experience indicates that this is often a moving target that seems to progress as the system evolves. It is further complicated by moving beyond a general definition of performance with a qualitative description of what constitutes good performance. This involves subjective decisions by subject matter experts who may well have differences of opinion. Researchers have recently demonstrated both the difficulty and possibilities of moving subject matter experts toward some sort of consensus so that their performance ratings have some reliability (Sollenberger, Stein, & Gromelski, 1997). Given the right training, supervisory controllers can learn to focus on the same behaviors and set their personal issues aside to develop reasonably reliable performance ratings. This can be a powerful tool for system evaluation.

In the process of evaluating a system, there are very few standards that apply consistently.   Simply meeting the minimum requirements under MIL–STD–1472D (Department Of Defense, 1981) may not be enough to adequately evaluate a specific application. For example, 1472D is purposefully vague when it comes to human workload. It says that we should not overload the operator. What constitutes an overload would vary from one application to another and, for that matter, from one operator to another.

For ground systems, researchers will have to expend considerable effort to identify and validate suitable metrics that not only meet the criteria necessary for good measurement but have obvious relationships to performance of the systems being evaluated.   The bottom line of any system should be how it really performs, either in prototype or, preferably, in high-fidelity simulation before prototype testing is begun in the field.

Testing requires empirical and high-quality measurement, and, to date, there is no general agreement on either the measures themselves or what constitutes acceptable performance. This is especially true when a system is under development to replace one that is currently operating. Systems will evolve, and change will be gradual rather than meteoric. This means that measures looking for added value based on performance or safety improvement will be looking for subtle anticipated differences. They will have to work harder to find them. The more subtle the anticipated differences, the more difficult it will be to demonstrate them using conventional measurement and statistical tools. Although empirical testing offers the evaluators an opportunity to reach out for what could be a better estimate of operational reality, it does have its drawbacks: time and cost. It takes longer and it costs more to follow a philosophy like MANPRINT in which human factors are integrated into the developmental cycle and evaluators accomplish empirical evaluation whenever possible (at least in the ideal model of the philosophy).

Even if human factors experts do not have all the answers today and all the tools for tomorrow, they can still be of more help than they were in the past if human factors enters the development cycle sooner. These professionals need to develop groundside and improve airside measurement. This will not happen overnight, despite the belief by some that we could do it very rapidly. It seems that developers invite human factors professionals' participation and advice late in the developmental cycle. Fortunately, we have a tool set that, if not perfect, at least is adequate to assess feasibility, usability, and user preferences. Preferences are becoming increasingly important as user groups are making their needs and concerns known to anyone who will listen and even to those who do not initially see the role of the user in system development.

The question remains concerning the role of human factors in the certification process. As indicated earlier in this chapter, certification is a legal term. This chapter does not take a position concerning whether or not we should be directly involved in the legal certification of systems. It does take a very clear position that human factors should be involved throughout the life cycle of systems. Those handling the legal aspects of putting systems in the field should acquire all the human factors data they can to make the best possible decisions on the degree to which a system can perform its function safely and expeditiously. We in the human factors community have the obligation to speak out and to do our best to provide the best measurement and data possible. When others make certification decisions they should base their conclusions on an accurate picture of how the system really works with people in the loop.

Human factors professionals must never promise more than they can deliver and must deliver all that they promise. To do this, they have the dual role of helping systems developers avoid the obvious errors of design and constantly trying to develop new and better measurement tools. Measurement tools must be reliable and valid against systems goals. Measures must also meet the test of face validity if stakeholders are to accept them. Human factors experts and the products they generate have to make sense to the customers they serve. These include the users and the developers. Human factors professionals try to bring them together based on mutual interest and need.

REFERENCES

Booher, H. R. (1990). *MANPRINT: An approach to systems integration.* New York: Van Nostrand Reinhold.

Federal Aviation Administration. (1991a, August 8). *Maintenance control center operations concept.* (Order No. 6000.39). Washington, DC: Author.

Federal Aviation Administration. (1991b). *The national plan for aviation human factors* (draft). Washington, DC: Author.

Federal Aviation Administration. (1995). *National plan for civil aviation human factors: An initiative for research and application.* Washington, DC: Author,

Fitts, P. M. & Jones, R. H. (1960). Analysis of factors contributing to 460 "pilot error" experiences in operating aircraft controls. in H. W. Sinaiko (Ed.); . *Selected papers on human factors in the design and use of control systems.* (pp. 332-358) New York: Dover. (Original work published 1947)

Department Of Defense. (1981). *Human engineering design criteria for military systems, equipment and facilities* (MIL-STD-1472). Washington, DC: Author.

Nagel, D. C. (1988). Human error in aviation operations. In E. L. Wiener and D. C. Nagel (Eds.), *Human factors in aviation.* (pp 263-303). San Diego, CA: Academic Press.

Sollenberger, R., Stein, E. S., & Gromelski, S. (1997). *The development and evaluation of a behaviorally based rating form for the assessment of air traffic controller performance* (DOT/FAA/CT–TN96/16). Atlantic City, NJ: DOT/FAA Technical Center.

Stein, E. S. (1985). *Air traffic controller workload: An evaluation of workload probe* (DOT/FAA/CT–TN84/24). Atlantic City, NJ: DOT/FAA Technical Center.

Thomas, D. D. (1985). ATC in transition, 1956-1963. *Journal of ATC,* 30–38.

Wagner, D., Birt, J. A., Snyder, M. A., & Duncanson, J. P. (1996). *Human factors design guide for acquisition of commercial-off-the-shelf subsystems, non-developmental items, and developmental systems* (DOT/FAA/CT–96/1). Atlantic City, NJ: DOT/FAA Technical Center.

Warm, D. D., & Dember, W. N. (1986). Awake at the switch. *Psychology Today, 20*(4), 46–53.

7

Certification of Tactics and Strategies in Aviation

Hartmut Koelman
European Organization for the Safety of Air Navigation, Belgium

This chapter takes a close look at *layered operational concepts* in Air Traffic management (ATM) and flight operations. The aim is to expose common principles behind the tactics and strategies governing the operation of complex goal-oriented (self-optimizing) systems such as the air transport system. That analysis is developed into a subject called *planning theory*, which is subsequently discussed in the context of certification and human factors.

TYPICAL OPERATIONAL CONCEPTS

An operational concept is a model that describes the dynamics of managing an operation. It is expressed in terms of planning and execution responsibilities, control loops, and operational procedures. The operation is described for the composite human–machine system.

Traditionally in a composite human–machine system, the human still carries ultimate responsibility for the satisfactory performance of the composite system. Hence, the planning and execution responsibilities, control loops and operational procedures—although supported by machines (automation)—are primarily a human factors concern.

Operational Concept for ATM

If we look at the way ATM is organized today and is foreseen to be organized in the coming 20 years, we see the use of planning layers with different planning horizons (Eurocontrol, 1992, 1999). The general objective for each layer is to deliver an acceptable situation to the next layer. Each layer works as a filter for the following one. This filtering strategy defines specific roles for each layer, with the higher (strategic) layers addressing a general scope and a long planning horizon, and the lower (tactical) layers concentrating on specifics and short planning horizons.

A typical layering for ATM is as follows:

- Development of operational concepts, standards and recommended practices on a worldwide and regional scale (10–15 years ahead): This serves to provide guidance for system procurement actions and aircraft mandatory carriage requirements, items that typically have lead times of 5 to 10 years.
- Planning the renewal and upgrade of infrastructure on regional and subregional scale (5–10 years ahead): deals with site construction, development and procurement of hardware/software, and strategic human resource planning.
- Strategic Airspace Management (ASM) on a regional scale (several years ahead): This establishes basic strategies for air traffic service route networks (ARN), airspace use (segregation or flexible), and traffic segregation.
- Strategic Air Traffic Flow Management (ATFM) on regional scale (up to 1 year ahead): Strategic ATFM activities are intended to resolve major demand–capacity imbalance problems and generally concern summer traffic flow. They result in a traffic orientation scheme based on published flight schedules, airspace structure and system structure and capacity. After processing, estimates of traffic loads over any navigation point or air traffic control (ATC) sector can be provided. Discussions are initiated with all the partners concerned (states and aircraft operators), starting each year in October and ending in February. The outcome of this planning is the traffic orientation scheme, which dictates the routes to be used by operators when planning flights from specific departure areas to specific destination areas during the coming summer (Martin, 1993).
- Pretactical ASM and ATFM on a regional scale (1 day ahead): Pre-tactical activity is directed at the specific situation 1 day ahead. On the basis of updated demand data (incorporating repetitive flight plans [RPLs] and last-minute changes notified by aircraft operators), archived data of traffic situations on a similar day in the recent past, and taking into account the latest information about temporary airspace reservations and capacity in the area control centers, a pretactical ATFM plan for the coming day is developed. This plan, which defines the restrictions to be applied to traffic flows on the day concerned, is

published every day around noon in the form of an ATFM notification message (ANM) and is dispatched to thousands of addressees (air traffic services and aircraft operators). The ANM describes in a single message all the tactical ATFM measures that will be in force on the following day (Martin, 1993).

- Tactical ATFM on regional and flight information region (FIR) scale (several hours ahead): On the day of operation itself the Central Flow Management Unit (CFMU) will apply the measures announced in the ANM and will monitor whether the pretactical plan is having the desired effect. At the present time, tactical ATFM is based on the application of acceptance rates, expressed in terms of the number of flights per unit time that will be allowed to enter a specific congested area from a particular area of origin. Aircraft that plan to fly through a congested area—as detailed in the ANM—are expected to request a slot from the CFMU. Slots are allocated mainly in the form of a revised departure time but sometimes as a time of arrival at an en-route point (Martin, 1993).

- ATC Area Management on sub-FIR scale (1 - 2 hours ahead): This corresponds to operational supervision and tactical ASM. This activity is responsible for dealing with events having a significant effect on traffic handling and throughput (such as notices to airmen (NOTAMs)), system failures, changes of airspace, airport and runway availability, meteorological hazard, or traffic overflow). It adapts the sectors and selects the overall strategy for dynamic routing, airspace use, traffic segregation, and runway usage to meet the strategic regulation plan (tactical ATFM plan) and the required capacity of the involved sectors. This includes assessing and smoothing the center's workload for the coming hours.

- Planning ATC on a single or multisector scale (20–30 minutes ahead): This serves to organize the traffic entering and leaving the planning area so as to avoid unmanageable situations inside the planning area (from a flight safety, flight economy, and ATC workload point of view). This involves aircraft sequencing, allocation of runways, routes, levels, delays and coordination with adjacent planning areas to establish agreed transfer conditions. The resulting plan must include a certain contingency to give executive ATC the "maneuvering liberties" needed to resolve unexpected problems.

- Executive (tactical) ATC on a single-sector scale (5–10 minutes ahead): This is responsible for implementing the plan established by planning ATC and maintaining satisfactory levels of safety (through separation assurance and aircraft guidance). The provided separation and aircraft guidance has to meet certain legal requirements (minimum separation values that depend on geographical, technical and institutional circumstances). Executive ATC has to monitor a highly dynamic system in which the nature of the problems at hand can significantly change in the course of a few minutes. The resources available for problem solution are limited, and there are hard real-time constraints for the control loop that must detect these problems, develop solutions, and implement those solutions.

- ATC Safety Net layer on single aircraft scale (2 minutes ahead): This complements Executive ATC with functions such as short-term conflict alert (STCA) and minimum safe altitude warning. The purpose of this layer is to catch those safety-threatening situations that were not resolved by the Executive ATC layer.

- This hierarchy of layers can be continued to shorter and even negative time horizons (in this case the term *planning layer* is not appropriate any more):

- *Real-time operations* layer on single aircraft scale (real-time): This is the physical act of communicating clearances, instructions, advice and requests to individual aircraft, executing procedures and changing the internal state variables of the ATM system as a reaction to the occurrence of triggering events (this progresses the chain of events). Real-time operations are what an incidental visitor sees happening when he or she observes a controller on duty. The incidental visitor would not see the meanings (all the tactical and strategical considerations) that are behind the observed real-time operations.

- *Forecasting and extrapolation* layer (past situations extrapolated to a target time; i.e., to real-time or to the future): This serves to produce assumptions about the state vector of the relevant objects (individual aircraft, weather phenomena, the traffic flow, etc.) based on a description of the situation in the past (obtained via the history data collection layer). The purpose of forecasting and extrapolation is to close the feed-forward loops by transforming history data into a state that matches the time horizons used by the various higher planning layers. Forecasts and extrapolations may be deterministic (with reduced accuracy of extrapolated state variables) or probabilistic (if the target time is too far ahead of the recording time). A track extrapolation 10 seconds ahead is an example of deterministic extrapolation; a 2 day weather forecast is an example of probabilistic forecasting. Note that forecasting and extrapolation are to be distinguished from prediction based on an object's intentions (such as the planned trajectory or the clearance of a flight).

- *History data collection* layer (negative time horizon; i.e., delayed): This is called *surveillance data acquisition* in the ATC context. It serves to create an accurate recording of the air traffic situation (or weather situation, etc.) over time. The obtained accuracy depends on the sensors used. The average age of the most recent data (the delay) depends on the sampling rate and the processing or communication delay. The data may be used in realtime for planning and control purposes (after extrapolation), or for offline applications (route charges, accident or incident investigations, statistics, etc.).

OPERATIONAL CONCEPT FOR FLIGHT OPERATIONS

The flight operations operational concept is a layered model quite similar to the ATM operational concept. Aircraft operators are faced with managing the three main phases of aircraft operations: flight time, taxi time, and turnaround time in preparation for the next flight. Each of these phases is managed in a number of planning layers and properly coordinated with the other phases. To illustrate the similarity with the ATM planning layers, here is a typical set of planning layers for the flight phase, as applicable to a scheduled airline:

- Long-term strategic planning to determine business opportunities and decide between fundamental options (10–15 years ahead).
- Aircraft fleet planning (5–10 years ahead).
- Acquisition and planning of commercial routes and destinations (several years ahead).
- Development of timetables for the coming season (6 months–1 year ahead).
- Negotiation of routes with strategic ATFM (6 months–1 year ahead).
- Filing of Repetitive Flight Plans (RPLs) (about 6 months ahead).
- Strategic flight planning: Determination of aircraft maintenance schedules and initial allocation of resources to individual flights (aircraft, crew, logistics, etc.; approx. 3 months ahead).
- Tactical flight planning for individual flights, based on pretactical ATFM (1 day ahead).
- Direct flight planning (offboard crew activities), based on latest NOTAMS, actual weather forecasts, cabin briefing and so on. (45–60 minutes before scheduled departure time, planning ahead until a few hours after scheduled arrival time).
- Direct flight planning (onboard crew activities), based on load sheet, ATIS, and so on. Results in completed fuel order, computed takeoff data, planned 4-D trajectory, estimated time enroute, and so on. (15–45 minutes before scheduled departure time, planning ahead until scheduled arrival time).
- Strategic flight management: coordination with ATC to obtain predeparture clearance, including a departure slot, and other flow restrictions. Adjust planned 4-D trajectory accordingly. This corresponds to decisions made by tactical ATFM. Subsequently obtain start-up approval (5–15 minutes before "off blocks").

- Pretactical flight management: Obtain expected clearances for departure, enroute, climb, descent, and approach, and adjust planned 4-D trajectory accordingly. This represents coordination with decisions made by the planning ATC layer (during taxi and flight, 30 minutes ahead).
- Tactical flight management: In a ground-based ATC environment, obtain actual clearances for push-back, taxi, takeoff, departure, enroute, climb, descent, approach, landing, and taxi, and adjust planned 4-D trajectory accordingly. This represents coordination with decisions made by the executive ATC layer. In a free-flight environment (Eurocontrol, 1999; Federal Aviation Administration, 1999), use the airborne separation assurance system (ASAS) to adjust the planned 4-D trajectory (from 5 minutes before "off-blocks" to docking and engine shut down, 5–10 minutes ahead).
- Prepare execution of tactical maneuvers, based on the latest 4-D trajectory plan and tactical ATC instructions (vectoring). This represents execution of cockpit procedures and coordination with decisions made by the executive ATC layer and the ATC safety net layer, or ASAS in a free-flight environment (during taxi and flight, 30 seconds–2 minutes ahead).
- Prepare execution of collision avoidance maneuveres, based on visual observations and/or ACAS (during taxi and flight, 5–30 seconds ahead).
- Real-time operations layer (realtime): The physical act of operating (controlling) the aircraft and maneuvering it in accordance with the most up-to-date 4-D trajectory plan, and the act of communicating with various partners such as ATC.
- Forecasting and extrapolation layer (past situations extrapolated to a target time; that is, to real-time or to the future): For example, dead-reckoning techniques, fuel burn prediction, and other estimations.
- History data collection layer (negative time horizon, i.e., delayed): For example, flight data recording and aircraft position determination.

PLANNING THEORY FOR OPERATIONS

The foregoing describes the functioning of "layered" ATM and flight operations in specific (operational) terms. The advantage of this approach is that it is pragmatic and permits the reader to relate the story to his or her own operational experience as opposed to being theoretical. The disadvantage is that the same basic operational problems are unknowingly solved over and over again, for different look-ahead time scales, in different terminology, and by people with different backgrounds.

This section of the chapter addresses issues such as:

- ˙ The different goals that may be set in an automation strategy (these goals are heavily influenced by human factors)
- The development and assessment of certification strategies for operations management.

Thus for certification purposes we need a kind of theoretical insight into this layered planning technique. What is called *planning theory* in this chapter, represents an attempt to identify some of the basic underlying principles in the operation of an Air Transport System. The subject of planning theory is now presented next.

The Players in Operations Management

The Air Transport System. The generic term *air transport system* refers to the aggregate of weather, airspace, aerodromes, aircraft, aircraft operators (commercial, military, general aviation, and aerial work), and air traffic management/communication navigation surveillance (ATM/CNS) Systems, all operating together in a particular geographical region.

Actors. In accordance with the preceding definition, the Air Transport System consists of a number of interacting elements, such as airlines, aircraft, pilots, airports, runways, airspace, routes, ATC units, controllers, systems, weather phenomena, aircraft separation, and so on. For the sake of generalization, I call these elements the *actors* of the air transport system. They are the real-life equivalent of what are called *business objects* in modern software development (Object Management Group, 1997).

Relationships Between Actors. The operation of the air transport system is far more than the sum of the operation of the individual actors. Indeed, these actors are in constantly changing interaction with each other. Managing the operation of the air transport system means managing these interactions. Some interactions are to be promoted because they are necessary ingredients of the proper operation of the air transport system. Others represent problems, and system management efforts are directed at avoiding or removing such interactions. An example of the latter is the separation conflict between two aircraft. Some of the most important general types of interactions (relationships) are:

- User–resource relationship: Actors in a resource role exist in limited supply with a variable number of actors in a user role competing to use that supply (establish a user–resource relationship). Examples are aircraft operators with respect to aircraft, aircraft with respect to runways, clients with respect to database servers, aircraft with respect to route capacity, aircraft with respect to mutual separation, flights with respect to ATM, and so on.

- Competitor relationship: Users competing for the same resource are in a competitor relationship. Examples are two aircraft approaching the same airport at the same time, two radios attempting to transmit on the same frequency at the same time, areas of severe weather with respect to aircraft wishing to use that same airspace, and so on.
- Collaborator relationship: Resources able to distribute the user load between them (i.e., to reduce bottlenecks) are said to be in a collaborator relationship. Examples include parallel runways, different flight levels, parallel routes, main versus reliever airports, etc.
- Buffer relationship: Resources able to temporarily absorb the user load of another resource are called buffers. Examples are holding patterns, route extensions, queues, contingency measures, and so on.
- Control relationships: Sometimes actors are responsible for the operation of another actor. This responsibility may range from defining an operating envelope (providing policy, guidance, operating constraints, or allocating workload) to assuming detailed control. Examples include pilots with respect to aircraft, controllers with respect to controlled flights, ATFM with respect to ATC, and so on.
- Target relationships: When actors are planning to reach a goal, actor and goal are said to be engaged in a target relationship. An example is a flight with respect to its destination airport.
- Part–of relationship: This indicates the (permanent or temporary) assembly of individual actors into a composite system, having a certain state vector in common. Examples: a runway is part of an airport, a pilot is part of an aircraft in flight.

The Environment. In simple terms, the environment of an actor is everything that surrounds that actor. However, only that part of the environment with which the actor interacts in one way or another is relevant. This leads to the following natural definition of environment: the total set of existing, expected, and planned relationships with other actors. The environment is dynamic because the membership of this set is subject to change as time elapses: Relationships disappear and new ones come into existence.

The Conduct of Operations

Operations in a World Without Planning. Object-oriented analysis (OOA) techniques (Coad & Yourdon, 1991; Shlaer and Mellor, 1992) describe the world in a mechanistic manner as a set of objects (actors) with predefined relationships, predefined life cycles (state transition diagrams), predefined event chains, and communication capabilities. This may perfectly suit the needs of system analysis for the purpose of developing static (nonadaptive) pieces of software or analyzing rigid organizations (mechanistic systems), but it seems a bit inadequate for documenting the operation of complex, goal-oriented systems such as the Air Transport System. OOA may correctly capture such a system on a syntactical, real-time operations level (existence of actors, possible relationships, state transitions, etc.), but it misses out on the semantics (the whole layered planning process preceding the physical conduct of operations).

Operations in a Goal Oriented World. In contrast, the operation of the Air Transport System is highly adaptive (i.e., it is governed by a set of constantly modified and [re]created scenarios, scripts, procedures, project plans, flight plans, story boards, event models, rules, regulations, strategies, tactics, philosophies, etc.). This modification and (re)creation is the previously mentioned layered planning process that precedes the physical conduct of operations. It happens this way because most of the actors in the Air Transport System are goal-oriented entities. To reach a goal in a world full of uncertainties and conflicting requirements, one needs to plan the future and reduce the amount of improvisation. In fact, all the actors of the Air Transport System spend a considerable part of their energy on such planning activities. So what are these planning activities all about, in a nutshell? They are about developing a scenario, refining it, and finally executing it in an environment of external and self-induced perturbations.

The external perturbations are the unforeseen events, interactions, and timing in an actor's environment. External perturbations occur because the environment may be inherently unpredictable, but can also be due to lack of overall coordination in the Air Transport System. The self-induced perturbations come from an actor's inability to accurately execute his or her own operational scenario. These are cases of mismanagement in the operational sense. On top of that, the actor may simply be following a bad scenario (e.g., with lack of feasibility and full of inconsistencies). This type of problem and the perturbations give rise to the need for constant situation assessment and revision of the scenario.

Fuzziness in the Planning Process. Actors deal with two types of scenarios: probabilistic and deterministic. The countdown toward the moment of physical execution of a particular operation is normally spent in different uncertainty phases:

- *Phase 0*: The need for the operation has not yet been identified.
- *Phase 1*: The need for the operation is identified, usually with an approximate target time, but no plan or scenario is available.
- *Phase 2*: The phase of fuzzy and probabilistic scenarios.

- *Phase 3*: The reduction of uncertainty, to transition from fuzzy and probabilistic scenarios to a very limited number of candidate scenarios (scripts).
- *Phase 4*: The phase in which one of these candidate scenarios has achieved a very high probability of occurrence, and has become a structurally stable script for the operation (*structurally stable* means that the sequence of events is stabilized, and partners for various types of relationships are known; e.g., "contractual" status of relationships is established).
- *Phase 5*: The phase in which the script does not structurally change, but the accuracy of its various parameters (e.g., timing, planned value of state vector, details of interactions with other actors) is improved.
- *Phase 6*: The actual execution of that script, resulting in a physical operation (the chain of events is progressed).
- *Phase 7*: The phase in which factual data on what has happened are not yet available.
- *Phase 8*: The availability of history data describing what actually happened.

To visualize the relation between these uncertainty phases and time, an uncertainty–time diagram is used in this chapter. The time axis is to be seen as absolute time; and the uncertainty axis lists the preceding phases to give a qualitative idea of the accuracy of a given scenario.   Figure 7.1 illustrates the production of different versions of a scenario. Scenario 1 is associated with Time t_1, scenario 2 is the same (but revised) scenario as it stands at Time t_2, and so on.   These reference times correspond to the real-time execution of what is spelled out in a particular scenario version.   In the example of Scenario 1, the part before t_1 is history, and the part after t_1 is the plan for the future.

An example will clarify this. A flight plan is to be seen as a scenario. At Time t_1, part of the flight has been completed already and the next leg is quite accurately planned, but the details of arrival are uncertain. For example, because of the unpredictability of the weather it is uncertain whether the alternate destination will have to be used or not.Fig. 7.1.

This story can also be looked at from the perspective so well known from space vehicle launches: the countdown view. While counting down, a particular time t_{target} in the future (e.g. the time of arrival of a flight) is associated with different scenario versions as time elapses. Each new version is more accurate with respect to time t_{target} because prediction is less of a factor.

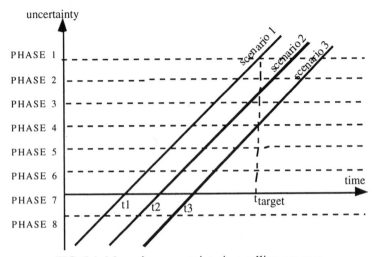

FIG. 7.1. Managing uncertainty in a rolling process.

All this explains why operations management works as a "rolling program" of scenario development. In this approach, a particular situation at Time t_{target} seems to be planned in an iterative fashion because it is associated with a number of different scenario versions. On the other hand, the planning process chases a moving time horizon because the subsequent scenario versions are time stamped differently, but must maintain the same outlook with respect to their time stamp.

This scenario revision process is implemented by a control loop. In project management, the control loop is called the *PDMA cycle* (plan–do–monitor–adjust). In ATM, one usually distinguishes the following phases (Eurocontrol, 1992):

- Acquire information.
- Monitor current situation.
- Predict evolution.
- Identify problems.
- Propose and evaluate solutions.
- Choose solution.
- Communicate and implement solution.

Each cycling through these control loop phases produces a new version of the scenario. But let us return to the previously mentioned hierarchy of uncertainty phases. In system terms, this hierarchy could be expressed as the following strategy:

- Determine the desired start state and end state (goal and target) of the operation.
- Plan the time of occurrence for start state and end state.
- Select intermediate states (subgoals).
- Determine the order of intermediate states (temporal organization of activity).
- Elaborate synchronization and coordination requirements for the operation (type and sequence of the relationships needed or expected during the operation).
- Determine the partners for these synchronization and coordination relationships (production of a structurally stable script, establishment of "contractual" relationships with partners).
- Work out the timing (start and end) of the synchronization/coordination with each of the above partners.
- Determine the detailed timing (accuracy) for the intermediate states.
- Operate in real-time, i.e. perform state transitions and interact (exchange events) with various synchronization/coordination partners.
- Collect history data.

Take any type of operation—ATFM, planning ATC, flight planning, flight management, project planning—and it is possible to express the operations planning in these terms.

A third way of expressing this strategy is reflected in the traditional what–how–where–when sequence:

- The *what* phase identifies the operation.
- The *how* phase is responsible for identifying the needed interactions.
- The *where* phase produces the stable script that identifies the partners for the individual interactions.
- The *when* phase puts on the accuracy by refining the timing.

One can continue repeating this strategy in different terminology disguises. In a systems development context this scenario development strategy is called a *life cycle*:

- User requirements definition.
- Operational concept definition (sometimes termed *requirements analysis*).
- Operational requirements definition (also called *system requirements definition*).
- Architectural design (alias technical concept definition).
- Detailed design.
- System procurement and installation.
- System operation.

In project management terms, the hierarchy looks as follows:

- State the overall mission of the project.
- Determine the completion date of the project.
- Develop the work breakdown structure.
- Identify the interdependencies between work packages (production of Pert Chart).
- Perform rough allocation of the total project duration to individual work packages (production of initial Gantt Chart).
- Allocate resources to work packages.
- Refine timing of work packages by eliminating resource overallocation. (production of Gantt Baseline Chart)
- Adjust the project plan based on plan deviations
- Execute the project plan
- Do progress tracking.

All these strategies are nothing more than variations on the same basic theme. Depending on the complexity of the operation and the expected number and magnitude of perturbations, the length of this countdown process—alias planning strategy—may take just a few seconds, and at the other extreme several years or even decades.

The Feasibility of a Scenario. The objective of each layer (or countdown and anticipation phase) is to deliver an acceptable situation to the next lower layer. This means maintaining a set of conditions (i.e., a solution space or operational performance envelope) in which one or more feasible action plans exist. If the higher layer fails to maintain those conditions, the lower layer may end up in a dead-end street, a situation in which the operation cannot be successfully completed. The expression "to pass the point of no return" emphasizes the timing and state transition aspects of this feasibility collapse.

Consequently, one of the responsibilities of a higher layer is to maintain a constant awareness of the operational performance envelope of the next lower layer. The ATC concept of minimum legal aircraft separation requirement is an example of such an operational performance envelope. The previously mentioned feasibility collapse may have internal and external (environmental) causes. A mistake in the calculation of aircraft endurance during flight planning is an example of an internal cause. So is the failure of a pilot to initiate the landing flare at the right moment, or the failure of a controller to detect a loss of separation between two aircraft. An unexpected weather change to instrument meteorological conditions during a visual flight rules flight is an example of an external cause.

The Impact of Perturbations. As mentioned, there is a constant need for situation assessment and revision of the scenario due to internal and external perturbations. In addition, the notion of the operational performance envelope has been introduced.

The impact of perturbations depends on the magnitude of those perturbations. In this context, *magnitude* can refer to size as well as duration. If the magnitude of the perturbation exceeds the operational performance envelope of the planning layer under consideration, then there is nothing this planning layer can do to solve the problem. It is up to a higher layer to take care of the situation. That, of course, cannot be done in a reactive way after the problem occurred. By virtue of its longer planning horizon, the higher layer is supposed to have prevented the problem.

If the magnitude of the perturbation does not exceed the operational performance envelope of the planning layer under consideration, but it exceeds the envelope of the next lower layer, then this planning layer is responsible. It has to modify the scenario within the possibilities of that particular planning horizon.

If the magnitude of the perturbation does not even exceed the operational performance envelope of the next lower layer, then this planning layer does not have to change the scenario with respect to that particular planning horizon.

Whatever layer is responsible, in a properly functioning system the problem is solved in anticipation (a certain time before it would actually occur). This can be seconds, minutes, hours, days, or even years in advance. Additionally, this revision of the scenario by a particular layer invalidates all the plans under the responsibility of lower layers. This imposes certain time constraints on the lower layers which that to re-create their part of the scenario from scratch. Indeed, imagine a situation in which there exists a sufficient number of possible solutions on the shorter planning horizons, but the responsible human or machine is unable to produce these solutions in the available time. An example of this is the situation in which the pilot does not keep up with the airplane: He or she is overtaken by events rather than staying abreast of them.

The Impact of the Environment. To plan the scenario of an actor with a certain accuracy to a certain time horizon, the predictability of the environment must be equal to or better than the fuzziness or accuracy of the desired scenario. This is illustrated in Fig. 7.2. The scenario labeled "Actor" can be implemented in Environment 1 but not in Environment 2.

Let us illustrate this with the example of planning a conflict free 4-D tube clearance (the actor scenario) in a given environment of other aircraft trajectories. The bottleneck of the planning process is the part with the greatest uncertainty. Uncertainty translates in this example into planning horizons, time windows (departure, overflight, climb, descent, arrival), positional accuracy, and confidence levels.

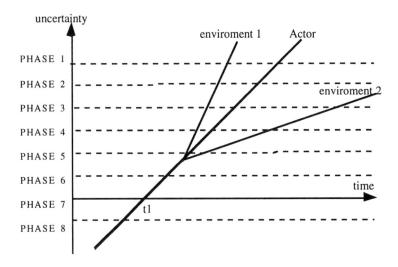

Fig. 7.2. Varying degrees of predictability of the environment

Assume the following operational goal for the actor: touchdown in exactly 2 hours, plus or minus 30 seconds. Of course the actor needs to have the capability to execute this scenario with the required accuracy. It is intuitively clear that it is feasible to plan this in environment 1 where the landing times of the other aircraft will occur with an accuracy of 10 seconds. It is equally obvious that such planning is pointless in environment 2 where the landing times of the other aircraft will occur with an uncertainty of 5 minutes, unless the minimum separation values (safety margins) are greatly increased, with a resulting reduction of control capacity.

The Impact of a Lack of Knowledge. The role of knowledge is quite similar to what was said about the environment. Note the choice of words in the previous example: "will occur with an accuracy of." If I replace this with the words "is known with an accuracy of," we see the impact of a lack of information.

Figure 7.3 illustrates that the environment is assumed to be less deterministic than it really is due to insufficient information. The actors in the system have a lack of situational awareness, which translates into reduced planning horizons (the horizontal delta in Fig. 7.3) and reduced certainty at a particular outlook time scale (the vertical delta in Fig. 7.3). In plain words, the planning can only be as accurate as the knowledge on which it is based.

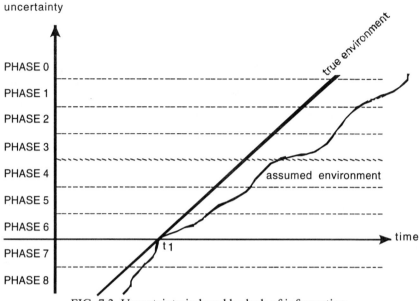

FIG. 7.3 Uncertainty induced by lack of information.

The attempt to approximate the inherent unpredictability as much as possible is the reason why future operational concepts strive to use better surveillance, better coordination, and higher levels of system integration—air–ground integration and use of data links in particular.

Lack of Correlation With Reality. The other extreme is a lack of correlation with reality: sophisticated models of the future that make the actors believe that the situation is very much under control.

That situation is represented in Fig. 7.4. The description of the future is too precise: It is a collection of unfounded assumptions that will probably turn out to be false; for example, an ATC system that "sells" the idea that the next separation conflict of Aircraft X will be with Aircraft Y, whereas at the given moment it is inherently impossible to say whether it will be with Aircraft Y or Z. A system acting like this will either fail or at least exhibit very poor performance because each such case of overconfidence probably creates a mistake; that is, an internal perturbation in the system's operation.

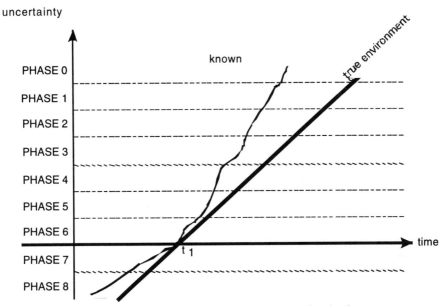

Fig. 7.4. Over confidence in the ability to predict the future.

The Need for Planning. Orville and Wilbur Wright did not need ATC in 1903. They certainly did not have any use for flow management. There seems to be a tendency that systems evolve to lose their simplicity over time (people call this "more advanced"): The environment changes into a more complex one, the system's internal complexity increases, the operation needs to be more optimized, uncertainty is less and less acceptable, and the operational performance envelopes are extended to enable the previously impossible. What used to be simple now requires advanced and accurate planning: no more "flying by the seat of the pants." This is now "follow the procedures" and "fly by the numbers." There was a time when aircraft flew but ATC did not exist, then a time with ATC but without planning controllers, and finally there was a need for ATFM (which is, of course, fairly recent).

At some stage, all this planning gets into conflict with the *flexibility requirements* of the airspace users. The idea of freeflight (Eurocontrol, 1999; Federal Aviation Administration, 1999) advocates a reversal of the "more planning is better" trend. However it is also argued (Eurocontrol, 1997) that more planning is needed in high-density traffic areas to achieve higher capacity—much like a parking lot needs white lines to accommodate more cars. Theoretically, a completely deterministic operational concept could provide 100% safety and nearly unlimited control capacity by virtue of highly accurate preplanned (booked) 4-D conflict-free flight trajectories from takeoff to landing. One negotiates (books) a flight plan (all previously booked flight trajectories remain unchanged), and from then on everything unfolds mechanistically clockwork.   Unfortunately, that would not lead to the best possible optimization. In addition, there are many uncertainties in real life that cannot be completely eliminated, not even by the best planning. Besides that, even if unexpected external events did not exist, operations managers on all planning levels might want to change their minds once in a while, instead of having to stick to plans that were cast in concrete long ago.

A few things need to be remembered from the preceding:

- The need for planning tends to increase as systems, organizations, and technology become more mature.
- Historically (i.e. during the evolution of a system) the stack of planning layers builds from the bottom up.
- Those higher planning layers always address the need for more global optimization of operations (the need for bird's-eye views and crystal balls).
- The need for planning depends on the complexity of the operating environment (e.g., traffic density and geographical context).
- The need for planning depends on the complexity of the system (organization and equipment) and the type of operation.

OPTIMIZATION STRATEGIES FOR OPERATIONS MANAGEMENT

What do goal-oriented entities (humans, machines, or composite systems) usually do to optimize their operations? A number of general strategies always return:

- Have a plan as early as possible.
- Have a plan, as feasible as possible (maximize contingency).
- Have a plan which includes a strategy for dealing with probabilistic situations.
- Re-assess and revise the plan as often as possible.
- Have a plan, as stable as possible (i.e., the revisions should be as small as possible).
- Have the ability to (re)create a plan quickly (improvise if necessary), to minimize the time delay between the last situation assessment and the availability of the new plan.
- Have the ability to stick to the plan (a minimum of internal perturbations).
- Solve problems (external perturbations) as early as possible.
- Solve problems (external perturbations) as thoroughly as possible (full impact analysis of solution).
- If there is a choice, operate in as stable an environment as possible.

- If there is a choice, operate in as predictable an environment as possible.
- If there is a choice, operate in an environment with the least number of interdependencies (low-complexity environment).
- Avoid the unknown; that is, operate in as well known an environment as possible (maximize the available information).
- Split the planning process into different concurrently operating layers with different responsibilities based on the dynamics of the possible perturbations (work with a hierarchy of plans).
- Know the true extent of all the performance envelopes (how far can you go on each layer without compromising safety).
- Devise a proper filtering strategy to dispatch perturbations to the responsible planning layer (full impact analysis of perturbation): A dispatch to a layer that is too high leads to unnecessary replaning; a dispatch to a layer that is too low leads to safety problems.

CERTIFICATION ISSUES

How should planning theory (i.e., the aforementioned considerations of planning layers), control loop phases, and performance envelopes be seen within the context of certification? Rather than trying to give a complete answer, this chapter attempts to give a number of useful indications. Before going into details, however, let us clarify the terminology used.

Definitions

For the purpose of this chapter, I assume the following definitions. *Verification* is a review process to check the system requirements against their source. The validity of the requirements themselves is verified to ensure that a system that would be built to satisfy these specifications would also be suitable. *Validation* is the checking of a system design against the requirements. It is the production of (formal or experimental) proof, serving to establish a measure of confidence in the correctness and effectiveness of important system features. Validation is performed after the fact, as, for example, during acceptance tests. *Feasibility study* is similar to validation, but different in the sense that it is done in the exploratory phase (before the fact) to select a suitable solution amongst different possible alternatives. A feasibility study never replaces validation. *Certification* is the administrative "rubber stamping" of a validation, an endorsement to give it an official status and level of authority.

The Limitations of Certification

There is one catch about certification: The correctness and effectiveness of the certified system features (the fitness for operation) is not endorsed under unlimited operational circumstances. Every certified system or person "carries" a piece of paper that lists these circumstances or limits of authority: System or person such-and-such is certified to deliver operational performance X during a period Y under operational circumstances Z. In fact, these limitations can be equated to the operational performance envelopes that were introduced earlier in this chapter.

Certification of the Planning Process

Normally, operational concepts are not certified in their totality, probably because that is beyond today's state of the art. However, there is a need to develop, validate, and certify the standards and recommended practices that are used in support of operational concepts.

Traditionally, certification was focused on systems (equipment or humans) to qualify their functional capabilities and operational performance. Emphasis always seemed to be on the real-time operational performance (uncertainty Phase 6 in the figures in this chapter), because that is the easiest to observe and, more important, because it represents the ultimate judgment on proper operation of the system. It is the stuff the whole operations planning process is finally all about.

Now, systems become more complex and rely more and more on planning (automated or human). This means that it is no longer sufficient to certify that the system works, regardless of how it achieves this goal. The planning process itself needs to be certified because it determines those situations in which a system will and will not work. As mentioned earlier, the documentation of these limitations is a crucial element in certification.

In other words, the time has come to consider planning layers and their individual control loop phases as objects for certification, instead of just physical people, equipment, functions, and procedures.

After having said this, it must be clearly stated that planning layers and control loop phases already exist in today's systems because all kinds of functions and responsibilities fulfill these roles. However, these functions and responsibilities have not been consciously designed based on sound planning theory. Instead, they historically evolved in a bottom-up fashion, over many system generations, just as human language is a product of history rather than a careful design. Thus the certification problem is considered to be twofold:

- Certify the operational principle (i.e., the effectiveness of a certain combination of planning strategies).
- Certify the implementation (i.e., the functions, responsibilities, etc.) of these planning strategies to certain performance standards.

Let us rephrase this in a bit more detail. Operations management uses a layered planning process that includes strategies for the reduction of uncertainty and for dealing with perturbations. The performance of this planning process can be expressed as specified previously and in terms of the uncertainty phases defined earlier in this chapter. The interoperability of these layers and phases depends on mutual awareness of operational performance envelopes (internal operation and expected range of external perturbations) plus proper matching of these layers and phases.

Certification needs to concentrate the effectiveness of these strategies (the quality of the produced scenarios) and on the interoperability of layers and control loop phases. The exact documentation of performance envelopes is a key issue in the certification process. This is the framework for certification of the implementation components of operational concepts; that is , the functions, responsibilities, and operational procedures.

Large systems have many players: groups of people and automated systems operate together in teams to make collective tactics and strategies (planning layers and control loops) happen. This task sharing defines the information flows between people, on the human–machine interfaces, and between automated systems. Different tactics and strategies require different information flows.

Functions, responsibilities, and operational procedures are the "bricks" for building the implementation of an operational concept. The previously mentioned information flows are the cement that keeps the bricks together. The existence of these bricks is to be justified in terms of the certified planning strategy, and the role of existing bricks is to be mapped on the planning layers and their control loop phases. New bricks should be designed to fit specific niches in that framework of layers and phases.

To make the implementation of an operational concept perform as intended, the individual bricks need to be certified to meet the requirements imposed by the previously certified interface and performance specifications of planning layers and control loop phases.

Human Factors

All the preceding considerations about planning layers, control loop phases, scenarios, and performance envelopes apply to any goal-oriented system. That, of course, includes composite human–machine systems.

In such a system, the planning responsibilities outlined in the operational concept are allocated to humans and machines. This can be done in a top-down fashion (in an arbitrary manner), or bottom-up; built around the human capabilities and the state of the art of technology.

Many studies have investigated the role of the human. The human factors field aims at determining the automation environment in which the human performs best. This chapter does not attempt to draw specific conclusions from the existing literature, but in the end various strategies are possible to integrate the human into an automated system or to support the human with automated functions. One strategy may be to give the complete responsibility for some planning layers to humans and automate the remaining layers. For example, in ATC, automate safety nets but keep executive control largely manual.

Another approach is to take the control loop of a planning layer, automate some phases, and leave responsibility for others to the human. To again use an ATC example; automate the monitoring and problem detection phases, but keep the proposal and evaluation of solutions and the decision-making phases manual.

No matter whether a layered or phased automation strategy is chosen, it is necessary to correctly use the strengths and weaknesses of humans and machines. In other words; within the framework of layers and phases, find out whether the human or the machine fits the requirements best, in terms of monitoring capabilities, problem-solving capabilities, memory, speed, assimilation, pattern recognition, span of attention, reliability, and so on.

Thus, the human gets certain modular chunks (layers, phases) of the overall operational concept (the planning strategy as described earlier). The problems of human factors certification can then be seen as the certification of interoperability of these chunks with the overall operational concept. In that sense, the human role is no different from an automated function taking the same responsibilities. The performance needs to meet the requirements as foreseen for that particular responsibility within the context of the total planning strategy.

CONCLUSIONS

General

This chapter suggests that the tactics and strategies notion is a highly suitable paradigm to describe the cognitive involvement of human operators in advanced aviation systems (far more suitable than classical function analysis), and that the workload and situational awareness of operators are intimately associated with the planning and execution of their tactics and strategies. If system designers have muddled views about the collective tactics and strategies to be used during operation, they will produce suboptimum designs. If operators use unproved or inappropriate tactics and strategies, the system may fail.

I want to make the point that, beyond certification of people or system designs, there may be a need to go into more detail and examine (certify?) the set of tactics and strategies (i.e., the operational concept) that makes the people and systems perform as expected.

The collective tactics and strategies determine the information sharing and situational awareness that must exist in organizations and composite human–machine systems.

The available infrastructure and equipment (automation) enable this information sharing and situational awareness, but are at the same time the constraining factor. Frequently, the tactics and strategies are driven by technology, whereas we would rather see a system designed to support an optimized operational concept; that is, to support a sufficiently coherent, cooperative and modular set of anticipation and planning mechanisms.

Again, in line with the view of MacLeod and Taylor (1994), this technology-driven situation may be caused by the system designer's and operator job designer's overemphasis on functional analysis (a mechanistic engineering concept), at the expense of a subject that does not seem to be well understood today: the role of the (human cognitive and automated) tactics and strategies that are embedded in composite human–machine systems. Research would be needed to arrive at a generally accepted planning theory that can elevate the analysis, description, and design of tactics and strategies from today's cottage industry methods to an engineering discipline.

Planning Theory

A theory based on planning layers, control loop phases, uncertainty phases, and performance envelopes would provide a modular framework for the task of designing and documenting operational concepts (i.e., sets of tactics and strategies). The second half of this chapter represents an initial attempt to highlight the key issues of such a theory. When such a framework is used, the benefits may spin off to the certification task. In addition, it will put the role and contribution of human factors into clear perspective.

OOA

A few references to OOA techniques have been made in this chapter. It is felt that is too mechanistic; that is, it misses some expressiveness when used to analyze and document systems consisting of goal-oriented entities. Planning theory could be a suitable candidate to remedy that problem.

Acknowledgment

The content of this chapter expresses the opinion of the author and does not necessarily reflect the official views or policy of the EUROCONTROL Agency.

REFERENCES

Coad, P. , & Yourdon, E. (1991). *Object-oriented analysis*(2nd ed.). Englewood Cliffs, NJ: Yourdon Press.

EUROCONTROL (1992). *Open ATM system integration strategy (OASIS).* Brussels, Belgium: European Organization for the Safety of Air Navigation.

EUROCONTROL (1999). *European air traffic management system (EATMS) — operational concept document (OCD)* (ed. 1.1) Brussels, Belgium: European Organization for the Safety of Air Navigation.

Federal Aviation Administration .(1999). *National airspace system architecture.* Washington, DC: FAA Office of System Architecture and Investment Analysis .

MacLeod, I. S., and Taylor, R. M. (1994). Does human cognition allow human factors (HF) certification of advanced aircrew systems? In J. A. Wise and V. D. Hopkin, & G. J. Garland (Eds.), *Human factors certification of advanced aviation technologies.* (pp 163–186). Daytona Beach, FL: Embry-Riddle Aeronautical University Press.

Martin, B (1993, February). *Progress through better air traffic flow management.* Paper presented at the ATC '93 Conference, Maastricht, UK: Jane's Information Group.

Object Management Group (1997). Business object DTF —Common business objects (OMG Document bom/97-12-04 Version 1.5). Framingham, MA: Object Management Group.

Shlaer, S., & Mellor, S. J. (1992). *Object lifecycles—Modelling the world in states.* Englewood Cliffs, NJ: Yourdon Press.

III

PRACTICAL APPROACHES TO HUMAN FACTORS CERTIFICATION

8

Achieving the Objectives of Certification Through Validation: Methodological Issues

Paul Stager
York University, Canada

The intent of this chapter is to review methodological issues surrounding the question of what constitutes the appropriate human factors certification of advanced aviation systems. In this context, an advanced system is assumed to be one in which there is a high level of system automation and to which leading-edge technologies and design concepts have been applied during its development. The argument presented herein is that the primary objectives of human factors certification must be accomplished within the design and validation phases of the human engineering program.

Examples of advanced aviation systems to which the question of human factors certification must certainly be applied would include the series of substantial air traffic control (ATC) subsystems that are intended to be part of the larger U.S. National Airspace System (cf., Federal Aviation Administration, 1994, 1996a, 1996b) and the Canadian Automated Air Traffic System (CAATS) system (Stager, 1991, 1996). The human factors issues associated with the current system as well as with the transition to greater levels of automation have been reviewed by a National Research Council Panel on Human Factors in Air Traffic Control Automation (Wickens, Mavor, & McGee, 1997). More recently, the same panel identified a continuing requirement for a robust human factors engineering program to support the development and integration of the different subsystems (Wickens et al., 1998).

The significance of the design and validation requirements for complex human–machine systems, such as these ATC systems, has been recognized (cf. Wise, Hopkin, & Stager, 1993; Wise, Hopkin, Stager, Gibson, & Stubler, 1993), although the inherent complexity of many current and planned operational systems presents a methodological challenge for the human factors engineering profession (Meister, 1997). The rapid evolution of technologies that are a part of contemporary systems makes an empirical and incremental development of any complex system that much more difficult. New technologies upset the traditions and practices that would otherwise facilitate the dialogue between the system designers and the users. The identification of behavior-shaping constraints and user preferences is more difficult and the process of system design becomes one of continuing experimental evaluation (Pejtersen & Rasmussen, 1997; Wickens, Mavor, Parasuraman, & McGee, 1998).

Presumably, in this context of iterative evaluation, human factors certification might apply to (a) the human engineering design and development (including evaluation) of a system, (b) the operational system itself, or (c) both the process and the product. However, it is most likely that the measurements taken as part of a human factors certification of an operating system will be determined by human engineering design issues that would normally be addressed in the design and development cycle. As Woods and Sarter (1993) proposed:

> Tests of human–machine systems are as much about design as evaluation—think of the designer as experimenter. The critical criterion for investigations of human-machine systems is whether the methods chosen are sensitive to detect design errors without waiting for the lessons of experience. Tests of human–machine systems are as much about discovery as about evaluation—discovering what would be effective; discovering the problem space that captures the range of variation and contextual factors that operate in a particular domain. (p. 156)

Human factors certification of more complex cognitive systems is tantamount to certification of the underlying design development methodology. As stated at the outset, the argument presented here is that the primary objectives of human factors certification must be accomplished *within* the design and validation phases of the human engineering program.

A PRELIMINARY EFFORT TO DEFINE THE ISSUES FOR HUMAN FACTORS

The process of validation can usefully be seen as a matter of measurement (Kantowitz, 1992). Validation issues concern not only *how* one measures the behavior of human–machine systems (cf., Hollnagel, 1993; Reason, 1993; Woods & Sarter, 1993), but also *what* one measures as criterion variables. In a discussion of criterion variables, Harwood (1993) distinguished between the human-centered issues of technical usability, domain (i.e., operational) suitability, and user acceptability. Small and Bass (chap.11) also emphasize human-centered measurements (based on the stakeholders' concerns) that are used to evaluate system performance from the operator's perspective. Validation of the system–operator interface must inevitably demonstrate that the computer–human interface, for example, will satisfy the criteria for domain suitability and user acceptance across the range of conditions and operational contexts. Several questions can be asked from the perspective of domain suitability: Does the system support relevant task situations? Can the user easily navigate among tasks or pursue several different task-related goals and maintain situation awareness? How is the safety of the system and its error tolerance to be measured?

If the objectives of human factors certification are to be accomplished within the design cycle itself, how is this concept of certification to be implemented or realized? Elsewhere in this volume, Hanes (chap.5) has described the approach to human factors certification taken by the U.S. Nuclear Regulatory Commission (NRC), an approach in which design certification is based on an implementation process. Just as for current aviation systems, the development of advanced control rooms for nuclear power plants, with their advanced computer-based human–interface technology, must proceed in the absence of years of lessons learned from previous design efforts. Consequently, system evaluation has to be such that it can achieve a convergent validation derived from several techniques (cf., O'Hara, 1994; O'Hara, Stubler, Brown, Wachtel, & Persensky, 1995). The implementation process for the NRC provides a programmatic approach, from the development of guidelines for advanced human–interface technology through to the completion of the verification and validation activity. Thus, human factors certification is based primarily on process.

Critical elements in such an approach to certification of advanced technologies, however, are the availability of (a) appropriate engineering standards and guidelines, and (b) objective operational criteria.

Existing Guidelines

Current engineering requirements, as outlined in MIL–H–46855B (Department of Defense, 1979), call for any contractor to establish and conduct a test and evaluation program to assure fulfillment of the applicable requirements. The objectives of the program are:

- To demonstrate conformance of system, equipment, and facility design to human engineering design criteria.
- To confirm compliance with performance requirements where personnel are a performance determinant.
- To secure quantitative measures of system performance that are a function of the human interaction with equipment.
- To determine whether undesirable design or procedural features have been introduced.

Section 3.2.3 of MIL–H–46885B states that even though these objectives may be addressed at various stages in system, subsystem, or equipment development, a final engineering verification of the complete system is still required. Human engineering testing is to be integrated into engineering design and development tests, contractor demonstrations, research and development acceptance tests, and other development tests, and is to be directed toward verifying that the system can be operated, maintained, supported, and controlled by user personnel in its intended operational environment. The testing is to include:

- Collection of task performance data in simulated or operational environments.
- Identification of criteria for acceptable performance of the test.
- Identification of discrepancies between required and obtained task performance.
- Analysis of failures to differentiate between failures due to human error, to equipment, and to operator–equipment incompatibilities.

The MIL–H–46885 was used to derive the NATO standard STANAG 3994 (*The Application of Human Engineering to Advanced Aircrew Systems*) and the U.K. Defence Standard DEF–STAN–00–25 (*Human Factors for the Designers of Equipment*). In the opinion of Taylor and MacLeod (1999),

STANAG 3994 is perceived as a potentially valuable aid both for maintaining HE [human engineering] quality assurance and for managing the HE risk in the procurement of complex mission systems. The risk for HE is perceived to be particularly important in the procurement of complex mission systems. In complex systems, situation assessment and mission performance effectiveness are functions of the integration and interaction between the operator and equipment information processing and cognitive decision-making capabilities (p. 116).

The engineering requirements that form MIL-H-46885 clearly lend emphasis to the issues for measurement, particularly criterion measures, in validation and acceptance.

The Criterion Problem

The operational requirements detailed in the system specifications do not present a problem for human engineering evaluation. It is those criteria that are left unspecified or are to be identified in the human engineering program that create the challenge. Often the unspecified criteria will relate to error rates, workload, maintenance of situational awareness, and the potential for performance decrement associated with prolonged operational stress or fatigue. But these questions also relate to typical usability issues as well as system safety and efficiency (Harwood, 1993).

*Identifying the Criteria.*Meister (1985, 1991, 1992) has long called attention to system design issues, including the questions of measurement and evaluation. In his discussion of the difficulties surrounding measurement, a distinction is made between *measures of performance* and *measures of effectiveness*, the latter requiring a standard of required performance for comparison (Meister, 1991). In their description of the Merlin helicopter program, for example, Taylor and MacLeod (chap.9) describe the use of measures of effectiveness to aid the design assessment, support progressive human engineering acceptance, and to anticipate future simulation requirements. The measures of effectiveness that were identified depicted specific performance characteristics that had to be demonstrated over a series of trials.

However, as Harwood (1993) argued, it is usually the deficiency of stated performance requirements that poses the greatest challenge to design validation. If it is to be determined that system goals (as opposed to design goals) can be achieved to some acceptable level or that personnel can perform tasks effectively without excessive stress, there needs to be some standard of what constitutes acceptable performance that is expressed, preferably, in quantitative terms. Meister (1991) suggested a standard might be developed "by creating a consensus of what would constitute *obviously* acceptable and non-acceptable performance, and by gradually narrowing the bounds or limits between them" (p. 506).

Frequently, new systems do not have testable requirements and criteria for human performance and human-machine interface issues. Moreover, in a systems context, the measures associated with human factors criteria are not always linked to higher task and/or mission requirements (Andre & Charlton, 1994). In the absence of salient links to system performance, there may be only a superficial examination of those human factors deficiencies that are identified in the course of operational testing and evaluation.

One of the advantages of early field testing during the development cycle is that the field test provides not only an insight into how the system elements will function in the operational environment but also an opportunity to capture and refine meaningful requirements for system certification (Harwood & Sanford, Chap. 19). Direct behavioral observations in field settings can sometimes provide the necessary data for the critical incident technique (Flanagan, 1954; Meister, 1985). Researchers have traditionally relied on the critical incident technique to determine measures of proficiency or the attributes relevant to successful performance in complex operations. With an increasing dependence on computer-based technology and the concomitant emphasis on cognitive activities in operational environments, behavioral observation has become more difficult, if not irrelevant, for understanding operator performance. As a consequence, the assumptions underlying the critical incident technique have to be reviewed to determine how they are impacted by this shift in emphasis (Shattuck & Woods, 1994). This is particularly the case in air traffic control, for example, where observable behaviors do not adequately reflect what the controller is doing at any given moment.

Human Performance Models as Sources of Criteria. System design reflects the designer's knowledge or assumptions about operator limitations, but the validity of the implemented design will depend on the validity of the behavioral models used in the requirements specification and in the design process (see Bainbridge, 1988; Burbank, 1994; Pew & Baron, 1983; Rouse & Cody, 1989). However, when there are insufficient grounds for extrapolation from an existing system, there must be greater reliance on models of human performance and cognition for system specification and design (Elkind, Card, Hochberg, & Huey, 1989; MacLeod & Taylor, Chap. 14 Wickens et al. , 1997).

Comprehensive overviews of several types of human performance models (including models for task allocation and workload analysis, individual tasks, multitask situations, and crew performance) are available (cf. McMillan et al., 1989; McMillan et al., 199), but many of the human performance models have not yet been systematically validated.

The Measurement Problem: Identifying the Measurement Strategy

Is there a prerequisite measurement strategy for certification? The particularly critical issues within the question of measurement strategy include the matter of whether (or when) process or outcome measures are the more appropriate—or are specifically required. Another concern relates to the challenge of matching performance criteria to the level of system activity being addressed. The higher the system level at which measurements are taken, the more encompassing the performance criteria must be, but at the same time, the more difficult to justify the measures, perhaps, for purposes of validation. Still another issue concerns the degree of experimental manipulation that is required to elicit potential (and perhaps adverse) interaction effects. For example, a comprehensive simulation of an ATC environment (cf. Graham, Young, Pichancourt, Marsden, & Irkiz, 1994; Wise, Hopkin, & Smith, 1991) can enable a variety of individual and system-level performance measures to be taken under controlled experimental conditions, and critical performance measures may be compared against operational data.

Can operational testing, using real-time simulations, realistically address the requirement for experimental manipulation in the certification process? It could be argued reasonably that a scenario-driven or phenomenon-driven approach (Carroll, 1995; Sarter & Woods, 1994) affords the opportunity to select and control indirectly the parameters of the critical variables in comprehensive simulations. The scenario-based approach addresses in large measure the need for representativeness in validation. Simply stated, representativeness of subjects, representativeness of variables, and representativeness of setting (or ecological validity) can be viewed as the major components of external validity (Kantowitz, 1992), where external validity is the generalizability of findings from a study (cf. Campbell & Stanley, 1966; Locke, 1988; Sherwood-Jones cited in Taylor & MacLeod, chap.9; Weimer, 1995). (See also Westrum, chap.16)

For example, in outlining the challenges in system validation, Hollnagel (1993) alluded to the questions of whether test cases are representative and whether the functions provided by the interface are sufficiently complete. Are they representative of situations likely to be encountered? Woods and Sarter (1993) asked whether the validation methods chosen are sufficiently sensitive to detect design errors and whether the problem space captures the range of variation and contextual factors that operate in a particular domain. "New technology introduces new error forms; new representations change the cognitive activities required to accomplish tasks and enable the development of new strategies; new technology creates new tasks and roles" (Woods, Patterson, Corban, & Watts, 1996, p. 970). Consequently, the scenarios used to validate system design must truly sample the operational space, including the complications and challenges that confront the end users (Carroll, 1995; Gawron & Bailey, 1995; Guidi & Merkle, 1993; Neiderman, 1993; Sarter & Woods, 1994). The utility of contemporary complex simulation studies is likely to be limited only by the failure either to include or to manipulate the more critical variables and the measurement process becomes a critical element in the design and interpretation of what Woods and Sarter (1993) referred to as "converging evidence on system performance" (p. 141)

PROGRESSIVE ACCEPTANCE TESTING AS HUMAN FACTORS CERTIFICATION

Unquestionably, the concept of validation carries an implicit notion of evaluation undertaken to substantiate the effectiveness and viability of a system design, and it could be argued that validation as certification ought to be a *cumulative* process. Taylor and MacLeod (Chap. 9) describe a concept of progressive acceptance testing and evaluation. Similarly, it has been argued that iterative evaluation activity—intended as validation—should be an integral part of system design rather than a "fig leaf" at the end of the process (Woods & Sarter, 1993) and that the objective of system evaluation should be to help the designer improve the system, not simply justify the resulting design. The process of iteration—between domain modeling, discovering requirements, and prototype refinement (i.e., what Woods, Patterson, Corban, and Watts, 1996, call *understanding*, *usefulness*, and *usability*)—is seen as fundamental to the design activity. System development, to be effective, requires a balanced investment across the iteration of modeling, innovation, and refinement of prototypes. A common practice in many instances, however, is for system designers to be overly dependent on prototyping to uncover design requirements.

For advanced complex automated systems, Taylor and MacLeod advocate that progressive acceptance testing and evaluation should be embodied throughout the different stages and levels of system design and development. There is a particular concern that as the integration of system functionality is achieved, an assessment of effectiveness, with the operator in the loop, may be deferred until final acceptance testing. The risks associated with this approach are significant for systems "requiring major engineering integration activity to avoid potentially high operator workload" (p. 128).

A key element in the description of the iterative process in system design for Woods et al. (1996) is the understanding and modeling of human error. Hollnagel (1993) and Reason (1993), among others, had considerable concern about how the question of error and system failure is to be addressed in system evaluation. For Hollnagel, erroneous actions can be seen either as context dependent and thus system induced or as context independent and thus residual. The first are associated with the interaction between the person and the system. The second are due to the inherent variability of human cognition and performance. Reason (1990, 1993, 1997) would move the focus beyond human reliability in operations *per se* to be cognizant, as well, of human reliability in system design and management. Reason provided a distinction between active failures and latent failures. Whereas active failures are unsafe acts committed by the operators, latent failures are usually fallible decisions taken at the higher echelons of the organization. The damaging consequences of the decisions become evident only "when they combine with local triggering factors (i.e., active failures, technical faults, atypical system states, etc.,) to breach the system's defences" (Reason, 1993, p. 224).

Obviously, these uncontrolled sources of variance, within the domain of human–machine reliability, pose significant challenges to the concepts of validation and progressive human engineering acceptance testing—with or without the added implication of certification.

IMPLICATIONS FOR THE CONCEPT OF CERTIFICATION

If we assume that need for certification could be met by an iterative validation process or progressive acceptance testing, how are we to reconcile the constraints that often accompany validation studies (cf. Baker & Marshall, 1988; Chapanis, 1988)? In addition to the frequent constraints on representativeness of variables, the failure to replicate studies, inadequate training, and long-term changes in the work environment, we can add the cautions for evaluation methodology that have been described in Wise, Hopkin, and Stager (1993).

Although the concept of validation can be argued to be inherent in the process of design and experimental investigation, the confidence that we can place in a program of iterative evaluation, as a certification process, will depend on our ability to identify and accommodate, through the design, construction, and management of systems, potential sources of human and system variance. In the final analysis, the most appropriate objective in system validation is to minimize the variance not accounted for in system design. The acceptable level of risk associated with the variance not accounted for that can be accepted will likely be a function of experience with similar systems, perhaps of societal expectations, and certainly of real costs. However, the act of certification—whether by an iterative process as advocated in this chapter or by an independent process at a terminal point in the development cycle—in essence, is saying that the level of risk is acceptable for the operational objectives of the system in question.

SUMMARY OBSERVATIONS

Relying only on final operational testing and evaluation for system acceptance constitutes a significant risk, but beyond the question of risk as it relates to the possible need for design change is the issue of whether the traditional objectives of certification can be met, at that point in the cycle, with reasonable validity.

Contemporary evidence and opinion would seem to argue for progressive and iterative human engineering evaluation throughout the design and development cycle. However, it is likely that the economics of the activity will require that the pursuit of this human engineering assessment be contractually specified (cf. Hanes, Chap. 5; Taylor & MacLeod, Chap. 9). Further, additional effort toward the development of human engineering standards and guidelines, particularly for system testing and evaluation, is an obvious prerequisite if a contractually bound evaluation process is to be fully effective.

Identifying and defining objective operational performance criteria will always present a challenge in the evaluation of advanced human–machine systems. Perhaps the most difficult challenge, however, lies in the requirement to articulate an appropriate measurement strategy for validation. As the functional complexity and sophistication of advanced operational systems continue to increase, it becomes particularly urgent in system validation studies to be able minimize the behavioral variance not accounted for during the design process and to address the known sources of variance associated with human performance.

Human factors certification of advanced systems has to do with approving the level of risk associated with a system design, and hence certification, ultimately, has to do with the measurement strategy applied in system validation.

REFERENCES

Andre, T. S., & Charlton, S. G. (1994). Strategy-to-task: Human factors operational test and evaluation at the task-level. In *Proceedings of the Human Factors and Ergonomics Society 38th Annual Meeting* (pp. 1085–1089). Santa Monica, CA: Human Factors and Ergonomics Society.

Bainbridge, L. (1988). Multiple representations or 'good' models. In J. Patrick & K. D. Duncan (Eds.), *Training, human decision making and control* (pp. 1-11). Amsterdam: North-Holland.

Baker, S., & Marshall, E. (1988). Evaluating the man–machine interface—The search for data. In J. Patrick & K. D. Duncan (Eds.), *Training, human decision making and control* (pp. 79-92). Amsterdam: North-Holland.

Burbank, N. S. (1994). *The development of a task network model of operator performance in a simulated air traffic control task*, (DCIEM Tech. Rep. No. 94-05. Toronto, : Defence and Civil Institute of Environmental Medicine.

Campbell, D. T., & Stanley, J. C. (1966). *Experimental and quasi-experimental designs for research*. Chicago: Rand McNally.

Carroll, J. M. (1995). *Scenario-based design: Envisioning work and technology in system development*. New York: Wiley .

Chapanis, A. (1988). Some generalizations about generalization. *Human Factors, 30*, 253–267.

Department of Defense. (1979, January 31). *Human engineering requirements for military systems (MIL–H–46855B)*. Washington, DC: Author.

Elkind, J. I., Card, S. K., Hochberg, J., & Huey, B. M. (Eds.) . (1989). *Human performance models for computer-aided engineering*. Washington, DC: National Academy Press.

Federal Aviation Administration (1994). *Automation strategic plan*. Washington, DC: U. S. Department of Transportation.

Federal Aviation Administration (1996a, January). *Aviation System Capital Investment Plan*. Washington, DC.: U. S. Department of Transportation.

Federal Aviation Administration. (1996b, October). *National Airspace System architecture version 2.0*. Washington, DC: U.S. Department of Transportation.

Flanagan, J. C. (1954). Critical incident technique. *Psychological Bulletin, 51*, 327–358.

Gawron, V., & Bailey, R. (1995). Lessons learned in applying simulators to crewstation evaluation. *The International Journal of Aviation Psychology, 5*, 277-290.

Graham, R. V., Young, D., Pichancourt, I., Marsden, A., & Irkiz, A. (1994). *ODID IV simulation report*. (EEC Rep. No. 269/94, Task AS08). Brétigny-sur-Orge, France: Eurocontrol Experimental Centre.

Guidi, M. A., & Merkle, M. (1993). A comparison of test methodologies for air traffic control systems. In *Proceedings of the Human Factors and Ergonomics Society 37th Annual Meeting* (pp. 1196–1200). Santa Monica, CA: Human Factors and Ergonomics Society.

Harwood, K. (1993) . Defining human-centered system issues for verifying and validating air traffic control systems. In J. A. Wise, V. D. Hopkin, & P. Stager (Eds.), *Verification and validation of complex systems: Human factors issues* (pp. 247–262). Berlin: Springer-Verlag.

Hollnagel, E. (1993). The reliability of interactive systems: Simulation based assessment. In J. A. Wise, V. D. Hopkin, & P. Stager (Eds.), *Verification and validation of complex systems: Human factors issues* (pp. 205–221). Berlin: Springer-Verlag.

Kantowitz, B. H. (1992). Selecting measures for human factors research. *Human Factors, 34,* 387–398.

Locke, E. A. (1988). *Generalizing from laboratory to field settings.* Lexington, MA: Lexington Books.

McMillan, G. R., Beevis, D., Salas, E., Strub, M. H., Sutton, R., & Van Breda, L.(Eds.). (1989) (Eds.), *Applications of human performance models to system design.* New York: Plenum Press.

McMillan, G. R., Beevis, D., Salas, E,Stein, W., Strub, M. H., Sutton, R., & Reynolds, K. C. (1992). *A directory of human performance models for system design.* (NATO Report AC/243 (Panel 8) TR/1.), Brussels: NATO Defense Research Group.

Meister, D. (1985). *Behavioral analysis and measurement methods.* New York: Wiley.

Meister, D. (1991). *Psychology of system design. Advances in human factors/ergonomics.* Amsterdam: Elsevier.

Meister, D. (1992, April). Validation in test and evaluation. *Test & Evaluation Technical Group Newsletter, 7*(2), 2–3. (Human Factors Society Test and Evaluation Technical Group).

Meister, D. (1997). *The practice of ergonomics: Reflections on a profession.* Bellingham, WA: Board of Certification in Professional Ergonomics.

Neiderman, E. C. (1993). Real-time simulation of multiple parallel approaches. In *Proceedings of the Human Factors and Ergonomics Society 37th Annual Meeting* (pp. 1191–1193). Santa Monica, CA: Human Factors and Ergonomics Society.

O'Hara, J. (1994). Evaluation of complex human–machine systems using HFE guidelines. In *Proceedings of the Human Factors and Ergonomics Society 38th Annual Meeting* (pp. 1008–1012). Santa Monica, CA: Human Factors and Ergonomics Society.

O'Hara, J., Stubler, W., Brown, W., Wachtel, J., & Persensky, J. (1995). Comprehensive guidance for the evaluation of human–systems interfaces in complex systems. In *Proceedings of the Human Factors and Ergonomics Society 39th Annual Meeting* (pp. 1160–1164). Santa Monica, CA: Human Factors and Ergonomics Society.

Pejtersen, A. M., & Rasmussen, J. (1997). Effectiveness testing of complex systems. In G. Salvendy (Ed.), *Handbook of human factors and ergonomics* (pp. 1514–1542). New York: Wiley.

Pew, R. W., & Baron, S. (1983). Perspectives on human performance modeling. In G. Johannsen,& J. E. Rijnsdorp (Eds.), *Analysis.design and evaluation of man-machine systems:Proceedings of the IFAC/IFIP/IFORS/IEA conference*, (pp. 1–14). Oxford UK: Pergamon.

Reason, J. (1990). *Human error.* Cambridge UK: Cambridge University Press.

Reason, J. (1993). The identification of latent organizational failures in complex systems. In J. A. Wise, V. D. Hopkin, & P. Stager (Eds.), *Verification and validation of complex systems: Human factors issues* (pp. 223–237). Berlin: Springer-Verlag.

Reason, J. (1997). *Managing the risks of organizational accidents.* Brookfield, VT: Ashgate.

Rouse, W. B., & Cody, W. J. (1989). Designers' criteria for choosing human performance models. In G. R. McMillan, D. Beevis, E. Salas, M. H. Strub, R. Sutton, & L. Van Breda (Eds.), *Applications of human performance models to system design* (pp. 7–14). New York: Plenum.

Sarter, N. B., & Woods, D. D. (1994). Pilot interaction with cockpit automation II: An experimental study of pilots' models and awareness of the flight management system. *The International Journal of Aviation Psychology, 4* 1–28.

Shattuck, L. G., & Woods, D. D. (1994). The critical incident technique: 40 years later. In *Proceedings of the Human Factors and Ergonomics Society 38th Annual Meeting* (pp. 1080–1084). Santa Monica, CA: Human Factors and Ergonomics Society.

Stager, P. (1991). The Canadian Automated Air Traffic Control System (CAATS): An overview. In J. A. Wise, V. D. Hopkin, & M. L. Smith (Eds.), *Automation and systems issues in air traffic control* (pp. 39–45). Berlin: Springer-Verlag.

Stager, P. (1996, December). Automation in the Canadian Automated Air Traffic System (CAATS). Briefing presented to the Panel on Human Factors in Air Traffic Control Automation, National Research Council, Somers Point, NJ.

Weimer, J. (1995). Developing a research project. In J. Weimer (Ed.). *Research techniques in human engineering.* (pp. 20–48). Englewood Cliffs, NJ: Prentice-Hall.

Wickens, C. D., Mavor, A. S., & McGee, J. P. (Eds.). (1997). *Flight to the future: Human factors in air traffic control.* Washington, DC: National Academy Press.

Wickens, C. D., Mavor, A. S., Parasuraman, R., & McGee, J. P. (Eds.). (1998). *The future of air traffic control: Human operators and automation.* Washington, DC: National Academy Press.

Wise, J. A., Hopkin, V. D., & Smith, M. L. (Eds.) (1991). *Automation and Systems Issues in Air Traffic Control .* Berlin: Springer-Verlag.

Wise, J. A., Hopkin, V. D., & Stager, P. (Eds.) (1993). *Verification and Validation of Complex Systems: Human Factors Issues.* Berlin: Springer-Verlag.

Wise, J. A., Hopkin, V. D., Stager, P., Gibson, R. S., & Stubler, W. F. (1993). Verification and validation of complex systems: Human factors issues. In *Proceedings of the Human Factors and Ergonomics Society 37th Annual Meeting* (pp. 1165–1169). Santa Monica, CA: Human Factors and Ergonomics Society.

Woods, D. D., Patterson, E. S., Corban, J. M., & Watts, J. C. (1996). Bridging the gap between user-centered intentions and actual design practice. In *Proceedings of the Human Factors and Ergonomics Society 40th Annual Meeting* (pp. 967–971). Santa Monica, CA: Human Factors and Ergonomics Society.

Woods, D. D., & Sarter, N. B. (1993). Evaluating the impact of new technology on human–machine cooperation. In J. A. Wise, V. D. Hopkin, & P. Stager (Eds.), *Verification and validation of complex systems: Human factors issues* (pp. 13–158). Berlin: Springer-Verlag.

9

Quality Assurance and Risk Management: Perspectives on Human Factors Certification of Advanced Aviation Systems

Robert M. Taylor
DERA Centre for Human Sciences , U.K

Iain S. MacLeod
Aerosystems International, UK

This chapter is based on the experience of engineering psychologists advising the U.K. Ministry of Defence (MoD) on the procurement of advanced aviation systems conforming to good human engineering (HE) practice. Traditional approaches to HE in systems procurement focus on the physical nature of the human–machine interface. Advanced aviation systems present increasingly complex design requirements for human functional integration, information processing, and cognitive task performance effectiveness. These developing requirements present new challenges for HE quality assurance (QA) and risk management, requiring focus on design processes as well as on the design content or product.

A new approach to the application of HE, recently adopted by NATO, provides more systematic ordering and control of HE processes and activities to meet the challenges of advanced aircrew systems design. This systematic approach to HE has been applied by MoD to the procurement of the mission systems for the Royal Navy Merlin helicopter. In MoD procurement, certification is a judicial function, essentially independent of the service customer and industry contractor. Certification decisions are based on advice from MoD's appointed Acceptance Agency. Test and evaluation (T & E) conducted by the contractor and by the acceptance agency provide evidence for certification. Certification identifies the limitations on release of systems to service. Evidence of compliance with HE standards traditionally forms the main basis of HE certification. Significant noncompliance could restrict the release to service.

The systems HE approach shows a concern with the quality of processes as well as with the content of the product. Human factors (HF) certification should be concerned with the quality of HE processes as well as products. Certification should require proof of process as well as proof of content and performance. QA criteria such as completeness, consistency, timeliness, and compatibility provide generic guidelines for progressive acceptance and certification of HE processes. Threats to the validity of certification arise from problems and assumptions in T & E methods. T & E should seek to reduce the risk of specification noncompliance and certification failure. This can be achieved by T & E being creative, informative, and an integrated component of the design process. T & E criteria for HE certification should be directly linked to agreed systems measures of effectiveness (MOE). HE risk should be managed principally through iterative T & E and progressive acceptance. Integrated and iterative HE T&E procedures linked to MOE criteria should feed progressive acceptance and provide confidence of compliance with specification and QA criteria. Certification should include human behavior as an integral part of the total systems functioning.

Traditionally, the risk for human performance in systems has been a customer responsibility. Recent initiatives in procurement policy seek to provide a more integrated approach in which human resources issues, including operator or maintainer capability and training, are considered at all stages of the procurement process. The success of this initiative will depend on the ability to measure and predict human competencies in systems operation. It may be possible to specify successfully the requirements for skill- and rule-based behavior. However, uncertainties inherent in the performance of knowledge-based behavior present difficulties for system specification and certification.

BACKGROUND

Experience in supporting HF aspects of various MoD air systems acquisition programs from the late 1970s through the 1980s revealed a number of general problems in procuring systems to conform to good HE practice (Taylor, 1987). These problems may be summarized as follows:

- HF requirements were poorly defined in system specifications.
- HE design standards focused on the physical characteristics of the human–machine interface and not on the design process, nor on the performance and effectiveness of functions, tasks, and operating procedures.
- Increasing systems complexity was increasing the impact of HF on operator performance and mission effectiveness.
- Poor systems integration increased human information processing and operator workload and reduced situational awareness.

- Responsibility for HF was shared between the customer and the supplier.
- The demand for HF advice was increasing beyond that which could be supplied by the customer HF advisors.
- Contracting policy (fixed prices) encourages rigid adherence to specifications and reduces the flexibility for changing HF requirements during system design and development.
- Acceptance procedures for HE QA based on ergonomic checklists and late demonstration evaluation were ineffective and not directly related to mission effectiveness criteria.
- Problems in operating complex systems were difficult and costly to resolve through in-service modification and rectification.
- Unacceptable risk in HF was carried by the customer.

THE HE APPROACH TO SYSTEMS DESIGN

In 1985, discussions with North American HF colleagues in the Air Standardization Coordinating Committee and NATO military aircrew systems and cockpit standardization fora revealed similar problems in HE procurement. The U.S. HF personnel had made substantial inroads into HE procurement problems during the Navy F/A 18 aircraft acquisition programme. This procurement was based on the extensive application of the principles of U.S. Department of Defense (DoD) Military Specification MIL–H–46855, *Human Engineering Requirements for Military Systems, Equipment and Facilities.* MIL–H–46855 concentrates on the importance of timeliness of key HE activities, traceability, and the performance of critical tasks. It highlights the importance of early "front-end" analysis techniques (mission and scenario analysis, function analysis, function allocation, task analysis, and performance prediction) in reducing subsequent system development costs and risks. The progressive nature of these stages in HE analysis is illustrated in Figure 9.1. The design and development process is iterative. Analyses are repeated several times during the course of design and development. MIL–H–46855 promotes the value of an agreed, tailored, and systematic human engineering program plan (HEPP) with tracing of the required HE effort from the initial analysis, design, and development to final system & E, including activities, responsibilities, time scales, products, and deliverables. The HEPP makes clear the detailed contractor HE responsibilities and requires full consideration of the resourcing, cost, and risk implications during contract tendering. Application of the HEPP is coupled with U.S. Military Standard MIL–STD–1472, *Human Engineering Design Criteria for Military Systems, Equipment and Facilities,* which provides detailed equipment design requirements for good HE practice. Canadian HF colleagues, using the same principles, verified that, used properly, MIL–H–46855 provided an excellent approach.

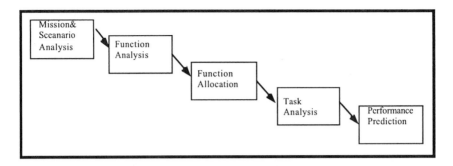

FIG. 9.1. Stages of Human Engineering Analysis (From Beevis, 1992)

In 1985, NATO and ASCC cockpit design standards were concerned with relatively specific technologies and equipment, and with individual controls, displays, layout, and lighting requirements. There was no statement of integrating policy. Based on the North American experience, it was decided to generate international standards similar to MIL–H–46855 and MIL–STD–1472 to specify HE activities during aircrew system acquisition. The derivative NATO and ASCC standards have been available since 1990. The sequence of NATO STANAG 3994 activities is illustrated in Fig. 9.2. Similar activities are identified in the triservice MoD Defence Standard DEF–STAN–00–25 *Human Factors for Designers of Equipment* (Part 12: Systems), published in 1989. This MoD standard provides *permissive guidelines* in accordance with the systems approach but without explicitly defining the requirement for a structured plan (i.e., no HEPP). Other initiatives aimed at a wider integration of human resources considerations in systems acquisition, including manpower, personnel, training, and safety requirements, such as the U.S. Army MANPRINT program recently adopted by the U.K. MoD Army, incorporate similar systems HEPP procedures based on MIL–H–46855. Detailed MANPRINT HE procedures are described in AMC–P 602–1, -*MANPRINT Handbook for RFP Development* (Barber, Jones, Chung, & Miles, 1987).

T & E IN SYSTEMS HE

According to STANAG 3994 and MIL–H–46855 philosophy, the aim of HE T & E is to verify that the human–machine interface and operating procedures are properly designed so that the system can be operated, maintained, supported, and controlled by user personnel in its intended environment. The following guidance is derived from the STANAG with extracts from the DoD Human Engineering Procedures Guide Military Handbook (DOD–HDBK–763; U.S. Department of Defense, 1987).

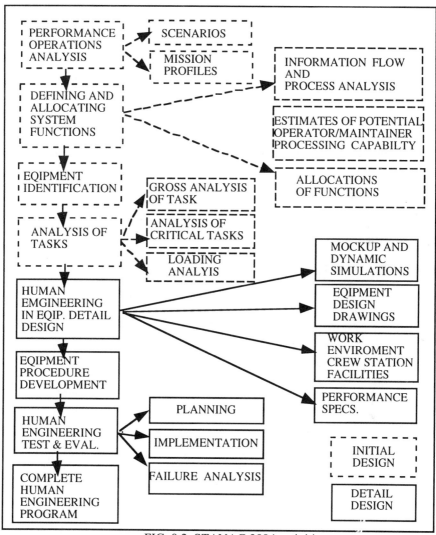

FIG. 9.2. STANAG 3994 activities

Identification of Test Parameters

System performance requirements need to be identified for verification during HE T &E . Identification of HE T & E parameters should be based on the mission analyses, in conjunction with the critical task analyses and the loading analyses. The criteria for selecting the system performance requirements should be the same as for the identification of critical tasks. These requirements should be used to develop an HE test plan for approval by the procuring agency.

Test Plan

The HE test plan (HETP) should specify the type of T & E techniques, rationale for their selection, procedures to be used, data to be gathered, number of trials, number and training of trial subjects, trial conditions, and the criteria for satisfactory performance. The relation with other T & E activities should also be indicated. The HETP should be specified to ensure that the human performance requirements of the system are met and reported to the customer. Areas of noncompliance and their consequences should be identified with justification. This information should enable the customer to determine the influences of the operators and maintainers, and their performance, on total system effectiveness and reliability. It should indicate how the test program results will influence the design and apply to follow-on equipment or similar systems.

QA Compliance

In indicating how the HETP data will be used, the plan should describe if the collected data are to be used as formal proof of QA compliance. Proof of compliance should be indicated as being either by analysis, inspection, demonstration or measurement. MIL–H–46855 reporting requirements call for data item descriptions (DIDs) which include a human engineering test report (HETR; DI–H–7058). Formal compliance may be provided by the HETR.

NATO Defense Research Group Endorsement

The systems approach to HE has been reviewed and endorsed recently by NATO Defense Research Group, Panel 8, RSG 14, Analysis Techniques For Man–Machine Systems Design. The report of RSG 14 (Beevis, 1992) offers the following observations:

- The concept of a system may well have been established prior to consideration of HF issues. As a result, designers and engineers have difficulty understanding the need for analyzing systems from a functional point of view. This makes HE analyses of function allocation of little value.
- The importance of the approach is that it permits engineers and designers to look at the system concept in new ways, by identifying the functions that must be performed, rather than by identifying the subsystems that may be required.
- This function-oriented point of view facilitates the development of novel system designs and encourages revolutionary as well as evolutionary changes.
- Increasing levels of automation and complexity in advanced mission systems make it more important that the roles and functions of the human operators are analyzed in detail.

- The effectiveness of HE analysis techniques is based on a decomposition of the system design problem into functions, subsystems, or states that are defined and validated.
- These items are then recombined to predict the system performance and operator and maintainer workload.
- It is generally assumed that the prediction of system performance is valid if it is based on the validated performance of subsystems.
- The QA aspects of the various techniques need to be better understood.
- The link from HE analyses to system performance requirements must be made explicit.
- In most analyses, particularly for function allocation, the link is indirect and it can only be provided by additional analyses of system performance.

MERLIN HE

In the United Kingdom we have experience in applying MIL–H–46855 principles by calling up STANAG 3994 as a mandatory document on several air systems acquisition programs. We have been particularly keen to raise the profile and effectiveness of HE and to export more of the HE risk in procurement to contractors, maintaining HE QA. STANAG 3994 is perceived as a potentially valuable aid both for maintaining HE QA and for managing the HE risk in the procurement of complex mission systems. The risk for HE is perceived to be particularly important in the procurement of complex mission systems. In complex systems, situation assessment and mission performance effectiveness are functions of the integration and interaction between the operator and equipment information processing and cognitive decision-making capabilities. The U.K. program that provides the most advanced example of STANAG 3994 application is the procurement of the prime contract for the Royal Navy Merlin (formerly EH101) antisubmarine warfare helicopter. This project is known as the Merlin Prime Contract (MPC). RAF IAM, DRA Farnborough and Aerosystems International have acted as HE technical advisors on this program. This chapter is largely based on the HEPP acceptance and compliance assurance issues that have arisen on the MPC program.

Merlin Specification Rationale

The development of the U.K. Royal Navy (RN) Merlin helicopter was progressed from the RN EH101 development program by giving responsibility for the RN EH101 helicopter to a prime contractor (IBM/ASIC). In the process the helicopter was renamed Merlin. To aid the submission and assessment of bids by the potential primes, the Merlin aircraft was specified with relation to design, functionality and its operational performance and acceptance specification (OPAS). the technical requirement specification (TRS) lists standards and rules governing design. The OPAS dictates the trials, their types and formats, and the methods to be used for the acceptance of Merlin by the RN. Figure 9.3 shows the basic contents of the Merlin specification.

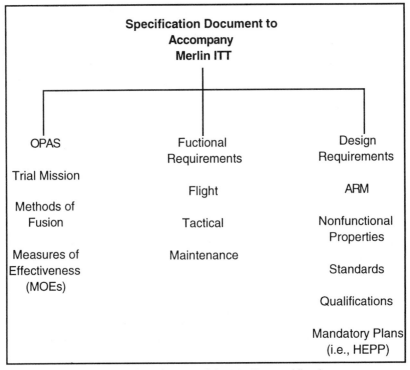

FIG. 9.3. The contents of the Merlin specification.

OPAS

The OPAS trials are in two forms. Single-task trials assess the operational performance of individual equipment. Stressing mission trials assess the operational performance of multiple systems within a realistic flight trial and operational scenario. The requirements for trial aircrews are specified. Where the need for trained service aircrews is identified, the need for appropriate qualifications, experience, and conversion training are specified. The means of assessing the performance of the trials is specified. One of the main criteria for assessment is the establishment of MOE. The MOE are based on specific high-level functions that are progressively decomposed to a level of MOE depicting specific performance characteristics that have to be demonstrably met over a series of trials. Pass–fail acceptance criteria are agreed for the deterministic Single

Task Trials. The operator-in-the-loop Stressing Missions will be performed on a test and declare basis (i.e., with no pass–fail criteria). The view has been taken that service crew competence is not a Contractor responsibility. Thus, crew performance is considered to be an uncontrolled and unpredictable variable. The contractor's intention is to reduce the risk in the stressing missions by additional operator-in-the-loop simulations prior to OPAS.

Merlin HEPP

The application of HE to the Merlin is governed by a mandated HEPP, in accordance with STANAG 3994. The HEPP is managed by Westlands Helicopters Ltd. (WHL) on behalf of IBM/ASIC. The agreed HEPP is a tailored implementation of STANAG 3994 applicable to all new or modified equipment and systems of the Merlin specification, compared with the earlier EH101 specification, namely active dipping sonar, data link, identification friend or foe, global positioning system, and digital map. Figure 9.4 illustrates the concept of the HEPP and T & E binding together the Merlin high-level functionality.

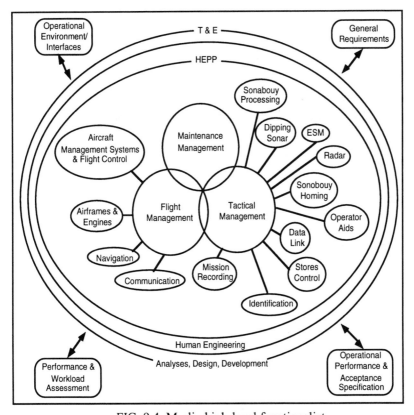

FIG. 9.4 Merlin high-level functionalist.

The weakness of the HEPP is the limited influence of equipment or systems that were developed for RN EH101 without a mandated HEPP, which will remain largely unmodified. The plan focuses on the extended mission systems human–machine interface (HMI) in the rear cabin where the Merlin specification has the main impact. Aircraft HE integration issues arising from the flight deck have little influence on the Merlin HEPP, having been covered under the RN EH101 development. OPAS is considered to fulfill the mission analysis requirement. Also, system functions are based largely on the existing EH101 definition and allocation, amplified by the Merlin functional requirements definition (FRD). This renders further function analysis either unnecessary or potentially ineffective. Notwithstanding the requirements of the new Merlin equipment, the HEPP largely concerns activities after equipment identification, from task analysis to equipment detail design, with the traditional emphasis on the HMI. The primary focus is to ensure that, as new features are added, the operator HMI workload is manageable. Early identification of workload and design challenges reduces the risk of future cost and scheduling problems. Consequently, the HEPP has a strong workload emphasis. It specifies the analyses, simulation assessments, workload measurement trials, and tools for HMI development. In summary, through the extended HMI, the HEPP and associated T & E linked to OPAS MOEs can be conceived as the means of delivering the HE of the required TRS and FRD high-level functionality. Figure 9.5 shows the HE testing sequence in relation to the system life cycle.

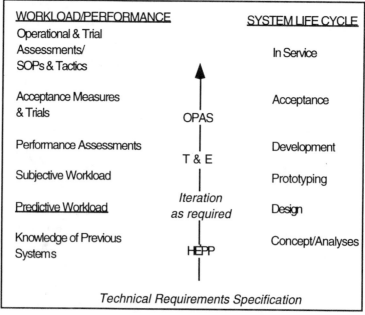

FIG. 9.5. HE testing sequence in Merlin life cycle.

Merlin Predictive Analysis

A key feature of the Merlin HEPP is the role of predictive analyses of workload and decision making to aid the design assessment, to support progressive HE acceptance, and to anticipate future simulation and flight trials (MacLeod, Biggen, Romans, & Kirby, 1993). Critical mission segments were selected from OPAS. Mission "storylines" were created for the segments from interviews with subject matter experts (SMEs). These storylines were transformed into operational sequence diagrams (OSDs) at the level of aircrew subtask activity. These OSDs were the basis for the workload and decision analyses. The sequencing and the relation of the analyses are depicted in Figure 9.6.

Workload Analysis

In the workload analysis, detailed task time lines were generated from empirical observation and published task time data. Attentional demand loadings were created from SME loading estimates using VACP (visual, auditory, cognitive, psychomotor) workload model criteria recommended by MoD (Taylor, 1990),and subsequently validated by the contractor (Biggen, 1992). The results are used to indicate the peaks and troughs of workload, to determine their causes, and to suggest solutions to the amelioration of unwanted workload. The data generated to date have indicated predicted task time overruns on critical mission segments, compared with the baseline intended times. These overruns are being addressed largely with reference to the efficiency of the proposed operating procedures. The predicted workload data obtained so far indicate some short transient areas of multitask conflict during continuous monitoring tasks, leading to reduced situational awareness, mostly due to the demands of simultaneous intercom tasks. There are also indications of imbalance in the workload distribution between the two rear-operator positions (observer and aircrewman).

On the whole, the predictions are judged by the contractor as indicating manageable workload problems with amelioration action through procedure examination during simulator workload validation. Maintaining and refining the workload prediction model, and keeping it up to date with new equipment and task requirements are recognized as important requirements for progressive HE acceptance. The initial analysis was static and deterministic. It is intended to conduct future analyses using dynamic and stochastic network simulation.

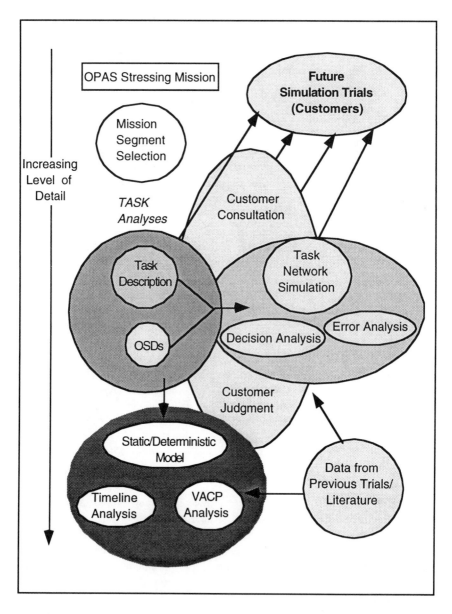

FIG. 9.6 Relationship of Merlin HE Predictive Analyses

Decision Error Analysis

The decision analysis uses a novel technique to examine task-related decision processes and their associated errors. The TRS called for particular attention to be given to the cognitive aspects of the Merlin HE. The quality of situation assessment and decision making are considered to be a key factors in determining operational effectiveness of the Merlin mission system. This consideration influenced the choice of the stressing missions for OPAS. Stiles and Hamilton (1987) pointed out that the interdependency of mission goals means that there are often decision points that permit the operator to modify his or her intent according to the assessment of the situation. The options associated with goals are controlled at these points. The designer must ensure that the option paths are clearly presented at these points in the context of the situation. Decision analysis could become the controlling activity for the design process, complementing the information analysis. It was necessary to develop a novel technique because decision analysis is a relatively new activity. Several attempts at developing a task analysis technique for decision making have been reported in the literature. But, as noted in a NATO research study group report, no single technique has emerged as the most promising approach (Beevis, 1992). The form of decision analysis used on Merlin is described in detail by MacLeod et al. (1993).

In summary, based on the OPAS mission storylines OSDs, human error probabilities associated with the performance of task segments were generated from the literature or from SMEs. The effects of errors on subsequent decision processes were estimated by SMEs in terms of error probability and error severity. These error influences on critical tactical decisions were then mapped against estimated task times through dynamic stochastic network simulation in MicroSAINT for Windows, which provides dynamic simulation of critical decisions and errors through various decision paths to operator task completion using Monte Carlo rules. The results provide traceable evidence of the efficacy of tactical decisions on the probability of mission success. The output identifies critical decision points affecting mission performance. These critical decision points can be correlated with the workload analysis. They can be used to guide design action through improved information provision, option clarification and highlighting, and procedure modification and training

Certification

By definition, to certify is to endorse or guarantee that certain required standards have been met. According to a dictionary, certification is "the act of certifying" or "the state of being certified". The word *certify* has its roots in the Latin *certus* (certain) and *facere* (to make). To be certain means to be positive and confident about the truth of something. In law, certification is a document attesting to the truth of a fact or a statement.

The requirements for the act of certification are that the system should fit its intended purpose and meet specific requirements of reliability, safety, and performance. Certification is more than endorsing compliance with the system specification, which is the concern of the contracting authority, because the specification may not include all the necessary requirements.

Government Functions

In the government and management of systems design, the role of certification can be considered as a judicial function, rather than as a legislative or executive function. Certification is a judgment on the design standard of the system; it carries with it major implications for program risk and costs. The following are further notions of how these functional distinctions can be applied:

- *Legislative functions*: Staff requirement generation, system technical requirements specification, design standards definition, acceptance standard definition, technical transfer agreement, contracting.
- *Executive functions*: Contract management, programme planning, concept analysis, prototyping, design, development, documentation, production.
- *Judicial functions*: T & E, compliance demonstration, acceptance, concession negotiation and agreement, audit, QA, certification.

Legislative functions are the responsibilities of the customer, task sponsor, the contracting authority (MoD) and its project or program office. Executive functions are largely the responsibilities of the contractor or manufacturer, in consultation with the customer authority. Separation from the legislature and the executive is essential to preserve the effectiveness of the judicial function. Failure to achieve certification has major implications for both the customer and the contractor. It follows that in the interests of independence and impartiality, HE certification needs to be independent from both the legislative and the executive functions. Certification of the overall testing and acceptance plan ultimately should be the responsibility of an independent agency appointed by the customer authority and recognized by the contractor or manufacturer.

Certification Authority

Certification is the end product of successful test and evaluation T & E. Logic dictates that T & E follows analysis and design. In the United Kingdom, the ultimate endorsement for military aircraft systems is the release to service granted by the MoD Controller Aircraft (CA), namely the CA release. Certification for civil aircraft is issued by the Civil Aviation Authority (CAA). CAA certification needs to be particularly stringent because of the responsibility for carrying civilian passengers. The object of CA release is to provide a statement to the service department that the aircraft will perform its intended in-service role with acceptable levels of safety and effectiveness. This statement includes any limitations or restrictions to be observed in operating the aircraft at the defined build standard. All systems should be safe to operate and fully effective under all specified environmental conditions. CA release covers the performance of the mission systems and the vehicle engineering systems, as well as the basic handling qualities of the aircraft. CA release is a progressive activity, beginning with an initial temperate functional CA release covering the temperature environment for the initial delivery aircraft for flight testing. Subsequent stages of release extend the scope of clearances for flight testing of early production aircraft through to the activities leading to the formation of the first operational squadron.

It is currently MoD policy to appoint an acceptance agency to advise that the system produced is adequately tested to prove that it meets the specification requirement. The acceptance agency liaises directly with the contractor on behalf of the MoD authority to endorse trial plans, monitor trials, and assess results against the agreed performance criteria, recommending acceptance or rejection by MoD. Responsibility for trials planning and control rests with the contractor. An MoD trials agency may be appointed to assist the contractor in aspects of trial planning and control involving MoD facilities and to advise on operational and support requirements. The MoD Aeroplane and Armament Experimental Establishment (A & AEE) at Royal Air Force Boscombe Down is the MoD agency for aircraft operational trials and acceptance testing. A & AEE provides the aircrew for the Merlin contractor T & E progressive acceptance demonstrations and flight trials. CA release is based on recommendations by A & AEE. A & AEE assessments are governed by the requirements of the aircraft technical specification and relevant MoD Defense Standards, MIL specs and MIL standards, in particular DEF–STAN–00–970 *Design and Airworthiness Requirements for Service Aircraft*. DEF–STAN–00–970 includes chapters on general HE requirements for cockpit vision, controls, displays, layout, and lighting. These are called up in the system specification and they are used by the manufacturer to guide design. The manufacturer is required to provide evidence of qualification for compliance to assist the certification process. Avionics systems rigs with representative human-machine interfaces are used by A&AEE to support the process of CA Release. Data generated by the Contractor during the development trials testing also contribute to CA Release. A & AEE does not employ HE specialists. This weakens the ability of A&AEE to act as an Acceptance Agency for HE. There is merit in having a single Acceptance Agency responsible for all aspects of aircraft acceptance. DRA and IAM provide A&AEE with technical advice and scientific support on HE Acceptance. As the demand for HE Acceptance increases and becomes more sophisticated, the need may arise for A&AEE to employ HE specialists as an integral part of its Acceptance function.

Certification Validity

The credibility or trustworthiness of certification depends on the validity of the evaluation on which it is based. This requires that attention be paid to the threats to validity for the particular evaluation and design decisions. Sherwood-Jones (1987) provided a summary of the threats to quality in evaluation using quasi-experimental designs, familiar to behavioral scientists and HE specialists. These include nine threats to internal validity, namely:

- History—Events other than those studied occurring between pretest and posttest that could provide an alternative explanation of effects.
- Maturation—Processes within the system producing changes as a function of the passage of time.
- Instability—Unreliability of measures, fluctuations in sampling.
- Testing—The effect of taking a test on the scores of a second test.
- Instrumentation—Changes in calibration, observers, or scores that produce changes in the obtained measurements.
- Regression artefacts—Pseudoshifts from selection of subjects or treatments based on extreme scores.
- Selection—Bias from differential recruitment of comparison groups leading to different mean levels on the measure of effects.
- Experimental mortality—Differential loss from comparison groups.
- Selection maturation interaction—Bias from different rates of maturation or autonomous change.

Six threats to external validity can be identified covering the problems in interpreting experimental results and of generalizing to other settings, other treatments, and other measures of the effect:

- Interaction effects of testing—For example, pretesting effects and sensitivity to variables.
- Interaction of selection and experimental treatment—Unrepresentative responsiveness of the treated population.
- Reactive effects of experimental arrangements—Artificiality in the experimental setting being atypical of the normal application environment.
- Multiple treatment interference—Effects of multiple treatments different to separate treatments.
- Irrelevant responsiveness of measures—All complex measures have irrelevant components that may produce apparently relevant effects.
- Irrelevant replicability of treatments—Complex replications failing to reproduce the components responsible for the effects.

QA

In accordance with the emphasis in MIL–H–46855 and STANAG 3994 on functional effectiveness, certification of criteria for HE acceptance should provide a broad endorsement of QA or fitness for purpose. The word *quality* is defined in British standards as "the totality of features and characteristics of a product or service that bear on its ability to satisfy a given need". The definition of QA is "all activities and functions concerned with the attainment of quality". MoD Defence Standard DEF–STAN–05–67, Guide to *Quality Assurance in Design,* emphasizes that all concerned with a project have a contribution to make to its quality and are involved in the assurance of quality. QA organizations undertake specific activities in measuring quality and ensuring that appropriate contributions are made by all to QA. But the responsibility for the quality of the final product rests with the line management responsible for design and production, including performance over the system life cycle. This is a basic tenet of *total quality management* (TQM).

HE can support the TQM approach by helping to identify the characteristics of the system users and their requirements, and the features of operator or maintainer performance, which contribute to variance in the system product or output. The RSG 14 Report notes that the distinction is made between *quality of design,* meaning "the process of task recognition and problem solving with the objective of creating a product or a service to fulfil given needs"; and *quality of conformance,* meaning " the fulfilment by a product or service of specified requirements". HE QA is a function of how well it contributes to the design of an effective system (quality of design) and how well it provides accurate, timely, and usable information for the design and development team (quality of conformance). The following indexes or criteria are proposed by RSG 14 as providing evidence for HE QA:

- Schedules that show that the analyses will be timely.
- Organization charts that indicate that the HE effort will be integrated with other systems engineering and integrated logistical support (ILS) activities.
- Use of metrics and measures of effectiveness that are compatible with each other and with other engineering activities.
- Compliance with a relevant specification.

The scheduling and charting of HE activities are key MIL–H–46855 and STANAG 3994 characteristics. On the basis of an analysis of HE analysis techniques, RSG 14 recommends the following QA criteria be considered in the development of an HEPP:

- Completeness.
- Consistency with preceding analyses.

- Timeliness.
- Compatibility with other engineering analyses.

Consideration of QA draws attention to the need for concern about both the design process and the content of the product. Advanced systems employ new interface technologies and concepts. Existing HE standards for detailed equipment design have decreasing relevance and influence as new technologies and concepts are introduced. With new ideas, what has increasing importance for the quality of the product is the nature of the design process. HE certification for advanced aviation systems needs to be concerned with proof of process more than proof of content. This is the philosophy of MIL–H–46855 and STANAG 3994.

Creative Evaluation

The certifying authority might wish to conduct some form of human factors audit or ergonomic checklist for QA certification purposes. Indeed, the U.S. General Accounting Office (U.S. GAO, 1981) provides guidelines for this purpose, identifying questions to assess whether human factors were considered during the weapon system acquisition process. However, such an audit would not serve to inform the design process. Evaluation should be useful, informative and preferably creative. The need for evaluation to be useful was addressed by Patton (1978). Evaluation can be either *formative* and aimed at improving the design, or *summative* and aimed at deciding whether to proceed with a design. There are two fundamental requirements for making evaluation useful, namely:

- Relevant decision makers and information users must be identified rather than having an abstract target audience.
- Evaluators must work actively, reactively, and adaptively with those identified decision makers to make all other decisions about the evaluation; that is, focus, design methods, analysis, interpretation, and dissemination.

Progressive Acceptance

Both in common engineering practice and in the formalized approach advocated by MIL–H–46855 and STANAG 3994, HE acceptance testing is embedded as an integral part of the design process. HE involves a logical sequence of mostly iterative activities, each involving the application and testing of design and performance criteria and associated standards. Like software QA, T & E for HE acceptance needs to be phased or progressive. Progressive acceptance T & E should be embodied in the different stages and levels of the system design and development process. This could be referred to as technical rather than operational T & E. The higher levels of HE QA concerned with functionality and effectiveness are the most significant and yet the most difficult to check. Consequently, there is a danger that checking integrated functional effectiveness of the total system, with the operator or maintainer in the loop, will be fully addressed only in final operational acceptance testing. Relying only on final operational T & E for full HE acceptance is risky, particularly with complex mission systems requiring major engineering integration activity to avoid potentially high operator workload. In theory, the system should be designed to pass operational T & E without uncertainty. Progressive HE acceptance testing is needed during integration on rigs, simulation facilities, and development aircraft to ensure that the lower level requirements are being dealt with correctly. Otherwise it is unlikely that the higher levels will be acceptable. But this process needs to go particularly deeply into the operational performance of complex mission systems to guarantee functional integrity and effectiveness. Progressive acceptance is a key contributor to proof of process.

CERTIFICATION OF HUMAN BEHAVIOR

The Government-Furnished Equipment Approach

Formal acknowledgment of human functioning as an integral component of systems, together with equipment functioning, is a relatively recent development. Certification of systems in which the human is considered as a system component presents new challenges for systems engineering. The traditional approach to systems engineering focuses on the equipment functioning. It treats the human operator or maintainer as a given quantity, over which the contractor has little or no control and responsibility, often "jokingly" referred to as government-furnished equipment (GFE). The traditional design aim is to provide a system that is fit for a purpose and can be operated reliably, safely, and effectively by the average operator or maintainer. Unfortunately, average is ill defined, and becomes a quantity left to the judgment of the MoD A & AEE test aircrew. The danger in the GFE approach to human capability is that it carries the implicit assumption that it is sufficient to treat the performance of the average operator or maintainer in a deterministic, predictable, and mechanistic manner, when of course the uniquely human characteristics in systems are flexibility, adaptability, and unpredictability. Consequently, traditional HE analyses have tended to be physicalistic (anthropometry, ingress–egress, workspace layout, visibility and reach, lighting, task timeline analysis) rather than cognitive (situation assessment, decision making, errors of judgment, expertise, intentions, application of knowledge, tactics, strategy, goals). The consequences of this physicalistic–cognitive distinction are discussed in detail in the second paper by the authors (MacLeod & Taylor, 2000). The GFE approach prevents the Merlin OPAS stressing missions from being more than a test-and-declare process. The customer still bears the risk of total integration failure because this can be attributed to GFE variables. MANPRINT procedures, introduced since the EH101 procurement, seek to address this problem on future programs by procuring manpower, personnel, training, and HE.

Cognitive Functions

The traditional HE assumptions about human design requirements are at best limited in scope, and at worst invalid, if they are based on inappropriate models of human functionality in systems. They may lead to inaccurate, unrealistic, and optimistic assessments of overall system capability and effectiveness. Recent U.K. procurement experience indicates a tendency to be overoptimistic in predictions of future operational performance with complex advanced systems under development. With the GFE approach, the risk for human functionality in system performance is carried by the customer rather than by the contractor. Failure to achieve system performance targets in T & E can be ascribed to human capability or performance variability. This then becomes a problem of the human not matching the machine rather than the converse, to be solved by improved customer-provided training or by raised customer selection standards, and not by in-service system upgrades. This is increasingly untenable in a procurement climate seeking to minimize the risk to the customer. It is particularly inadmissible in the procurement of complex advanced mission systems in which system performance effectiveness is increasingly a function of operator–equipment integration and cognitive–level interactions in information processing, situation assessment, and decision making. The RSG 14 Report concludes that although it is generally assumed that new advanced systems place increasingly high demands on the cognitive aspects of operator or maintainer behavior, most HE techniques lend themselves to the description of skilled behavior, not cognitive behavior. It seems that certification of HF in advanced systems will require better resolution, analysis, and engineering of cognitive functions in future systems than HE techniques presently provide. Stiles and Hamilton (1987) described how a cognitive engineering approach to function analysis will be needed to identify what the intention of the pilot will be during his or her interface with the system, and then to provide a design (information and/or control) that helps achieve that intention. The requirement for improved resolution of cognitive functionality is discussed further in the second paper by the authors (MacLeod & Taylor, 2000).

Aircrew Certification

Certification procedures in aircrew selection and training might provide some of the missing human cognitive functional concepts and behavioral parameters needed in advanced aircrew systems HF certification. However, aircrew selection and training criteria are not yet firmly based on an understanding of the theory of cognition and behavior. Criteria for certifying aircrew ability as "adequate" for civil flying, or "above average, and not requiring further training" for military flying are largely based on performance on instrument flying tasks and on knowledge of rules and procedures for air safety. The required standards of *airmanship* are still highly subjective and judgmental, and largely the responsibility of experienced assessors and flying instructors. However, it is possible that the mystery surrounding airmanship will not remain much longer. MIL–H–46855 and STANAG 3994 call for a potential operator capability analysis to provide data for defining and allocating functions. Also, MANPRINT requirements for target audience description demand a more explicit, objective, and theoretically consistent approach to the definition of aviator performance.

The problems of measuring and developing competence in the cockpit are major concerns of training technologists. Brown (1992) noted the increasing concern with cognitive decision-making competencies for combat aircrew, in addition to the requirements for traditional skills and knowledge of flying. In the systems approach to training, competency is viewed as an outcome of a system, and that competency occurs within a system, as part of system functioning. Recent procurement policy for "turn-key" training systems has created the need for more functional and performance-based specifications, rather than for the formerly equipment-based specifications (Brown & Rolfe, 1993). This requires the customer to define the operating constraints and the training outcomes required, including the activities to be learned on the device, the rate of learning, and the performance standard. This places increased importance on the quality of the task and training analysis performed by the supplier in determining that the equipment will meet the task. Also, it focuses attention on the role of evaluation in acceptance testing, which may need to extend the evaluation into the system lifecycle to demonstrate that the device actually trains.

A review of the requirements for operator and automation capability analysis, in the context of advanced aircrew system design and human–electronic crew teamwork, points to the key role of human performance modeling for the prediction of human system performance (Taylor & Selcon, 1993). The embedded human performance model for cockpit performance prediction and pilot intent inferencing incorporated in the U.S. Air Force, Pilot's Associate, indicates some of the necessary HE elements (Lizza, Rouse, Small, & Zenyuth, 1992). What is needed is a common performance-resource model and associated taxonomy for systematically linking human resource capabilities to mission performance task demands, incorporating the features required for HE analysis, with the addition of relevant human competence parameters (Taylor, 1991).

Skill-, Rule-, and Knowledge-Based Taxonomy

The taxonomy of skill-, rule-, and knowledge-based behavior provides a potentially useful way of thinking about HF certification issues. In *skill-based behavior*, exemplified by the performance of controlling tasks, performance is relatively easily measured, demand is relatively easily predicted, and the capability requirement can be specified and verified. Hence, skill-based behavior is a strong candidate for HF certification. More or less the same can be said for *rule-based behavior*, exemplified by supervisory and monitoring tasks. Difficulty arises with the certification of *knowledge-based behavior*, exemplified by planning and decision-making tasks. By definition, knowledge-based behavior is novel, measurement of performance is qualitative and at best nominal (e.g., correct or incorrect decision), and demand is stochastic and probabilistic rather than predictable and deterministic. The capability requirement for knowledge-based behavior is the most difficult to anticipate, specify, and verify.

It is difficult to conceive of a contractor being prepared to guarantee, say, that incorrect decisions in uncertainty would be made less than 5% of the time. Traditionally, analysis of the decision points, where the operator changes goals and alters the information and controls requirements, is left out of the design process. Some progress can be made toward this kind of aim through decision analysis (MacLeod et al., 1987; Stiles & Hamilton). Metzler and Lewis (1989) reported procuring the Airborne Target Handover System/Avionics Integration for the Apache aircraft by specifying a 30% reduction in crew task time for each task (60% overall), with 90% mission reliability and with no more than 5% of the mission aborts attributed to human errors. The Merlin decision analysis explores the impact of decisions on the probability of mission success, but the findings are considered indicative rather than definitive.

Ideally, the design aim is to provide systems that are totally predictable and reliable. This must mean avoiding, if possible, the need for knowledge-based behavior, and probably the provision of totally automated systems. However, it is in the nature of the military environment that human situation assessment, hostile intent inferencing, and unbounded knowledge-based behavior, applied through the flexible appreciation of goals, tactics, and strategy, often provide the combat winning edge. Systems that are intended to operate in uncertain environments need to provide the unrestricted scope for appropriate knowledge-based behavior. The recent debate about providing situational awareness in highly automated systems is an example of this problem. Arguably, for certain military systems in which effectiveness depends on flexibility, adaptability, and unpredictability, it is this limitless capacity for knowledge-based behavior that needs to be certified.

CONCLUSIONS

Notwithstanding system life cycle considerations (i.e., maintenance, in-service modification, updating), certification marks a formal end to the system design, development, and production process. It is the last operational endorsement of the proof of concept, proof of process, and proof of product. It is the final sanction of the solution to the design problem. The threat of noncertification and a severely restricted release to service is a potentially powerful device. It could help ensure that HF considerations maintain their rightful place at the center of the design process. Consideration of the ability to certify HF aspects of system design is a sign of the maturation and acceptance of HF methodologies and standards. But, realistically, most HF issues are a long way from being assigned sufficient importance to become potential "show stoppers" for certification. With power comes risk of abuse. This could be a problem if certification is seen as an end in itself. What happens if in assessing novel technology and a revolutionary new system concept, existing certification criteria are wrongly focused, invalid, and fail to measure the true impact on health and safety? There should be a duty of care on the certification authority that necessitates a continuing process of self-evaluation. Care must be taken not to assign blind trust to existing certification procedures. Certification alone is not generative or creative. Front-end analysis, iterative design and testing, and progressive acceptance provide the methods and tools for generating confidence and HE QA necessary for certification. There is a danger of certification encouraging rear-end analysis. As such, it carries many of the characteristics and weaknesses of traditional, 1970s late ergonomic assessments, as identified at the start of this chapter. Certification is not a panacea, capable of remedying the ills of poor design methodology. It can be only as good as the front-end analysis and T & E that feeds it. It is probably essential to ensure that HF considerations, HE processes, and HE standards are contractually mandated as an integral part of the design process using MIL–H–46855 and STANAG 3994 procedures. HF certification then can be added to endorse compliance with these contractually binding requirements.

The uncertainty of human reliability is a fundamental problem for HF certification. Certification concerns matters that are certain and true in fact. Obviously, one cannot be certain about matters that are uncertain. Certification cannot be obtained for design concepts or prototypes tested only in the abstract or by simulation. Certification can only be valid for the real product tested in the real operational environment. Progressive acceptance rather than certainty is all that can be obtained for concepts and prototypes. Certification can guarantee that specific absolute HF design standards are met, and that necessary design and test processes and activities have taken place. However, when there is a human in the loop as an integral system component, it is difficult to conceive of contractually meaningful expressions of certainty about total system fitness for purpose, system performance and functional effectiveness. Human performance, whether based on skill, rule, or knowledge, is inherently uncertain. All that can be expected with certainty is an endorsement or guarantee that sometimes the required standards of human–systems performance will not be met. Levels of confidence in human systems performance could be provided in probabilistic rather than absolute terms. Probabilistic certification of human–systems functioning might provide the basis for a form of limited release to service, perhaps associated with additional supervisory, performance monitoring, and training safeguards. In advanced systems, the role of humans is increasingly one of dealing with the uncertainty that cannot be handled automatically, or the variability that can not be predicted and controlled. The human component is responsible for generating the required system performance, and for achieving the intended system effectiveness goals, under circumstances that cannot be entirely predicted and anticipated. Probabilistic descriptions of the intended and expected system functioning, performance, and effectiveness are likely to become more common as specification goals and certification norms. Certainty is perhaps too absolute a term for many HF certification requirements. Confidence, acceptance, and perhaps certitude, may be more appropriate terms for describing the relative uncertainties of human–machine systems performance.

REFERENCES

Barber, J. L., Jones, R. E., Ching, H. L., & Miles, J. L. (1987, September). *MANPRINT handbook for RFP development.*(AMC–P 602–1) Alexandria, VA: H.Q. U.S. Army Material Command.

Beevis, D. (1992, July). *Analysis techniques for man-machine system design.* (AC/243, Panel 8 TR/7).DCIEM, Toronto.:NATO.

Biggen, K. (1992, March). *EH101 mission workload simulation validation trials report.* (ER02Q002W) Yeovil., Summerset: Westlands Helicopters.

Brown, H. M. (1992). Competency in the cockpit. In D. Saunders & P. Price (Eds.), *Developing and measuring competence: Aspects of educational and training Technology.*(Vol. 25). London: Page.

Brown, H. M. and Rolfe, J. M. (1993). *Training requirements or technical requirements.* Unpublished manuscript.

DEF-STAN-00-25, (1989) *Human Factors for the Designers of Equipment.* (Part 12: Systems). The Ministry of Defense, Directorate of Standardization. Glasgow, UK.

DEF-STAN-00-970, (1983) *Design and Airworthiness Requirements for Service Aircraft.*(Volume 1, Issue 1) The Ministry of Defense, Directorate of Standardization. Glasgow, UK.

DEF-STAN-05-67, (1967)*Guide To Quality Assurance In Design.*Her Majesty's Stationary Office: London.

Lizza, C. S., Rouse, D. M., Small, R. L. & Zenyuth, J. P. (1992, July). Pilot's associate: An evolving philosophy. In T. E. Emerson, M. Rienecke, J. Riesing, & R. M. Taylor (Eds.), *The human electronic crew: Is the team maturing?* (WL–TR–92–3078,) Wright-Patterson Air Force Base: Air Force Material Command.

MacLeod, I. S., Biggen, K., Romans, J., & Kirby, K. (1993). Predictive workload analysis- RNEH101 Helicopter. In E. J. Lovesay (Ed.) *Contemporary ergonomics 1993,* London: Taylor & Francis.

MacLeod, I. S.& Taylor, R. M. (1993). Does human cognition allow human factors (HF) certification of advanced aircrew systems? In J. A. Wise, V. D. Hopkin, D .J. Garland ,(Eds.) *Human factors certification of advanced aviation* Daytona Beach, FL: Embry-Riddle Aeronautical University Press.

Metzler, T. R. & Lewis, H. V. (1989). *Making MANPRINT count in the acquisition process.* US Army Research Institute, ARI Research Note 89-37, June.

MIL-H-46855, *Human Engineering Requirements for Military Systems, Equipment and Facilities.*

MIL-STD-1472, *Human Engineering Design Criteria for Military Systems, Equipment and Facilities.*

Patton, M.Q. (1978). *Utilization-focused Evaluation.* Beverley Hills : Sage.

Sherwood-Jones, B. (1987). Human-factors audits and fitness for purpose. Proceedings of the CAP Scientific Conference, 1987.

STANAG 3994 AI, *The Application of Human Engineering to Advanced Aircrew Systems.*

Stiles, L. and Hamilton, B.E. (1987). Cognitive engineering applied to new cockpit designs. *Proceedings of the American Helicopter Society National Specialists Meeting. Rotorcraft Flight Controls and Avionics.* October 13-15, 1987, Cherry Hill, PA.

Taylor, R.M. (1987). Some thoughts on the future of engineering psychology in Defence. RAF Institute of Aviation Medicine,. Position Paper for the British Psychological Society Conference on the Future of the Psychological Sciences, Harrogate, 16-18 January. .

Taylor, R.M. (1990). *Merlin MPC Workload Acceptance Criteria.* RAF Institute of Aviation Medicine, IAM Letter Report 016/90, 2nd May.

Taylor, R.M. (1991). Human Operator Capability Analysis for Aircrew Systems Design :Proceedings of a Panel Session at the British Psychological Society 1991 Occupational Psychology Conference, RAF Institute of Aviation Medicine, Letter Report No 004/91, March.

Taylor, R.M., & Selcon, S.J. (1993). Operator and automation capability analysis: Picking the right team. In: *Combat Automation for Aircraft Weapon Systems: Man/Machine Interface Trends and Technologies.* AGARD CP 520. Neuilly Sur Seine : NATO.

US Department of Defence, (1987). *Human Engineering Procedures Guide.* DOD-HDBK-763, 27th February.

US General Accounting Office, (1981). *Guidelines for Assessing Whether Human Factors Were Considered in the Weapon System Acquisition Process,* GAO FPCD-82-5, 8th December.

10

Certification of Flight Crews for Advanced Flight Control Systems

Richard D. Gilson
David W. Abbott
University of Central Florida, USA

Pilots in command of turbo jet-powered airplanes are required by federal aviation regulations (FARs) to obtain additional certification of capability to operate each specific type of aircraft or family of aircraft. Although by and large these pilot operators already know how to fly airplanes in general, *type certificates* are required because turbo jets typically have highly sophisticated capabilities and complex systems far exceeding those of small simple airplanes. The assumption then is that the operation of one large turbine aircraft differs markedly from that of another type to a degree that skill in safe operation of one does not guarantee skill in operating another (Billings, 1996).

In advanced transport air carrier aircraft being flown today, reconfigurable video instrumentation, glass cockpits, and updateable flight management systems (FMSs) are the most apparent and distinctive features to their crews. These interfaces, along with the associated flight automation, literally isolate pilot operators from the airplane itself, theoretically enough to allow the safe flight of quite different airplanes that utilize the same display and control system. However, the chief advantage of such systems, flexibility, has become its weakness due to the loss of standardization of input–output techniques: symbology, format, sequencing, priorities, and the like. Different designs and selected customizations have been implemented by systems manufacturers, by airplane manufacturers, and by air carrier organizations themselves. This has created such variations among cockpits that pilots type certified for the same airplane may find flight safety compromised by their unfamiliarity with the particular interface system available to them. A few years ago, Gilson and

Abbott (1994) proposed, in effect, *interface certificates* to require pilots to demonstrate a high level of skill with the FMS, guidance system(s), and automation installed in the aircraft, in all possible modes and phases of flight not just a demonstration of basic flight capability with the aircraft type itself. Such skill could be assured by adding automated flight tasks to the current type certificate standards required by the FARs. In fact, New Zealand now requires type-specific global positioning system (GPS) avionics ratings for any pilot flying IFR GPS approaches. Given the rash of recent accidents attributed to FMS mismanagement—for example, in new technology airplanes (Sarter & Woods, 1997)—this type of skill demonstration may be more essential today than it was in 1994.

VARIETY OF AUTOMATED SYSTEMS IN AIRPLANE OPERATIONS

Pilots flying different route lengths (e.g., regional, domestic, or international) may encounter several automation types of aircraft in current operation. At the regional airline level, commuter twin-engine turbo–props such as the Beechcraft 1900 may not be equipped even with a basic autopilot. They are flown by direct mechanical manipulation)by cables, push–pull rods, hydraulic actuation, etc.) of flight control surfaces, engine power, and wing lift devices. It is actually more convenient to fly by hand on frequent, short, and varied commuter trips than it is to set up an autopilot to work for only a few minutes—just as most drivers leave the car's cruise control off in town. Also, many younger pilots, those who usually fly the commuters, prefer to hand-fly anyway for the experience. Thus the expense and complexity of autopilots may be bypassed.

Major airlines generally fly longer trips than the regional airlines in jet transport aircraft. Some of these advanced large transports may be fly-by-wire, electronic flight instrument system (EFIS; also know as a glass panel), fully automated digital electronic control (also known as auto-throttles), and FMS, literally capable of automated flight from takeoff to touchdown. In the Airbus 320 or Boeing 777, the airplane is flown by inputs submitted to a computer and filtered through a complex set of software rules that limit pilot input or initiate flight control changes in response to the software creators' ideas about safe combinations of power, flight attitude, control settings, altitude, and so on. The computer almost instantly considers the pilot request, the software "envelope protection" rules, the packaged automated mode instructions, and interactions with other flight guidance and power constraints, and then it initiates flight guidance action. Sometimes, however, such automated action can surprise the pilot (Sarter, Woods, & Billings, 1997).

Although different large transport types have different systems and limitations, experienced pilots frequently report that they fly alike. Is it also true that after considerable experience, all automated flight control systems and computers fly alike?" The answer to this is clearly no. Automation systems vary widely in their operation, even for the same series of airplanes, often acting differently from what is expected. Sarter and Woods (1997) reported that among line pilots flying Airbus-320 (A-320) aircraft, previous experience with automation provides little advantage in their performance on the latest highly automated systems. Specifically, pilots who had no previous glass cockpit or FMS experience and those who did have such experience did not differ when it came to automation surprises in the A-320. Eighty percent of both groups reported such surprises, where the automation behavior is different from what is expected by the pilot.

CRITICALITY OF AUTOMATED SYSTEM KNOWLEDGE FOR PILOT OPERATORS

Automated controls frequently take action without letting the pilots know, in any very explicit way, what the computer's vision of appropriate action might be (Sarter & Woods, 1997). Certainly, not notifying the pilots of intended automated action(s) can create undesirable consequences, ranging from hidden conflicts with their human crews to an incident or accident if the action(s) are radically inappropriate. The pilot in command, charged with responsibility for the safety of flight, must have an opportunity to input positive feedback that the planned action is desired, agreed with, and thus approved for implementation. Alternatively, the crews must have sufficient knowledge to avoid situations that might activate unwanted automated envelope protection features. To do so, however, the crew also must have sufficient proficiency in understanding the automated systems and their interactions, so that they can project a mental model (see Rouse & Morris, 1986) of current and next actions from the automation. Then they can competently judge those projected operations for appropriateness, including all downstream implications, before entering their request to the computer.

Considering all eventualities of automation is a daunting task even for the most knowledgeable pilot. In 1994, the chief test pilot for Airbus died in an A-300 accident while trying to evaluate envelope protection features in certain combinations of flight modes and procedures.

Even years after automated systems have been in routine operation, unknown surprises have surfaced. For example, in 1997 an Airbus A-300 crew discovered that a 40^0/second rate of bank causes the symbol generator units to shut off and automatically reset through a self-test mode. This blanks the EFIS screens and leaves the crew with no flight information except for the emergency backup instruments. An accident was avoided only by very careful monitoring that allowed the crew to keep up with the systems as they continue to take action. Other examples do not have such a favorable outcome and are not limited to Airbus. The crew's realization of an icing problem in an Aerospatiale

ATR 72 was masked and exacerbated by the automation until it was too late. The altitude hold feature continued to trim nose up (unnoticed) as the icing reduced the wing and tail efficiency until a stall and spin served to alert crew to the problem. Unfortunately, the first hint of impending trouble may be trouble itself, even for relatively low-level automated systems.

INDUSTRY THREATS TO OPERATOR PROFICIENCY

The airline industry is highly competitive, with frequent fare wars that doom the inefficient. Profitability depends on slimming expense margins for each and every operation. No operations are scrutinized more than those involved in flight, not only for safety, but because huge savings can be had for the smallest of efficiencies. Investments for long-term savings can have a short payback time, particularly with economies of scale. Flying safely is certainly a strong industry concern, but one must survive financially to fly at all.

Cockpit automation was initially introduced to replace the third person on the flight deck with a nonsalaried electronic crewmember, thereby leading the transition to our present day two-person flight crew (McLucas, 1981). It was not instituted with the expectation to improve safety or to reduce human errors, although that was the hope. Once implemented, however, designers quickly realized that aircraft could be operated far more efficiently with automated systems than when flown by hand, raising the prospects for additional savings. Since then efforts have been directed at honing efficiencies through smoothing flight control, optimizing flight profiles for fuel burn, obtaining direct routes, managing arrivals in poor weather, and so on. All of these can be aided by automation.

Training of pilots for the initial type rating and for maintaining proficiency is a high cost to the industry. Automation provides trade-offs by eliminating tasks. However, increasingly, the use of complex automation along with a reduced training footprint is a threat to operator proficiency (Abbott et al., 1996). Training frequently covers only basic operations of the automation, leaving much to learn on the job while flying the line (Sarter, Woods & Billings, 1998). It is true that different carriers may define basic differently, but deep system(s) understanding may not be attained through current training. Federal Aviation Administration (FAA) aircraft type certification does not now demand it, and thus type training does not now teach it.

NEW SYSTEMS FOR AUTOMATED COCKPITS

Several new technologies are coming online for automated cockpits. Concerns with controlled flight into terrain (CFIT) have led to two new developments.

First, advanced ground proximity warning systems are currently being installed throughout the industry. A digital database of terrain along with current position, altitude, velocity, and track allows the automation to issue warnings and advisories regarding potential flight into the ground ahead. Several suppliers manufacture such systems. Another technology to enhance positional awareness and reduce the CFIT accident is the use of the same digital terrain databases to provide a synthetic three-dimensional depiction of the out-the-window views on navigational display(s) during instrument flight. The notion is that synthetic terrain seen ahead compels awareness.

In another area, the strong interest in technology to support Free Flight[21] is driving the development of improved traffic avoidance systems beyond the current traffic collision avoidance system implemented as a cockpit display. Data-linked position reports of a highly accurate GPS position, velocity vector, altitude, time to next waypoint, estimated time of arrival, and so forth, will provide the basis for onboard cockpit display and automated avoidance of other aircraft traffic augmenting (ATC; Gilson, 1993). Direct routes, selected by the airline or flight crews themselves, offer substantial fuel savings for the industry. Such operations, however, will in turn necessitate a new type of automated system for the crew to manage, and of course will involve substantial differences in implementation by various manufacturers.

Finally, some manufacturers of automated flight controls are developing entire new interfaces for pilot computer input. One manufacturer has a cockpit control language interface under development, that will enable the pilot to talk to the flight control computer in the same language as pilots use with ATC. This has great potential for reducing hands-on workload but not necessarily mental workload, ambiguity, or surprises arising from current or proposed interfaces in the industry. It also probably means that the variety of conventions for computer interface design will only increase in the future.

NEW ROLE FOR PILOTS OF ADVANCED AUTOMATION

Today, pilots of advanced automation aircraft rarely fly the airplane; instead they monitor it. Crews enter or simply download preprogrammed flight plans into the automated system and then monitor or supervise the actions of the automation. Gilson and Abbott (1994) noted that automation or even semiautomation is an asset for crews who are highly FMS proficient, by eliminating the routine drain of attention by direct flight control. However, for less proficient crews, the mental workload of the automation itself may be greater. A computer analogy works well here. People highly proficient with a program application such as a mail merge program may find it convenient to automatically address even a few letters, relieving them of repetitive inputs and

1 Free Flight is the monitor for direct routing, by passing traditional airways

possible typing errors. On the other hand, those less proficient may find it far easier to manually address a few letters because mastering the complexity of the mail merge may take longer and be fraught with potential software surprises. A similar dilemma is inherent in working with flightdeck automation and may be the reason why "current FAA testing is for type ratings with the FMS fully functional and fully operational [presumably the highest level of difficulty]. In prior years, examiners turned *off* equipment to increase the difficulty level" (Gilson & Abbott, 1994 p.122). The new role is very different for different automated systems. Without a detailed understanding of the particular automatic systems, it is difficult to always avoid surprises, leaving the potential for incidents and accidents (Sarter, Wood,s & Billings 1997).

PROPOSAL

It is proposed that flight crew supervisors should be certified for mastery of specific FMS control systems, in addition to aircraft [airframe] type ratings, perhaps with FMS type ratings. The superb psychomotor skills of pilots of yesteryear have been upstaged now by needs for superior assessment and cognitive skills, particularly as applied to monitoring and proficiency with the automated systems installed in aircraft. Safety demands this refocus from airframe training to automation training. Training for airline pilots is designed by the airline, paid for by the airline, and designed to meet FAA standards for pilot certification. The need for further training must be addressed by revised FAA standards requiring pilots to achieve high proficiency with the automation by introducing a type certificate" in the airframe–automation combination. This even may extend beyond automated flight control to the possible certification of crews to use software-specific navigational systems that provide guidance, especially during critical instrument flight operations (Gilson & Abbott, 1994).

JUSTIFICATION

Airlines train only to required standards to remain competitive financially. As such the standards must include aircraft type and FMS rating to ensure compliance by all:

> An advantage to FMS certification (beyond the demonstration of competency) is that it provides hardware, software, and instructional system designers with human performance benchmarks to guide the design and training system process to optimize accommodation of the human component. It also begins to insure that the total system (hardware, software, and people) will meet the expected standard.s (Gilson & Abbott, 1994 p. 121)

EVALUATION FOR FMS TYPE RATINGS

Computer-based training and testing is now as commonplace in aviation as elsewhere, for written examinations, for part task trainers, and for full flight simulation. Substitution of high-fidelity flight simulators has been an option for portions of or even all of the in-flight check for certain type ratings. Simulators have proven themselves particularly valuable to test crew performance safely with emergencies and by simulating difficult instrument approaches under demanding weather conditions. However, full-flight simulators may not be needed for testing of an FMS or for advanced avionics proficiency. Appropriate software for personal computers could be developed to train and test flight guidance and avionics proficiency without expensive high fidelity simulators or the aircraft. Software packages could be developed to be manufacturer specific FMSs with appropriate simulated couplings to particular aircraft types. With such software, pilots could safely and inexpensively demonstrate proficient use of all modes and levels of automation in a variety of standard flight tasks, such as achieving clearance limit targets (waypoint, altitude, airspeed, time, etc.), holds, nonprecision approaches, unexpected approach runway changes, and so on..

The depth of testing proficiency in using the FMS should go well beyond just how to use the system into knowledge structure of how the system works, at least down to a level at which pilot intervention is possible. Specifically, pilot operators should have the ability to track what the system is doing at any given time, and to accurately predict what the system will do next, allowing for cross-checking of programming, mode awareness, and for hardware and software errors. Proficiency testing should also include checks for mistaken inputs, skill and speed of programming or reprogramming changes, and the ability to cope with failures or unexpected actions. For example, written and oral testing should include accurate explanations and expectations for a variety of flight tasks, including anticipating action that will be taken by the automation, and actions to take if such a sequence or information is absent. Intrinsic testing is also possible. Sarter and Woods (1994) tested deeper proficiency in their experimental work by embedding changes in ongoing scenarios and observing diagnostic searches and response planning. Similar tests could be developed for certification.

SUMMARY

Unsafe flying has always been more mental than physical. Physical performance errors can lead to incidents or injury, but it is usually the mental errors that kill. Today, FMS computers can perform physical tasks far more accurately and reliably than their human counterparts, thus minimizing those types of errors. But mental errors may go unchecked. FMS computers rely on programmers and pilots to be told what to do, and thus are subject to their respective mental errors and conflicts. With the current variations in interfaces, logic, and databases, appropriate pilot operator FMS proficiency testing is both reasonable and prudent. We suggest FMS type certification similar to aircraft type certification.

REFERENCES

Abbott, K., Slotte, S., Stimson, D., Bollin, E., Hecht, S., Imrich, T., Lalley, R., Lyddane, G., Thiel, G., Amalberti, R., Fabre, F., Newman, T., Pearson, R., Tigchelaar, H., Sarter, N., Helmreich, R., & Woods, D. (1996, June). *The interfaces between flightcrews and modern flight deck systems* (FAA Human Factors Team Rep.). Seattle WA: Federal Aviation Administration.

Billings, C. E. (1996). *Aviation automation: The search for a human-centered approach.* Mahwah, N.J.: Lawrence Erlbaum Associates.

Gilson, R. D. (1993, September). Electronic aerospace and the avionics engine. *The Aviation Consumer, 23*(17 –18), 19-25.

Gilson, R. D., & Abbott, D. W. (1994). User "type" certification for advanced flight control systems. In J. A. Wise, V. D. Hopkin, D.. J. Garland (Eds.), *Human factors certification of advanced aviation technologies* (pp. 119–123). Daytona Beach, FL: Embry-Riddle Aeronautical University Press.

McLucas, J. L. (chair). (1981, July). *Report of the PResident's Task Force on aircraft crew compliment,* Washington, DC.

Rouse, W. B. and Morris, N. M. (1986). On looking into the black box: Prospects and limits in the search for mental models. *Psychological Bulletin, 100,* 349-363.

Sarter, N. B. and Woods, D. D. (1994). Pilot interaction with cockpit automation II: An experimental study of pilot's model and awareness of the flight management system. *International Journal of Aviation Psychology, 4,* 1-28.

Sarter, N. B. and Woods, D. D. (1995). "How in the world did we get into that mode?" Mode error and awareness in supervisory control. *Human Factors, 37,* 5-19.

Sarter, N. B. & Woods, D. D. (1997). Team play with a powerful and independent agent: An operational experiences and automation surprises on the Airbus A-320. *Human Factors 39* (4), 553-569.

Sarter, N. B., Woods, D. D., & Billings, C. E. (1997). Automation surprises. In G. Salvendy (Ed.), *Handbook of human factors and ergonomics* (2nd ed., pp. 1926-1943), New York: Wiley.

11

Certify For Success: A Methodology For Human-Centered Certification Of Advanced Aviation Systems

Ronald L. Small
Simauthor, Inc. USA

Ellen J. Bass
Search Technology, USA

The methodology described herein uses *Design for Success* (Rouse, 1991) as a basis for a human factors certification program. The *Design for Success* methodology espouses a multistep and iterative process for designing and developing systems in a human-centered fashion. These steps are:

1. Naturalizing—Understand stakeholders and their concerns
2. Marketing—Understand market-oriented alternatives to meeting stakeholder concerns.
3. Engineering—Detailed design and development of the system considering trade offs between technology, cost, schedule, certification requirements, and the like.
4. System evaluation—Determining if the system meets its goal(s).
5. Sales and Service—Delivering and maintaining the system.

Because the present focus is on certification, we elaborate on Step 4, system evaluation, as the natural precursor to certification. Evaluation involves testing the system and its parts for their correct behaviors. Certification focuses not only on ensuring that the system exhibits the correct behaviors, but only the correct behaviors. Of course, many excellent suggestions exist for evaluating complex aviation systems (cf. O'Brien & Charlton, 1996; Wise, Hopkin, & Stager, 1993); but, our purpose here is to provide thought-provoking suggestions for improving certification by using a human-centered systems design perspective.

Before we delve into evaluation and certification issues, however, a brief explanation of the first step, naturalizing, is necessary to provide context for the subsequent steps. However, in the interest of brevity, we then jump from step 1 to step 4 because they are the most germane to our suggested certification improvements.

Naturalizing

The main purpose for naturalizing is to understand the purpose of the system to be certified and to understand the concerns of the system's various stakeholders. From a human-centered perspective, the system's purpose should be described in a way that explains why and how the system supports humans in accomplishing their goals. For example, if we define the airline pilot's job as safely and efficiently moving passengers from origin to destination, then the purpose of the airliner and all of its parts is to support the pilot. (By parts we mean electric, hydraulic, and engine subsystems; flight management and other software modules; and individual components such as radios, circuit breakers, throttle levers, and switches.) Note that we are not stating that the pilot's job is to fly the airplane. Nor are we stating that the airplane transports people. Rather, the emphasis is on the human, the pilot, whose job it is to transport passengers by using the airplane. The subtle distinction of such a statement of system purpose is a key to thoroughly understanding and properly executing human-centered design, development and certification of aviation systems. This distinction becomes clearer with practice and is at the heart of naturalizing.

Defining the system's purpose requires understanding the history of the domain and the environment in which the to-be-certified system is to operate. Questions for identifying these issues include:

- Is this a new system or upgrade?
- If new, what was done previously? Why?
- What is the purpose of the system? (Answers should be stated in a human-centered format, as in the preceding airplane example.)
- What problems are there with the existing system?

(Note: If the system is completely new with no predecessor, the risk is probably too great and the system is probably not suitable for certification.)

The reasons for asking these questions are to understand the system's purpose and operational goals from the human perspective and to begin defining the set of metrics for evaluation and certification. Other metrics surface during discussions with stakeholders that must be recorded for use during the evaluation step. Also, during this naturalizing stage, it is important to use general and specific lessons learned from similar systems (or predecessors) so that "typical" problems can be planned for, and avoided or minimized (Rouse, 1994, 1998).

Who are typical system stakeholders and what are their concerns? The following section briefly answers these questions.

Stakeholders and Their Concerns

Typically system stakeholders are designers, developers, users, maintainers, purchasers, and certifiers of the system. Groups of stakeholders as well as individuals should be identified so that questionnaires can be devised and interviews can be scheduled. Because our focus here is certification, it is important to pay special attention to groups or individuals experienced with the current certification processes of similar systems. Questions asked of stakeholders include:

- What is the purpose of the system from your perspective?
- How should the system support humans to achieve their goals?
- What behaviors are expected during normal operations?
- What behaviors are expected during abnormal (degraded system) situations?
- What are the expected roles of the humans in both of these conditions?
- Who are the other system stakeholders from your perspective? Why?

The purpose for asking these questions is to understand the various stakeholder concerns so that the eventual certification can proceed along a well-defined path; after all, typical certification budget and schedule resources are limited. This well-defined path is derived from the metrics identified during this naturalist phase; therefore, stakeholder concerns must be expressed in quantifiable and measurable terms. These stakeholder-defined metrics then combine with the system metrics to become the set of measurement issues that form the basis of the system evaluation and the human-centered certification. In other words, the stakeholders' concerns and the derived metrics form the basis for the system requirements and specifications.

Stakeholder groups should have representatives on the certification team that actually conducts the system evaluation. This team concept ensures that all relevant stakeholder concerns are properly addressed during the evaluation and certification process. System evaluation is the subject of the next section.

SYSTEM EVALUATION

The first step in system evaluation is to define human-centered metrics based on the system's goals and purpose and based on the stakeholders' concerns gathered during naturalizing. Human-centered measurements are those that evaluate system performance and behavior from the humans' perspectives (e.g., operator, maintainer, trainer). For example, a software function may be able to execute in 5 milliseconds; but the system operator may only be able to comprehend that function's outputs at a 1Hz rate. From a human-centered certification standpoint, then, there is no reason to test the software function at an execution speed faster than 1Hz (however other reasons may exist for testing the function at the 200Hz rate).

Quantifiable metrics must be defined not only for the whole system, but for subsystems, modules, and components to evaluate their performance and behavior as the system is constructed. Although the certification authority is concerned with the system-level performance and behavior of the completed system, it is also important that the certification team have confidence in the underlying parts of the system. Therefore, this team should have access to developmental testing metrics, methods and results (e.g., task and failure analyses); and the team should independently verify a subset of those earlier tests.

Also, for human-centered certification purposes, the parts of the system should be evaluated as they interact to form operator-observable behaviors. These threads of interaction allow an operator representative on the certification team to focus on specific behaviors under specific circumstances—something that is difficult to do when evaluating the entire system because repeatable conditions are harder to generate as the system grows in complexity.

Another consideration for the certification team is to evaluate subjective as well as objective metrics. Subjective metrics include those that measure operator workload, situational awareness, tendencies to commit errors (due to memory overload, operational stresses, mode confusion, a faulty mental model of the system, etc.), and the appropriate task mixes between automation and the operator. Methods for objective and subjective evaluation are presented in the next section.

Evaluation Methodology

The guiding principle for the system evaluation is to test the system and its parts in such a manner as to yield results that can be compared against the metrics determined during naturalizing. Some analyses and evaluation methods include:

- Paper-and-pencil (mathematical) analyses.
- Modeling of the system or its parts from a human-centered view.
- Operator-in-the-loop experiments for even greater fidelity.

There are many other methods (cf. Gawron, Dennison, & Biferno, 1996), but in the interest of brevity, we describe only these three, with examples from personal experiences, in the following sections.

Mathematical Analyses. An envisioned airport safety system is being designed to detect and help prevent runway incursions and have minimal false alarms—a typical engineering trade-off between increasing system sensitivity and minimizing false alarms. Airport tower controllers are also responsible for detecting and preventing runway incursions (among their many other duties), so we performed a signal detection comparison between the automated safety system's specified detection performance and the historical human controller detection performance. Because runway incursions happen so infrequently, and because controllers detect and act to prevent most impending runway incursion accidents, we wanted to know if an automated runway incursion prevention system would boost the overall detection and prevention of incursions.

Using a statistical distribution analysis, we found that the automated safety system is not likely to improve the overall detection and prevention of runway incursions (Small, Rouse, Frey, & Hammer, 1993). This result is mainly due to the fact that controllers are already very good detectors of impending incursions, so their signal detection performance distribution vastly dominates the specified signal detection performance of the automated system. Obviously, we made some very broad assumption, but even with this fairly easy and inexpensive evaluation method, we were able to recommend that the automated system's detection rate should be modified somewhat. Another recommendation was to further analyze the result using higher fidelity analysis methods, such as modeling, to verify this preliminary result and to explore other issues, especially other performance metrics that could not otherwise be analyzed.

Modeling. Modeling is useful for testing hypotheses about the real system, but under conditions that the real system cannot be exposed to—for cost, safety, or other reasons. Digital models also allow for testing system behaviors in faster-than-real-time, thus enabling the evaluation of many replications under specified conditions that may yield statistically significant results.

For example, we developed a digital simulation of Atlanta's Hartsfield International Airport to test hypotheses about the effects of various features of the airport automation system described earlier. There were many simplifying assumptions needed to develop a model of this complex environment in a reasonable amount of time, but we were able to make some recommendations about controller communication workload under varying conditions. We could never have done such an analysis on the real system because it would have interfered with airport operations. Plus, we ran the model for replications of 40 simulated days in just a few minutes, thus enabling us to quickly obtain statistically significant results (Small et al., 1993).

Another benefit of system modeling is that analytical results help focus higher fidelity analyses on issues that cannot be studied using lower fidelity methods. Such a progressive methodology makes the more expensive evaluation methods, such as human-in-the-loop simulation studies, more cost-effective.

*Simulation Experiments.*System simulations are the next step increase in fidelity over digital modeling. Simulation experiments, with real system operators participating in testing, are useful when human operator interactions are required to evaluate the system (or some part of it) and yet the real system cannot be used because it does not exist yet, or because safety, cost, or operational reasons preclude using the actual system.

For example, we were involved in the design, development, and evaluation of the Pilot's Associate (PA) for a next-generation single-seat tactical fighter (Aldern et al., 1993). A simulation of the fighter's cockpit was needed to conduct utility testing of PA. This testing compared PA and non-PA conditions and used performance metrics ranging from fuel consumption to kill ratios to situational awareness. A method chosen for evaluating this range of metrics was pilot-in-the-loop simulation experiments because pilot opinion and performance comparisons were of vital importance to many of PA's stakeholders (Cody, 1992). The PA evaluation also used digital models to focus the piloted simulation studies on the metrics and conditions in which the greatest performance differences were expected (as in the airport study already described).

Although operator-in-the-loop simulation studies cost more than other evaluation methods, their results are more credible. It is usually the case that the higher fidelity (and more expensive) evaluation methods are also more credible; but that does not usually detract from the conclusions reached by the less expensive methods, as long as assumptions are made explicit. Furthermore, the less expensive analyses serve to focus the more expensive ones on metrics that may not otherwise be evaluated.

Methodology Summary

The goals for system evaluation are to analyze the system's performance (and all earlier intermediate results) relative to the set of metrics defined during naturalizing, and then to formulate conclusions and recommendations for system modification. In accomplishing these goals, the evaluation team must define follow-up analyses and tests when the performance results do not meet expectations and the system requires modification. The team also determines if new metrics are needed. If so they refine the set of metrics, as appropriate, and then conduct additional analyses and tests, iterating as needed until all metrics are satisfied.

During the system's design and development, the evaluation team should be the system's designers and developers. All test results are then made available to the final evaluation team. It is important to emphasize that each analysis method helps define the following higher fidelity evaluations. That is, the results from each method must be analyzed relative to previously defined metrics, and then used to refine any subsequent evaluation methods, or the next iterations of previous methods.

Now that we have described human-centered system evaluation, we highlight the distinctions between it and certification.

CERTIFICATION ISSUES

Although certification can be described as a more formalized evaluation process, it is distinct from the evaluation process described earlier in that it must independently analyze the system. This independent analysis should be very structured in the sense that separate subsystems or components have to pass different levels of scrutiny during certification.

For example, Requirements and Technical Concepts for Aviation (RTCA), an industry group that devises standards for aviation systems advocates different categories for certifying an airplane and its parts. The categories are based on the criticality of failure conditions, namely:

- Catastrophic—Failure conditions which would prevent continued safe flight and landing.
- Hazardous—Failure conditions which reduce safety margins, cause physical distress and such high air crew workload that tasks may not be completed accurately.
- Major—Failure conditions which increase crew workload thereby impairing crew efficiency.
- Minor—Failure conditions which slightly increase crew workload.
- No effect—Failure conditions which do not affect the operational capability of the aircraft or increase pilot workload.(Struck, 1992, p. 5).

These categories guide the human-centered certification process, described next.

Certification Process

How should a human-centered certification be conducted? The RTCA seems to emphasize crew workload levels in its definitions, and so should a human-centered certification methodology. Of course, workload levels are not the only human-centered metric. A certification team must also address the following concerns:

- What are the error conditions and the likelihood of the human operators committing those errors?
- What are the normal and abnormal operator procedures, and their likelihood of being performed correctly under varying conditions?
- What training is required for the system operators and maintainers?
- What types of screening for skills, and physical or physiological attributes are required?
- What is the tendency for the system's human–machine interface to promote the development of accurate mental models by operators in typical operational environments?

Similar questions have been posed by others (cf. Abbott, Slotte, & Stimson, 1996). Answering these questions is a nontrivial exercise, but the methodology for answering them is similar to the evaluation methodology described earlier. The gist of the distinctions between evaluation and certification is that certification analyzes failure conditions and their consequences, whereas evaluation examines correct or expected system behaviors.

Other differences between evaluation and certification relate to rules of development that are designed to minimize the system's dynamic response to conditions. Certifiable systems should not have unpredictable failure conditions. For example, when we built a certifiable knowledge base development tool, we had to pay special attention to some specific software engineering issues, including:

- Pointers—Introduce the potential for directing software execution to places in computer memory that may not be available for normal computations.
- Dynamic memory allocation—Introduces the potential for allocating memory that is already being used for other purposes.
- Compilers—The compiler used for development must be the same as that used for creating the actual executable code and for certification. The effect of this rule is that it inhibits the use of software development environments that typically have debuggers or other enhancements that enable more efficient software development, but that also greatly increase the amount of executable code loaded into a mission computer, for example. Consequently, either a sparse environment must be used for development (which is bad for software development efficiency), or two compilers must be used—one for development and one for precertification compilation—an expensive proposition (Hammer, Skidmore, & Rouse, 1993).

Another major difference between evaluation and certification is the composition of the certification team. As mentioned earlier, the evaluation team should initially be the system's designers and developers, whereas, the human-centered certification team must be independent (although it should examine the metrics, tests and analyses used by the evaluation team to ensure that the metrics are suitable and provide the necessary and sufficient coverage for the entire system and its parts).

Certification Team

One last set of questions concerns the nature of the human factors certification team:

- Do the members of this team need to be certified in the human-centered certification of systems?
- If so, what should be done to determine the certification team members' qualifications?

Abbott et al. (1996) posed similar questions. To answer all the previous questions during the certification process, the certification team must be competent in a wide range of human-centered design issues. In fact, we think that the certification team members should be certified by the certification authority in accordance with some professional standards and formal training (the training curriculum also requires certification then).

Determining a person's or group's competency in human-centered system design was one project's task that we accomplished (Hammer, Small, Frey, Edwards, Resnick, Skidmore & Zenyuh, 1994; Small, Hammer , Resnick & Reuse, 1995, Small et al., 1993). We devised a set of questions with answers that could be weighted and scored according to the needs of the system's stakeholders; we also recommended scoring guidelines. The questions are too numerous to present here, but they are based on the decomposition of human-centered system design competencies into four major topics and 20 specific issues (Fig. 11.1). A human-centered certification team should have individuals competent in, and certified for, evaluating a system in terms of these 20 issues.

CONCLUSION

As implied by the RTCA categories listed earlier, not every system or component should be certified to the same level. The extent of certification should relate to the component's or system's criticality. The extent of human-centered certification should relate to the component's or system's level of interaction with the human operators (or maintainers). Criticality and level of interaction also affect which system stakeholders and issues require the mostattention during the certification process, which brings us back full circle to

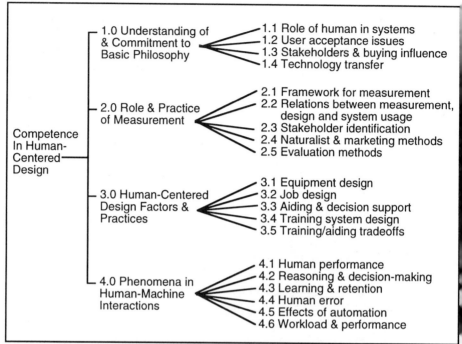

FIG. 11.1. Competencies in human-centered system design
(after Cody, 1993, p.C-3)

our initial naturalizing step and the analysis of stakeholder concerns.Typical human-centered stakeholder concerns are reflected in the topics and issues contained in our list of human-centered system design competencies.

Taken in its broadest meaning, human-centered issues comprise the set of issues to consider when certifying aviation systems. Because all system stakeholders have been surveyed and are represented on the system evaluation and certification teams, then, it is not only operator issues that are evaluated for certification. Indeed, human-centered systems design examines the role and responsibilities of all the humans involved in the system: operators, maintainers, trainers, managers, investors, designers and others. A thorough and iterative stakeholder analysis will uncover all the relevant issues and lead to a successful system certification.

ACKNOWLEDGMENTS

We gratefully acknowledge the inputs of Bill Rouse, John Hammer, Paul Frey, and Monica Skidmore, and the many Chateau de Bonas workshop attendees, especially Rene Amalberti, Vince Galotti, Dick Gilson, Lew Hanes, Hartmut

Koelman, Paul Stager and Ron Westrum, all of whom helped focus the ideas presented via thoughtful discussions and their individual perspectives. The opinions expressed are those of the authors and not necessarily those of any government agency, organization or other person mentioned.

REFERENCES

Abbott, K. A., Slotte, S. M., & Stimson, D. K. (Eds.). (1996). *The interfaces between flightcrews and modern flight deck systems.* Washington, DC: Federal Aviation Administration.

Aldern, T. D., Baker, H. G., Ball, J. W., Friedlander, C., Geddes, N. , Glover, M. R.., Hammer, J. M., Holmes, et al. (1993). *Final report of the pilot's associate program* (WL–TR–93–3090). Wright-Patterson Air Force Base, OH: Wright Laboratory:

Cody, W .J. (1992). *Test report for the pilot's associate program manned system evaluation*(contract F33615–85–C–3804). Norcross, GA: Search Technology.

Cody, W. J. (1993). *Competencies in human-centered system design.*In R. L. Small, W. B. Rouse, P. R. Frey & J. M. Hammer *Phase I report: Understanding the airspace manager's role in advanced air traffic control system concepts* (FAA contract DTFA01–92–C–00028). Norcross, GA: Search Technology.

Gawron, V.J., Dennison, T.W., & Biferno, M.A. (1996). Mockups, Physical and Electronic Human Models, and Simulations. In T.G. O'Brien. & S.G. Charlton (Eds.), *Handbook of Human Factors Testing and Evaluation* (pp. 43–80). Mahwah, NJ: Lawrence Erlbaum Associates.

Hammer, J. M., Skidmore, M. D., & Rouse, W. B. (1993). *Limits identification and testing environment* (NASA contract NAS1-19308). Norcross, GA: Search Technology.

Hammer, J. M., Small, R. L., Frey, P.R., Edwards, S. L., Resnick, D. E., Skidmore, M. D. & Zenyuh, J. P. (1994). *Phase II report: Understanding the airspace manager's role in advanced air traffic control system concepts* (FAA contract DTFA01–92–C–00028). Norcross, GA: Search Technology.

O'Brien, T. G. & Charlton, S. G. (Eds.). (1996). *Handbook of human factors testing and evaluation.* Mahwah, NJ: Lawrence Erlbaum Associates.

Rouse, W. B. (1991). *Design for success: A human-centered approach to designing successful products and systems.* New York: Wiley.

Rouse, W. B. (1994). *Best laid plans: How to create products that delight the marketplace; Business strategies to assure success; Life plans that are fulfilling and satisfying.* Englewood Cliffs, NJ: Prentice-Hall.

Rouse, W. B. (1998). *Don't jump to solutions: Thirteen delusions that undermine strategic thinking.* San Francisco: Jossey-Bass.

Small, R. L., Hammer, J. M., Resnick, D. E., & Rouse, W. B. (1995). *Phase III final report: Understanding the airspace manager's role in advanced air traffic control system concepts* (FAA contract DTFA01–92–C–00028). Norcross, GA: Search Technology.

Small, R. L., Rouse, W. B. Frey, P. R., & Hammer, J. M. (1993). *Phase I final report: Understanding the airspace manager's role in advanced air traffic control system concepts* (FAA contract DTFA01–92–C–00028). Norcross, GA: Search Technology.

Struck, W. F. (1992). *Software considerations in airborne systems and equipment certification* (Draft 7 of DO–178A/ED–12A, RTCA Paper number 548–92/SC167–177). Washington, DC: Requirements and Technical Concepts for Aviation.

Wise, J. A., Hopkin, V. D., & Stager, P. (Eds.). (1993). *Verification and validation of complex systems: Human factors issues.*(NATO ASI series; Series F: Computer and Systems Sciences, Vol. 110). New York: Springer-Verlag.

IV

SELECTION AND TRAINING

12

Certification of Training

Richard S. Gibson
Embry-Riddle Aeronautical University, USA

Training has been around as an informal process for countless years. Most higher order animals require some level of training in hunting, social skills, or other survival-related skills to continue their existence beyond early infancy. Much of the training is accomplished through imitation, trial and error, and good luck. In some ways the essentials of training in aviation have not deviated from this original formula a great deal. One of the major changes in aviation and other technical areas is that more complex response chains based on a broader base of knowledge are now required.

To certify means many things according to the *American Heritage Dictionary of the English Language* (Morris, 1969). These meanings range from "to guarantee as meeting a standard" to "to declare legally insane" (p.220). For this discussion, I use the definition "an action taken by some authoritative body that essentially guarantees that the instruction meets some defined standard" (p. 220). To make this certification, the responsible body subjects the educational process, training, training device, or simulator to some type of examination to determine its adequacy or validity.

ACADEMIC ACCREDITATION

In the academic community, the certification process is called accreditation. This refers to the granting of approval to an institution of learning by an official review board after the school has met specific requirements. In the United States, most universities and colleges are accredited through regional associations, which are voluntary associations of educational institutions. For example, Embry-Riddle Aeronautical University is accredited by the Southern

Association of Colleges and Schools (SACS), which is the recognized accrediting body in the 11 U.S. Southern states (Alabama, Florida, Georgia, Kentucky, Louisiana, Mississippi, North Carolina, South Carolina, Tennessee, Texas, and Virginia). SACS and other regional associations establish a set of criteria that the members must meet. These criteria address areas that are considered important to the effective operation of a college or school. In the case of SACS (1989) these include institutional purpose, institutional effectiveness, educational program, educational support services, and administrative processes. The accreditation process is a personnel-intensive procedure involving an internal review conducted by the university's faculty, followed up by a formal review by a visiting team from SACS composed of faculty from other universities and colleges. The process takes over a year and consumes thousands of personnel hours and many thousands of dollars. Generally a satisfactory accreditation is valid for a period of ten years before the accreditation must be reaffirmed. The reward to the university is that other universities will recognize the credits awarded to their students, and also that the university qualifies for many government loan and grant programs. Failure to win or retain accreditation can have catastrophic consequences.

Traditionally this rather complex process has relied upon the expert judgment of subject matter experts for both the self-study and the visiting review team. More recently, as the result of pressures from state legislatures interested in proof of the value of various college programs, there has been an increasing emphasis in the use of more objective, verifiable measures, such as the pre- and post testing of students (Did they learn anything?), performance of graduates on licensure examinations (Did they learn enough?), to surveys of employers of graduates (Did they learn anything useful?), etc. And, as importantly, asking how the institution has used this information to improve its programs. As this process of using objective evaluations continues to grow, the accreditation process shifts from using construct validity, based upon a systematic review by experts, to using empirical validity based upon observable results.

In addition to the regional accrediting associations, there are many specialized accrediting bodies based on specific academic disciplines, such as engineering, business administration, computer science, and psychology. Their procedures are similar to the regional associations. While not as important as the regional accreditation, the specialized accreditations demonstrate that the programs accredited meet the specialized requirements of various professional associations. Since all of these accreditations are paid for by the requesting institution, the cost in both time and money is significant.

Since all of the accreditation processes are pass/fail procedures, the outcome is not to guarantee academic excellence but to set the level of minimally acceptable academic mediocrity. The primary effect is to bring the weaker institutions up to a level of defined acceptability. This assures the consumers (the students and their parents) that they will get some reasonable value for their investment. However, for our purposes, the use of independent associations to establish and regulate accreditation or certification criteria can serve as one type of possible model for the certification of training, training devices, or simulators.

PROFESSIONAL LICENSURE/CERTIFICATION

Another approach to the certification problem can be found in the process of licensure for selected professions. The responsibility can be divided between a government regulatory body and a professional society. For example, for the licensure of clinical psychologists in the State of Virginia the applicant must have graduated from an American Psychological Association (APA) approved graduate program, must have passed an APA national licensing examination, must pass a state written examination, and finally a state administered oral examination (Regulations of the Virginia Board of Behavioral Science). As in the case of accreditation, the full costs are borne by the applicant. Again, the license does not mean that high quality services will be provided by the licensed individual. It does mean that sufficient minimum standards have been met so that the licensee is not considered to be an undue risk to the public. This joint relationship between a professional society and a government regulatory body provides another type of possible model for the certification of training, training devices, and simulators.

Interestingly, the APA has another level of recognition called the Diplomate. An individual with a Ph.D. and appropriate experience can apply for Diplomate status, and after a favorable review of credentials and the passing of a special examination, be awarded Diplomate status. This means that the association is essentially certifying the individual for private practice of the profession. Unfortunately, state licensing procedures do not give special recognition for the Diplomate status: the licensing process is the same for an individual with or without the Diplomate.

This brings up additional issues with respect to certification. The issues are, "who will recognize the certification" and "what is its economic value." With both academic accreditation and professional licensing, there is significant economic value for being certified. However, with the case of the Diplomate, the certification may have intrinsic value to the recipient of the recognition or certification, but have little or no real economic value.

TRAINING, TRAINING DEVICES, AND SIMULATION

Early aircraft simulators tended to look like miniature aircraft with stubby wings and tails. Their design gave them face validity. If they looked like airplanes and the instruments and controls appeared to be the same, they should be useful in teaching flying skills. Buyers of aircraft simulators have consistently had a strong bias toward purchasing devices that looked and acted like the real thing without actually becoming airborne. Researchers have tended to follow behind the development curve with questions such as: Does the training transfer? How much fidelity is enough? What is the cost effectiveness of simulator versus aircraft training?

A study by Provenmire and Roscoe (1973) used the Link Gat-1 simulators to train pilots to pass their final flight check in the Piper Cherokee Aircraft. Student pilots were given either 0, 3, 7, or 11 hours of training in the simulator before continuing their training in the aircraft. The results showed that larger amounts of time in the simulator led to larger amounts of time saved in the aircraft; however, the amounts of additional flight time saved diminished in the familiar shape of a learning curve. These results were important for two main reasons. First, they provided a basis for calculating the marginal utility of the simulator. Training in the simulator was cost-effective until the Incremental Transfer Effectiveness Ratio dropped below the simulator/aircraft operating cost ratio – in this case, about 4 hours in the GAT-1 for training student pilots to pass the final flight check for a private pilot's license. Second, the data also indicated that there was an upper limit to the transferability of simulator time to improved performance in the aircraft. Beyond a certain level of practice, about 8 hours in this case, the students were not showing increased benefits to their aircraft performance. This suggests that any within-simulator improvements were simulator peculiar without additional transfer value.

In an extensive review of the use of maintenance simulators for military training, Orlansky and String (1981) concluded that student achievement in courses that used maintenance simulators was the same as or better than that in comparable courses that used actual equipment trainers. In fact, not only was the training as good, it was cheaper. In one case that they cited, the total costs for the same student load over a 15-year period were estimated to be $1.5 million for the simulator and $3.9 million for the actual equipment trainer; that is, the simulator would cost 38 percent as much to buy and use as would the actual equipment trainer. In a subsequent study (Gibson and Orlansky, 1986) it was noted that student confidence and performance closely paralleled instructor ratings of simulator fidelity. They concluded that to make any generalizations about the effectiveness of simulator-based training without considering the fidelity of the simulators would be unwarranted.

Simulators that offer very high fidelity do not represent a serious problem for certification. The problem becomes more difficult as training devices depart in various ways from being faithful replicas of the aircraft and aircraft systems they represent. While initial students can benefit from a variety of relatively low fidelity training devices and simulators, experienced pilots receiving refresher training tend to need high fidelity simulators. The FAA Advisory Circular (AC) 120-45A specifies the evaluation and qualification requirements for six of a possible seven-level-of-flight-training devices. Level 1 is currently reserved and could possibly include PC-based training devices. A flight training device is defined by the FAA as:

> A full scale replica of an airplane's instruments, equipment, panels, land controls in an open flight deck area or an enclosed airplane cockpit, including the assemblage of equipment and programs necessary to represent the airplane in ground and flight conditions to the extent of the systems installed in the device does not require a force (motion) cueing or visual system; is found to meet the criteria outline in this Advisory Circular for a specific flight training device; and in which any flight training event or checking event is accomplished.

PC-based training devices do not meet these criteria, but many are offering some fairly impressive approximations. There will be a steadily increasing pressure for some type of certification of some of these hardware–software combinations for currency or refresher training.

TRAINER CERTIFICATION AND TRAINING VERIFICATION

Assuming that this will eventually happen, there are two problems to be addressed: one is the extent that any simulator training transfers and the other is to have a system that will verify the amount of flight experience with the training device and the quality of the individual's performance. The Provenmire and Roscoe (1973) model provides one way to establish the transfer effectiveness and to establish a metric for the upper limits of substitution for using the PC-based devices. This may be too costly and it may be necessary to assess performance relative to accepted reference simulators. This would be similar to the field practice of tests and measures in which paper-and-pencil intelligence tests are generally judged by how well they correlate with the individually administered intelligence tests, such as the Wechler Adult Intelligence Scale (WAIS).

The other problem will be that of verification of the actual amounts of practice. Regulators are wary of accepting unconfirmed self-reporting. Emerging technology may offer some assistance. It is currently possible to log onto networked games and play other opponents interactively, and the network charges for the time used. In the future, it may be possible to access approved (certified) networked software provided that you have the right PC hardware configuration and practice flying. The network could keep the necessary records. Another option would be to use a "smart" card system that would use the PC and attached "smart" card hardware to provide a record of the training hours. Obviously, there would have to be periodic checks in higher fidelity systems to provide the training not available on the PC, and check for possible abuses of the system. A pilot who had high PC training time but who performed poorly on the "check rides" would lose PC privileges.

CONCLUSIONS

Numerous models for training certification exist. All models require either construct validation based on expert opinion or some form of empirical validation that examines the results of the training. To be effective, the certification needs to be recognized by the appropriate regulatory agencies.

Techniques exist to assess the training effectiveness of training, training devices, and simulators. However, because of the cost and effort required, there is a need to examine the relation between performance on low-fidelity devices as a predictor of performance on higher fidelity intermediate devices that could be used as reference standards. If PC-based systems win certification, there will also be a need to establish a reporting and verification system based either on network usage or some type of "smart" card.

REFERENCES

Morris, W. (Ed.). (1969). *The American heritage dictionary of the English Language.* Boston: Houghton-Mifflin.

Provenmire, H K., & Roscoe, S. N. (1973). Incremental transfer effectiveness of a ground-based aviation trainer. *Human Factors, 15,* 534–542.

Southern Association of Colleges and Schools. (1989). *Criteria for accreditation commission on colleges* (1989–1990 ed.). Atlanta, GA: Author.

13

Presentation of a Swedish Study Program Concerning Recruitment, Selection and Training of Student Air Traffic Controllers: The MRU Project Phase 1

Rune Haglund
Consultant, Sweden

BACKGROUND PHASE 1

The Director of the Air Navigation Services (ANS) Department has set up an objective for the efficiency of screening and training procedures for air traffic controller students that implies that all students admitted shall be considered to have the qualification for—and be given the means of—completing the training.

As a consequence, a study project has been established. It is run by the ANS Department with members from the Swedish Civil Aviation Administration (CAA), in close cooperation with Uppsala University.

The task force of the MRU project consists of following members:

- Mr. Rune Haglund, project manager, Swedish CAA.
- Mr. Bertil Andersson, air traffic controller, Swedish CAA.
- Mr. Björn Backman, industrial psychologist, Swedish CAA.
- Mr. Olle Sundin, Manager Arlanda ATS, Swedish CAA.
- Professor Berndt Brehmer, Department of Psychology, Uppsala University.

Graduation Rate

On the first of January 1978, the military and civil ATS systems in Sweden were totally integrated into the Swedish CAA. As a preparation for this alteration a new ATS Academy was created and a new integrated air traffic controller training program was implemented in 1974. One of the aims for this training program was to decrease the failure rate to a maximum of 20%.

This objective has not been reached. However, since the start 1974, the average failure rate has been reduced by almost 20%. This improvement cannot be described as a steady curve. Instead, there is a great deal of unpredictable fluctuation around an average figure for successful training results:

- During the 1970s   Average 54%   Range 27%–71%
- During the 1980s   Average 66%   Range 57%–86%
- During the 1990s   Average 74%   Range 63%–90%

Conclusions about the success rate and trends regarding the present recruitment, selection, and training procedures are based on simple analysis of variance. Each decade was considered a group and the success rate of every class in that decade is the dependent variable. McNemar's (1969) formula for groups of unequal size has been used to test the significance of differences between the means.

$$F = s^2{}_b/s^2{}_w = .082/.015 = 5.467 \ p < .05$$

The conclusion of the task force is that there are systematic differences between the decade due to the greater experience of the people involved instead of the systematic changes in recruitment, selection, and training procedures.

Interestingly enough, the failure rate has decreased by one percentage unit per year since the start of the new integrated ATS Academy.

The mean for the years 1990 to 1993 has been calculated on the basis of eight completed classes with a total of 190 accepted students, of which 140 graduated (74%). This outcome can be compared with the rate of 80% graduating from the Federal Aviation Administration (FAA) Academy that FAA reports (Haglund &Backman, 1993).

ECONOMIC REVIEW

As a key figure for reviewing the costs of the recruitment, selection, and training system, one can calculate the total costs per graduated student. The total cost to the Swedish CAA for providing a new licenced tower control and terminal control (TWR/TMC) air traffic controller is $ 205,000 (U.S.) For an area control center (ACC) controller, the cost increases to a total of $ 255,000 (U.S.)

This total cost can be divided with the total amount of weeks in training as shown later (currency in Swedish Krona). Figure 13.1 shows the costs accumulated over weeks.

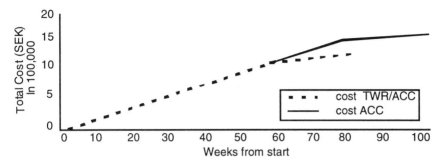

FIG. 13.1. Cost for licensing TWR/TMC air traffic controllers

Today, the Swedish CAA has achieved a balance between supply and demand with respect to air traffic controllers. This leads to an acute planning problem that can be described as follows.

Today CAA has to employ 27 students to be able to deliver 20 air traffic controllers into the ATS system. This is due to the unpredictable span between accepted and graduated students in the present screening and training system: The current system will provide an outcome of qualified licenced air traffic controllers that varies, by chance, from 17 up to 24. This uncertainty has great negative effects on both planning and economics (lack or surplus of personnel).

The outcome of today's system of recruitment, selection, and training of controller students is not satisfactory because it generates both a costly fluctuation around the mean and allows students who do not have the necessary abilities to remain in training for too long before they are expelled. An efficient system with a more predictable outcome and a higher success rate (i.e., a deviation of not more than 10% around the mean and a 90% success rate) would save the Swedish CAA at least $ 520,000 (U.S.) per class of 20 students.

GOAL SETTING

It is the opinion of the MRU task force that money and other resources invested in developing procedures for recruitment, selection, and training of controller students, so that the outcome is less affected by random errors, will be a good investment.

The Research

To pursue the causes of today's random errors, the MRU task force issues followed a two step procedure. Step 1 involved a job analysis, and Step 2 was a study of the correlations between tests and training results. These two steps were taken to validate current recruitment, selection, and training methods in use by the Swedish CAA.

JOB ANALYSIS TO DETERMINE THE JOB CRITERIA

A number of acceptable procedures exist for conducting a job analysis. One way is to interview observers who are aware of the aims and objectives of the air traffic controller's profession and who see the controllers perform their professional duties on a frequent basis. Thus supervisors, peers, and instructors may be interviewed about their observations of the critical requirements of the air traffic controller profession.

Current international research and analyses of the controller's job show that the air traffic controller profession is a very complex occupation in which the tasks are performed in a very special work environment.

THE SELECTION PROCEDURE

Brehmer (1993) noted that the current selection procedure is based on a series of tests and interviews. The tests have been chosen by ABAR, a consulting company specializing in the psychology of work and organizations. The choice of tests seems to have been made on the basis of a general analysis of the air traffic controller's job. However, there has been no standardization or statistical evaluation of the effectiveness of the selection procedure, except for a later addition: the use of percept-genetic (PG) techniques.

The paper-and-pencil tests are described in terms of four factors (ABAR, 1978):

- *Flexibility and ability to find new solutions.* The aspect is measured by means of three tests: *Skeppsdestination* (ship's destination), *Instruktionsprov II* (instruction test II) and *Kravatt* (neck tie) and concerns the ability to improvise and make decisions in unexpected situations.
- *Logical ability.* Logical ability is measured by means of two tests: Raven's matrices and number series, which are designed to measure logical ability
- *Spatial ability.* The aim is to measure the ability to construct a three-dimensional picture of the airspace from two-dimensional information. Three tests are used for this: *Klossar* (blocks), *Platmodeller* (metal sheet models) and *WIT Puzzles*.
- *Attention to detail, carefulness, and short-term memory.* This factor is measured with five tests: *Korrektur ABAR* (proofreading ABAR), *Sifferkorrektur* (proofreading of numbers), *Namnminne* (memory for names), *Sifferminne* (Memory for numbers), and *Figuridentifikation* (identification of figures).

In the final test battery, memory is treated as a separate factor and in addition to the tests already mentioned, two additional tests are used: *Uppskattning* (estimation) and *Felletning* (error search). The motives for including these are not given, and it is not clear what they are supposed to measure. In addition, a PG test is included together with an interview that aims to assess the applicant's motivation for the job. A test of capacity to process different information simultaneously is also included, as is an interview by personnel from ABAR that assesses ability to cope with stress, ability to cooperate, ability to take initiative, and motivation for the air traffic controller job (i.e., many of the factors also covered by the interview by the consultants in charge of the PG test). Finally, there is an interview by personnel from the CAA.

MRU Hypothesis 1: International research and job analyses regarding the air traffic controller's profession show that a majority of the work behaviors can be described in terms of cognitive skills. The first hypothesis is that a job analysis in the Swedish work environment will replicate these results.

MRU Hypothesis 2: The second hypothesis is that self-confidence plays an important role in coping with the critical job factors.

MRU Hypothesis 3: The third hypothesis is that interpersonal skills play a significant part in being a skilled air traffic controller or student.

MRU Hypothesis 4: The fourth hypothesis is that there is a significant difference in test results between those who successfully complete their controller training and those who fail.

MRU Hypothesis 5: The fifth and final hypothesis is that training based on cognitive skills training, coaching and mentoring the students will be more effective than traditional training methods (e.g., on the job training).

METHODS AND RESEARCH PROCEDURES

Job Analysis Procedures

The interviews were conducted as focused group interviews with a representative sample of ATS units. A total of 11 ATS units and 2 training units were visited. One hundred twenty-seven air traffic controllers participated in the focused group interviews. The interviewers, who were experienced in using this method, worked two by two. Each ATS unit had been contacted in advance about the purpose of the interviews. The interviewers used interview guides prepared in advance.

The interviewees were asked about how skillful air traffic controllers coped with stressful situations or events. Both the stressful situations or events and the effective work behaviors were recorded. The interviewers compared notes afterward and only notes that agreed were accepted. Almost 400 different measures to cope with stressful situations were recorded.

All responses noted were thereafter recorded and tabulated according to frequency. Thus the content of the job analysis was a frequency table of stressful events and corresponding key behaviors (the effective way to cope with a stressful event).

Control of Validity

The second step of the job analysis was to transform the responses of the interviewees into different questionnaires for different types of ATS units (i.e., TWR, TMC, and ACC). The same procedures as those already described were applied to the students' working situation.

The different questionnaires were distributed to a representative sample of 158 air traffic controllers and instructors working at TWR, TMC, or ACC (radar and nonradar). Their tasks were to list, on a 7-point scale, the importance of the behaviors and how often the related situations occurred in their daily work life. Step 2 was taken as a measure of the relevance or content validity of the results.

The recorded events and behaviors were compared with the causes of failures for students undergoing training. This was a test of the predictive validity of the job analysis (see Table 13.1 and 13.2).

The final step was to compare the results of the job analysis with data from the air traffic incident and information report system that exists at the ANS department. This step was taken to check the construct validity of the job analysis.

Analysis of the Test Battery Material

The material (Brehmer, 1993) consists of 145 students who were admitted to air traffic controller training in 1990 and 1991. The students came from courses 9007 (26 students), 9008 (24 students), 9009 (1 student), 9011 (26 students), 9107 (28 students), 9108 (24 students), and 9111 (16 students). There were 58 women and 87 men. Thirty-seven had failed and 104 succeeded, or at least not failed at the time when the evaluation of the test battery was done. Data with respect to success was missing for 4 students, who had taken a leave of absence from the training.

Complete data are available for only 134 of the 145 students. The number of students for whom data are available varies from 134 for the selection variable with the lowest number of students to 141 for that with the highest number. It is not likely that this had any important effect on the conclusions.

Analysis of Relations Among Variables

Two different kinds of analyses have been performed: regression analyses and discriminant analyses. Both of these aim at assessing the extent to which it is possible to predict success in training from the various predictor variables.

Regression analysis, which shows how well success can be predicted from the predictor variables, as well as the relative importance of different predictor variables, is the standard method for this purpose. However, some objections can be directed at this method in this case where the outcome variable is binary. Therefore, we also made discriminant analyses that show the extent to which it is possible to classify the students into two groups: those who pass and those who fail the training. As we shall see, these two methods give the same results.

In these analyses, the sex of the applicant has been entered as a predictor variable in addition to the test and interview variables. The analyses are based on the 134 students for whom complete data were available.

RESULTS

Introduction

The demanding tasks accounted for in the charts that follow have been compiled into five categories describing the nature of those stressful tasks. The terms used can be explained as follows.

Traffic Processing

This term is used to describe the actions taken and the decisions made to establish a safe and well-organized flow of traffic by use of clearances, separations, applicable working methods, and planning.

Coordination

This term describes the communication between air traffic controllers used to exchange information, obtain clearances, revise previous information, or hand over the control of an aircraft to establish a safe and well-organized flow of traffic.

Disturbances and Irregularities

This term is used to describe situations and duties when normal working methods cannot be used (e.g., technical malfunction, irregular behavior of an aircraft, etc.)

Fluctuating Workload

This is a description of the events and situations connected with uneven flow of traffic (e.g,. high traffic intensity with a variety of performance characteristics, followed by low traffic intensity, different flight status, and a mix of military and civil aviation).

Personalities and Social Skills

These terms describe how the persons interviewed perceive air traffic controllers, their personalities, and social behaviors.

The results of the studies show that the reported actions and behaviors are involved with information gathering, decision making, and communication in connection with traffic processing and coordination. Actions and behaviors caused by a (high) level of ambition and (high) demands on performance, account for a large portion of the strain appearing with irregular flow and varying traffic.

ACC

The most significant behaviors in the categories information gathering and decision making are found in the ACC function. Five of the 10 most common behaviors involve information gathering. The problem area social relations was not awarded the same significance as behaviors more closely connected to traffic processing, which is shown by the number of behaviors that came up in the group interviews. In the ACC function the highest importance was given to behaviors dealing with coordination and traffic processing:

- Accurate, short, and precise coordination with proper prioritization.
- Identifying conflicts early and following up on traffic.

TMC

In the TMC function, the majority of the behaviors that are ranked high in importance or frequency appear in the areas decision making and communication. The reasons for these behaviors can mostly be found in the straining tasks in connection with coordination and traffic processing.

A similar division of work behaviors as in the ACC function appears also in the TMC function, with the difference that by comparison, it is more important to:

- Dare to say "no."
- To be, and be perceived as determined.

TWR

In the TWR function, the most significant behaviors are found in the category decision making, followed by information gathering and demands on ambition and performance. The category communication was awarded lower significance than in the ACC and TMC functions.

In the TWR function, the importance of behaviors categorized under coordination decreases in favor of behaviors in the area of traffic processing. In the TWR function, the highest importance is awarded to:

- Making decisions, looking out, and following up on traffic.
- Working with confidence in one's ability and maintaining concentration also during periods of low traffic intensity.

TABLE 13.1.
Attribution and rank: Reasons for Failures During Basic Training
Between 1990 and 1993.

Attribution	*Ranking*
Slow starter, unprogressive learning curve	1
Rigid and uniform working methods	2
Passivity, lack of initiative, inactivity, late decisions	3
Low stress tolerance, makes mistakes in complex situations	3
Lack of theoretical knowledge	3
Slow worker, slow in decision making	4
Inequality of performance	4
Tense and nervous personality	4
Inadequate coordination	5
Insecure when working, doubts own decisions	5
Insufficient understanding of the ATC system	5
Excessively dependant on instructor	5
Inability to switch from low to high workload	5
Lack of concentration.	5
Constant inability to maintain separation	5
Insufficient planning skill.	5
Careless, not following instruction.	5
Lack of motivation, discontinuance due to other education	5
Total number of attributions	33
Average number of attributions per student	5

Note. The attributions come from 17 randomly selected students who failed to complete the training program.

CONCLUSIONS

As a description of the air traffic controller's profession and of air traffic control services, the survey largely corresponds with what the persons interviewed reported as significant behaviors in maintaining a safe and well-organized flow of traffic. We can therefore conclude that the job analysis is valid as well as reliable.

Movements and changes occur in the air and on the ground, and the air traffic controller is expected to handle these processes in a safe and orderly manner from her or his position in the tower or control center. The air traffic controller is thus not physically situated in the surroundings where these changes occur and cannot experience the movements and changes with her or his senses. Instead, the air traffic controller must create a mental picture of the present situation or of what the situation will be like within a limited time frame using the fragmentary information provided. As a support in constructing this mental picture, the air traffic controller has a number of technical aids: radio systems, direction finder, radar systems, data displays and monitors, telephones, telefax, telex, and so on.

TABLE 13.2.

A Comparison Between Air Traffic Controllers and Failed Trainees, Regarding Ranking of Problem Areas

Problem Area	Controller	Failed Trainee
Decision making	1	1
Ambition	2	2
Information gathering	3	5
Relations	4	3
Communication	5	5
Irregularities	6	4
Technical environment	7	6
Theoretical facts	8	5

Spearman's correlation of rank (Runyon &Haber, 1971) describes the statistical connection in ranking of problem areas between air traffic controllers and failed students. The correlation is .78.

The Predictive Validity of the Job Analysis A useful basis for studying success and failure is Heider's (Hastorf, Schneider, & Polefka, 1970) theory of attribution. The central issue of Heider's theory is viewing behaviors as caused either by environmental factors or by the individual himself or herself. The conception of reasons also leads to predictions about future behaviors and consequences. The attribution itself becomes a deterministic prediction of the future: Chance factors are not considered.

The Construct Validity of the Analysis

One noticeable discrepancy between the findings of Mattson (1979) and the MRU project is that Mattson only found two separate activities in air traffic control services: decision making and communication. The MRU project has found that, in traffic processing and coordination, information gathering and processing are highly important as a preparatory stage and that self-confidence is an important characteristic of the controller.

The interpretation made by Air Navigation Services/Headquarters (ANS/HQ) of the irregularity reports, taken from the ANS department's air traffic incident and information report system from 1991, is: "The air traffic controller assumes or expects, often as a result of indistinct or incomplete phraseology, that a pilot will act in a certain manner. The controller therefore neglects to take measures that would ensure the pilot to perform in the manner assumed by the controller."

The executives in charge of the ANS/HQ judged that all of those incidents could have been avoided if the controller had taken action to ensure that the pilots performed in the way intended by the controller.

The importance of following up on one's decisions and measures is regarded as a key behavior in the annual analysis, published by the ANS/HQ, and also in the MRU project's job analysis.

One interpretation in the job analysis is that important behaviors in the TWR function are decision making, to look out and follow up on traffic, to work with confidence in one's ability, and maintaining concentration during periods of lesser workload. This corresponds with the summary presented in the annual statistics, published by the ANS/HQ, concerning irregularity reports.The interpretation of those reports shows that the most serious incidents occur in the immediate vicinity of the airport. The final evaluation of the incidents is that operators and supervisors have not sufficiently emphasised methods of working and phraseology.

The importance of a distinct and fixed phraseology to verbally express one's decisions and measures to the party or parties concerned constitutes a key behavior in all sources accounted for.

Relation Between Test Results and Training Outcome

Table 13.3 shows each variable, and these computations are based on data for those students for whom the result in that variable was known; that is, the number varies from 134 to 141. In these analyses, the results for the PG test, which are reported only in terms of two categories, + (a positive value) and +/- (doubtful), dummy coded with 1 for the +/- category and 2 for the + category. No students with a pure negative value on PG (i.e., students for which the prognosis according to this test was clearly negative) had been admitted.

TABLE 13.3.
Means (M) and Standard Deviations (S) for the Different Predictor Variables.

	INST (R)	INST (F)	INST (S)	SERIER (R)	SERIER (F)
M	33.46	4.48	6.43	19.32	3.02
S	2.83	2.47	1.62	3.19	2.04

	SERIER(S)	KLOSS (R)	KLOSS (F)	KLOSS (S)	KORR (R)
M	6.39	39.41	8.11	6.51	21.49
S	1.56	3.39	3.74	1.65	3.20

	KORR (F)	KORR (S)	SKEPP (R)	SKEPP (F)	SKEPP (S)
M	2.44	6.18	41.70	5.14	6.12
S	1.76	1.70	5.21	4.47	1.61

	UPPSK (R)	UPPSK (F)	UPSK (S)	WIT (R)	WIT (F)
M	5.97	127.28	2.38	5.55	12.13
S	1.29	20.86	2.26	1.82	2.20

	KRAVAT (F)	KRAVAT (S)	PLÅTMO (R)	PLÅTMO (F)	PLÅTMO (S)
M	2.18	6.26	28.11	2.64	5.59
S	1.71	1.58	4.11	2.53	1.63

	PS IF (R)	PS IF (F)	PS IF (S)	FELLET (R)	FELLET (F)
M	51.24	1.44	6.22	16.39	1.08
S	5.61	1.44	1.74	2.92	1.20

	FELLET (S)	MATRIS (R)	MATRIS (F)	MATRIS (S)	VAR 1 (R)
M	6.77	38.18	8.09	6.31	18.86
S	1.45	4.91	3.67	1.43	3.24

	VAR 1 (S)	VAR 2 (R)	VAR 2 (S)	VAR 3 (R)	VAR 3 (S)
M	6.59	12.68	6.60	18.06	6.32
S	1.36	2.39	1.46	3.32	1.40

	VAR 4 (R)	VAR 4 (S)	VAR 5 (R)	VAR 5 (S)	SUMMA
M	17.97	6.26	13.32	6.99	32.76
S	1.15	1.29	1.51	0.55	0.67

	SUMMA (S)	SIMULTAN	MINNE	ABAR	LFV
M	7.07	5.71	6.04	5.55	6.01
S	1.15	1.29	1.51	0.55	0.67

	PG
M	1.92
S	0.27

It is important to note that the means in the predictor variables in terms of standard scores (stanine scores in this case, with $M = 5$ and $SD = 2$) are generally about one unit above the mean, and that the standard deviations do not differ very much from 2. This is likely to result from the compensatory effect of summing the scores for the individual tests into scales, as is done in the selection procedure. A result of such a summation is that high scores in one variable will compensate for low scores in another variable in the scale. It is therefore possible to find minimum scores of 2, and even 1, in most of the tests. This means that the scores for the tests for those who have been admitted to air traffic controller training will not deviate too much from the scores of those who apply, as is also shown by the fact that the standard deviations do not differ very much from 2. This means that correlational analyses, the results of which are affected by the standard deviations (but not the means) of the variables, will be meaningful, and that the effects of the possible restriction of range as a result of the fact that the students have been selected on the basis of the tests being evaluated will not be too serious. We certainly do not have to expect that the restriction is so serious that it will be impossible to detect the relations that might exist between success in training and the predictor variables

For some of the variables, the restriction is, however, considerable. This is especially true of the interview variables and the PG test. For these variables, it is clear that the whole range of scores is not represented in the present sample. Concerning PG, only two categories are found, in that no student with a clearly negative prognosis has been admitted, and the distribution of PG scores is quite skewed, with very few +/- values. In the ABAR interviews all of the scores are between 5 and 7, and in the interviews by the CAA personnel all scores are between 5 and 8. This means that it is difficult to say very much about these variables from the results of the regression and discriminant analyses.

Regression Analyses

In the regression analyses, training outcome, sex, and PG have been dummy coded. The results of a regression analysis with all predictor variables are shown in Table 13.4.

The multiple correlation, R, is 0.413. This represents an overestimation of the strength of the relations between the predictors and the outcome in that it capitalizes on sampling error. Moreover, the ratio of predictors to observations is high (19 to 134). After correction for such errors, the adjusted squared multiple correlation is 0.032 ($p < 0.25$). That is, there is no significant relation between the predictors and the training outcome and the whole set of predictors explain only 3% of the variance in the training outcome.

The best predictors are *Skeppsdestination* (B = 0.057, $p = 0.032$), *Korrektur ABAR* (B = 0.055, $p = 0.05$), *Felletning* (B = 0.054, $p = 0.073$) and the interview made by the personnel from the CAA (B = −0.121, $p =$

0.073). The latter variable has a negative weight; however, applicants given a high rating on the basis of this interview are less likely to succeed. Further support for the conclusion that these are the most powerful variables for predicting the training outcome is given by the results of a stepwise regression analysis that selected three of these variables (*Skeppsdestination, Korrektur ABAR* and the interview by the CAA personnel). The multiple correlation for this stepwise regression was $R = 0.334$, R_2 adjusted $= 0.091$, F (3,130) $=$ 5.437, $p < .01$.

TABLE 13.4.
The Results of the Regression Analysis With All Predictor
Variables

Variable	Coefficient	p
Intercept	0.125	0.869
ABAR-intervju	0.061	0.430
Skeppsdestination	0.057	0.032
Felletning	0.054	0.073
Instruktionsprov	0.007	0.791
Klossar	-0.026	0.328
Korrektur ABAR	0.055	0.050
Kravatt S	0.005	0.849
Kön	-0.019	0.844
LFV-intervju	-0.121	0.073
Matriser	-0.013	0.687
Minne	0.002	0.930
PG	0.124	0.416
Plåtmodeller	-0.011	0.694
PS IF figurer	0.026	0.291
Serier	0.019	0.504
Sifferkorrektur	-0.027	0.302
Simultankapacitet	0.016	0.645
Uppskattning	-0.033	0.307
WIT Pussel	-0.009	0.806

Discriminant Analysis

The discriminant analysis was performed with the same predictor variables as the earlier regression analysis. The results from the initial F tests for these variables agreed with those from the regression analysis (as would be expected) in that significant F values were obtained for *Skeppsdestination* F (1,132) = 18.885, p < 0.01 and *Korrektur ABAR* F (1,132) = 4.248, p < 0.05. As in the regression analysis, the results for the interview by personnel from the CAA F (1,132) = 3.429, p < 0.07 and *Felletning* F (1,132) = 3.586, p < 0.07 were close to significance. The discriminant function correctly identified 12 out of the 36 who failed and 89 out of the 98 who succeeded in the training; that is, 101 (75%) applicants were correctly identified. This should be compared with the number expected if the predictor variables are ignored and only the base rates are considered; that is, the number of correct classifications that would be expected randomly. This yields an expected rate of correct classifications of 62% or 83 students. Thus, the discriminant function improves the selection by 18 cases, compared to no selection procedure at all. Thirty-three students are incorrectly classified, compared to 47 that would be expected on a random basis. This agrees with what would be expected on the basis of the uncorrected multiple correlation between the predictors and the training outcome of R = 0.413. As noted in the discussion of the regression results, the sample estimate represents an overestimation of the possibilities of predicting the outcome; this is true also of the results of the discriminant analysis. Unfortunately, there is no procedure for estimating a discriminant function corrected for sampling errors comparable to the procedure for the multiple correlation. However, in this case with a binary outcome variable, multiple regression and discriminant analysis are basically the same, and we should therefore expect that after correction for sampling errors, we should have the same decrease in effectiveness; that is, we should expect that the ability to make correct classifications using the discriminant function should decrease by about 80% after sampling errors have been taken into account. A reasonable estimate of the improvement in the number of correct classifications of training outcomes from using the current set of tests and interviews is three to four cases (about 3%). This is the same estimate as that obtained from the regression equation.

DISCUSSION

The Main Task

The main task for this phase of the MRU project was to evaluate current screening and training procedures and create recommendations aimed at a reduction of the present span between intake of students and output of examined new air traffic controllers.

The Analysis

The task force chose to conduct a job analysis based on the critical incident technique. The result shows that 300 reported key behaviors could be cataloged into five groups:

- Decision making.
- Self confidence.
- Information gathering and processing.
- Social relations.
- Communication.

This result verifies the three hypothesis stated about the air traffic controller profession: Behaviors that are related to self-confidence are mostly reported in connection with unexpected events and variabilities.

The results from the job analysis have been compared to attributions for failure in the basic controller training. It has also been compared to incidents that have occurred in actual operations according to the current official report system. Both students and controllers fail to perform the key behaviors at a sufficient level.

The training process requires the students to practice key behaviors from the very first day, aiming at minimizing the number of errors to reach a full performance level, and finally to reach a mastery level. Today an uneven learning curve is the most frequent cause of failure during basic controller training.

At Stockholm ATS and Arlanda ATS units in Sweden, attempts have been made to improve on current methods of basic training. The results from this attempt to apply modern training techniques, for example to use programmed skill training and to transform the instructor into a mentor and a coach, are now the most promising measure taken to improve the outcome of basic air traffic controller training.

To quote one of the members of the task force, Professor Berndt Brehmer, "It is astonishing how little effort is made in general to train and develop an operator in a high-tech environment by modern training technology, and how much one still relies on an old-fashioned on-the-job training provided by a more experienced fellow worker."

Present rate and variation in span, in the outcome of the Swedish basic air traffic controller training, can only partly be explained by inadequate psychological tests and screening procedures. To reduce the uncertainty in the outcome, it is important to improve training and learning of key behaviors for the air traffic controller work, as well as to develop screening methods with high reliability and validity. This will give a prompt and positive result.

Efforts must also be made to create a continuing job analysis to keep up with changing technology and maintain screening and training methods with the highest possible effectiveness.

An important prerequisite for a successful training result is an efficient selection procedure based on a sophisticated chain consisting of information introduction and skill tests assessing the substance of the most important groups of key behaviors.

The Relation Between Test Results and Training Outcome

The results of the analyses just presented show that it is not possible to predict the outcome of the training on the basis of the variables used in the selection procedure.

One possible reason for these depressing results, and this is true both for the regression results and those from the discriminant analysis, is that these results are based on data only for those who were admitted to the training, that is, we have a classical case of restriction of range. To ascertain the effects of this, we need to look at the standard deviations of the various predictors for the sample used in the calculations. The relevant results are shown in 13.4. As already noted in the discussion of these results, the restriction is not as severe as might have been expected. The standard deviations for the predictor variables are between 71% and 91% of those for the unselected sample used to determine the stanine scores, and for most of the variables, the standard deviations for our sample are about 80% of those in the unselected sample. Moreover, we have the full range of the predictor variables for many of the variables; the lowest values are 1 and 2 for many of the variables. There is therefore little doubt that we would have been able to detect the relations that might exist between the predictor variables and the training outcome. The fact that we find very few significant relations, and that the correlations that we have found are very low, can therefore hardly be explained in terms of restriction of range. Instead, it seems more reasonable to assume that the results express real deficiencies in the selection procedure. That is, the predictor variables are not very powerful predictors of the training outcome. This is hardly surprising in view of the fact that these variables have been selected on the basis of a very general job analysis without real standardization and statistical evaluation; that is, the tests have not been chosen on the basis of an empirical evaluation of the actual predictive validity of the tests. This means that there was no reason to expect that the selection procedure would be very effective.

One could, of course, argue that the selection procedure concerns the job as an air traffic controller and not the training. It may well be that the training makes demands that differ from those of the job and that an evaluation of the selection procedure in terms of the training outcomes is not quite relevant. To answer this question, we need a more penetrating analysis of the demands that the training courses actually make compared to those that the job makes. At the present time, we do not have the data required for such an analysis.

Another objection is that the analysis may rely on the wrong model. The present analysis is based on a model in which the probability of success in training is assumed to be a monotone function of performance in the selection variables (see Fig. 13.1). That is, this model makes the reasonable assumption that if some ability is required, more of that ability leads to a higher probability of success than less of the ability.

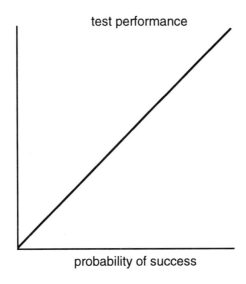

FIG. 13.1. Basic model for the analysis in this study. This model assumes that there is a monotone relation between training outcome and test performance.

An alternative model is illustrated in Figure 13.2. This alternative model assumes that the training only requires some minimum ability, and that all students having at least this minimum ability will have the same probability of succeeding in the training course.

If this model is true, the possibilities of detecting relations between the training outcome and the selection variables would be limited, especially if the students in the training course had been selected so that all of them had values exceeding the critical value. In the case here, this does not seem to be a very serious problem, however, because the full range of values is represented for most variables in this sample. Thus, it should have been possible to detect whatever relations might have existed between the test variables and the training outcome, even if Model 2, rather than Model 1, would have been valid for the present data. Moreover, when measurement error is added, Model 2 will generally be impossible to distinguish from Model 1.

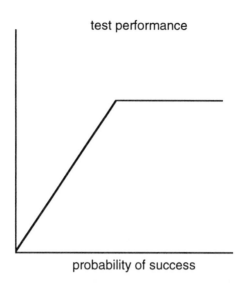

FIG. 13.2. Alternative model. This model assumes that the probability of succeeding in training increases up to some critical values and that it then stays constant at the same level.

The analyses have been based on the individual tests rather than the scales used by ABAR in the actual selection. The reason for that is that our analyses yielded no support for these scales in that we found that these scales were intercorrelated, and the tests included in the scales were not intercorrelated as they should have been. Moreover, a principal components analysis failed to yield the scales as components. Thus, there was little support for the usefulness of these factors. Additionally, regression and discriminant analyses based on the scales used by ABAR did not give better results than the analyses based on the individual predictors.

The results with respect to PG deserve special comment. This is the only variable included in the selection procedure on the basis of an empirical evaluation procedure. In this procedure, the PG test was given to the applicants, but not used for the selection. That is, the evaluation concerned the extent to which this test could improve the selection over and above what could be achieved with the original test battery. The results were quite encouraging, but we must now conclude that the conclusions from the original evaluation were overly optimistic. Thus, Svensson and Trygg (1991) concluded that it should be possible to decrease the proportion of students failing the air traffic controller training to less than 10% if the PG test was used. As shown in this analysis, this has not been the case. Even when the PG procedure is used, the proportion of students who fail is 26%.

It was, however, not realistic to expect that one would have as good results with this sample as with the standardization sample used to determine what PG variables should be used for the selection. First, the initial evaluation did not take into account the total effectiveness of the selection procedure with PG as one of many selection variables. The value of PG in such a procedure is dependent not on the correlation between this variable and the training outcome (which is what was reported in Svensson & Trygg, 1991) but on its unique contribution, which is dependent on the partial correlation between PG and training outcome, after its correlations with other selection variables have been taken into account. That such intercorrelations exist is demonstrated in this sample, despite the severely restricted variation in the PG scores. Such intercorrelations decrease the weight that the PG results will receive in the final selection.

Second, one must expect a certain shrinkage in correlation when the test is used for a new sample because the values obtained for the first sample capitalized on sampling errors. In this sample, the unique contribution from PG is far from significant. However, the extremely skewed distribution of PG values makes this correlation suspect.

A possibility of evaluating the effectiveness of the PG procedure is to compare the failure rates before and after the introduction of this test. The relevant comparison here should be with the failure rate for the 1980s, when the mean failure rate was 33.9% (with considerable variation among courses). In the 1990s, after the introduction of the PG procedure, the mean failure rate so far has been 26%, although this may well be an underestimation because not all students have completed their courses. That is, not all students have yet had a chance to fail. With this in mind, the maximum estimate of the improvement from PG would be 7.9 percentage units, but this would assume that all of the decrease in the failure rate from the 1980s to the 1990s can be attributed to the introduction of the PG procedure. This seems unlikely, especially in view of the fact that the decrease in the failure rate from the 1970s to the 1980s was about 6 percentage units (from a failure rate of 40% in the 1970s to a rate of about 34% in the 1980s) without any new selection procedures.

In the regression and discriminant analyses, three variables stand out. One of these, the results from the interview conducted by personnel from the CAA, receives a negative weight. That is, they are systematically wrong: Students with a high rating in these interviews perform systematically worse than those with a low rating. This suggests that this procedure must be improved.

Only two of the selection variables have systematic relations with training outcome: *Skeppsdestination* and *Korrektur ABAR*. The former of these is supposed to measure flexibility, the latter carefulness. The correlations are low, however, and they may well have been produced by chance. Therefore, one should not rely too much on these results until they have proven valid also for other samples.

CONCLUSIONS

This evaluation of the selection procedure is clearly limited, first because it is based on a limited sample, and second because it is based on the results for a group that has been admitted to the training on the basis of the selection procedure that is being evaluated. The restriction of range problem does not seem to be as severe as one might have suspected, however. It should therefore have been possible to detect whatever relations might exist between the selection variables and the training outcome. We must therefore conclude that the fact that it has been hard to find such relations probably means that they do not exist. There are therefore good reasons to reconsider the present selection procedure. It is not possible to decide whether it is possible to design a better selection procedure on the basis of the data we have today. For this, we need a careful analysis of the air traffic controller's job to determine what demands this job makes and how these demands can be met by means of selection and training.

Future ATS Systems

Coming automated ATS systems cannot replace the human controller. But manual repetitive work can be eliminated, thus facilitating information seeking and information collection.

If a new technology or a new system is to be introduced, it is fundamentally important to be assured that the operators accept the new technique and that the new technique will create opportunities for them to improve their performance. The controllers must also be informed in advance in what way they will be trained to achieve this new standard of performance.

In our view, a continued automation of the air traffic controller's work will only further emphasize the importance of adequate training to execute new key behaviors.

REFERENCES

ABAR. (1978). *Förslag till bedömningsvariabler vid psykologisk lämplighetsprövning av flygledarspiranter, våren* [Proposal for assessment of variables concerning psychological aptitude tests for air traffic controller applicants]. Arbetspsytcologist Radgivning, Stokholm.

Brehmer, B. (1993). Prestation vid urvalet och utbildningsresultat i flygledarutbildning. MRU-projektet [Performance in selection and training results in air traffic controller training]. (Rep. No. 7). Psykologiska Institutionen, Uppsala University.

Cullen, J., & Hollingum, J. (1987). *Infr total kvalitet* [Introduction of total quality]. Konsultfölaget AB.

Flannagan, J. C. (1954). The critical incident technique. *Psychological Bulletin,*
51, 327–358.

Haglund, R., & Backman, B. (1993). *Studiebesök på FAA Huvudkontor och Mike*
Monroney Aeronautical Center [Study visits at FAA Washington Headquarters
and Mike Monroney Aeronautical Center]. (MRU-Project Rep. No. 3).
Luftfartsverket.

Hastorf, A. H., Schneider, D. J., & Polefka, J. (1970). *Person perception.* Reading,
MA: Addison-Wesley.

Mattson, J. (1979). *Kognitiva Funktioner vid flygtrafikledning* [Cognitive
functions in air traffic control] Lund, Sweden:. Lund University Department of
Psychology.

McNemar, Q. (1969). *Psychological statistics.* New York:Wiley.

Runyon, R., & Haber, A. (1971). *Fundamentals of behavioral statistics.*Reading,
MA Addison-Wesley.

Svensson, B., & Trygg, L. (1991). *Personlighetskarakteristika hos elever i*
flygledarutbildningen [Personality characteristics of candidates for air traffic
controller training]. Rapport från Lunds universitet, Institutionen för tillämpad
psykologi.

Wanous, J. (1980). *Organizational entry. recruitment, selection and socialization*
of newcomers. Reading, MA: Addison-Wesley.

14

Does Human Cognition Allow Human Factors (HF) Certification of Advanced Aircrew Systems?

Iain S. MacLeod
Aerosystems International Ltd, UK

Robert M. Taylor
DERA Centre for Human Sciences, UK

A *system* may be defined as a set of parts with the output of the whole greater than the sum of the output from the individual parts. The systems approach to design is considered a formal and systematic set of procedures for systems development. Within the systems approach, systems certification is defined as the result of an applied examination process devised to formally test and affirm that the system being inspected satisfies certain accepted criteria. More formally, certification can be defined as the procedure by which a third party gives written assurance that a product, process or service conforms to specified requirements.

If certification is achieved by a system, it indicates that the system should fit its intended purpose and that it meets specific requirements of reliability, safety, and performance. In U.K. military terms there are certain types of design activities that must be performed to the satisfaction of the certification authority. An example of a software standard that defines such activities is the U.K. DEF STAN 00–55. Currently there are no similar standards for human factors (HF) design-related activities.

Within the definition of a system, it is obvious that a system may contain a human component. However, it should be noted that only recently have there been formal acknowledgments that a system is made up of human and equipment components (e.g., in the U.S. military U.S.A. Department of Defense I, 5000.2, 1991;. U.K. Def. Stan. 00–25 1989; NATO STANAG 3994AI, 1990–91). DoDI 5000.2 defines a total system as including the humans that will operate and maintain the equipment. Def. Stan. 00–25 states, "This Standard should be viewed as a permissive guideline, rather than a mandatory piece of technological law"(Part 2-12, p. 1). A *System* is defined as: "A purposeful organization of equipment (hardware and software), personnel and procedures all of which interact and thus influence each other to produce some specific result or goal" (Part 12, p. 65). STANAG 3994AI. This STANAG does not define a system though its list of definitions strongly implies that the human is an integral part of an advanced aircrew system.

HF certification should be an integral part of systems certification. HF certification implies that HF specification, testing, and evaluation has a secure foundation; therefore, the process of certification follows as a matter of carefully progressed HF appraisal of the system. Where possible, such appraisal should be an integral part of the overall system test and evaluation plan. The final HF certification tests should be the culmination of a planned process of allowing orderly and conditional HF certification to be progressed throughout the duration of the project. HF certification is ultimately concerned with how efficiently the human element(s) of a system can perform through their use of the system and how human performance affects that system's performance capabilities and the safe achievement of system-related goals.

Specification and certification of engineered systems can be conducted under any of several well-documented and accepted methods. Further, HF analysis and measurement has attracted a great deal of attention since World War II and can be performed. (A description of the methods generally used by 'behavioral specialists, including HF engineers, is contained in Meister, 1985.) However, the human role in complex human–machine systems is recognized as becoming predominantly one of system management, supervision, use of knowledge in uncertain environments, situation analysis, system control, and system direction. Thus, the main emphasis of the contribution of the human to human–machine systems has changed in nature from physical to cognitive. Cognition is normally taken to refer to the processes of the mind involved in human knowing such as use of knowledge, reasoned thought, understanding, and situation awareness. These properties are normally relegated solely to the domain of human expertise and are not considered within system requirements and performance specification. In contrast to human cognition, mental or psychological processes are more global and cover the total remit of mind functions and processes including such as emotion and long term memory.

If the human contribution is largely cognitive, the HF specification and certification of human complex interactions through complex systems should require a sound knowledge of the contribution to system performance of human cognition or of any form of cognition within the system. It is our contention in this chapter that a viable approach to the consideration of cognition within system design resides within the context of system performance as system cognitive functionality (SCF). By system cognition (or SCF) we refer to system goal-related properties of work concerned with system direction and control, system situation analysis, system management & supervision, system wide knowledge application, and system performance-associated teamwork (MacLeod, 1996). Such cognition has a focus on system performance, including human performance as applied through the system, rather than on human performance with or without the system. However, applicable knowledge of cognitive processes at the required quality does not currently exist in the realms of HF, engineering, or psychology. Therefore, with such a gap in knowledge, any HF-related certification of complex or advanced aircrew systems must be carefully qualified.

We start this chapter by considering conceptual issues associated with any considered use of cognition within system design. Problems inherent in traditional system design processes are then discussed, followed by some illustration of these problems through consideration of HF as applied during the U.K. RN Merlin program. Finally, the importance of encapsulating SCF within system specification is discussed with relation to its necessary contribution to both system performance and system certification.

COGNITION AND OPERATION OF ADVANCED AIRCRAFT SYSTEM

Conceptual Issues

It can be argued that engineering ideas are about as far removed conceptually from the ideas of human psychology as any ideas can be. This is because engineering ideas are mechanistic and physicalistic and based in the natural sciences, whereas psychology can be termed the science of mental life, both of its phenomena and its conditions. *Physicalistic* is taken as pertaining to the physical, mechanistic, or observable. Many tenets of HF and psychology are founded on physicalistic approaches, for example, anthropometrics, biomechanics, human manual control of systems, many approaches to human–machine interface (HMI) design, and behaviorism, to name but a few. Disciplines such as HF, cognitive psychology, and industrial psychology attempt to bridge the conceptual gap. In addition, disciplines such as engineering psychology and cognitive engineering are attempting to adopt more teleological approaches to the appreciation of human work. However, with the rapid advances in computer-based systems and automation and their burgeoning complexity, it appears that the conceptual gap remains wide. *Teleological approaches,* as applied to the human, refer to human mental goal seeking and purposeful behavior. For an interesting and teleological associated discourse on paradigm shifts in science see Ackoff (1972). To give examples from diverse HF viewpoints over the last decade, the following five quotes show a general HF and psychological concern on engineering approaches to system design and automation, approaches that are biased to only considering observable manifestations of human behavior.

- "Physicalistic descriptions can only capture those aspects of man, which submit to the metaphor of the machine, and must fail to account for the rest. This inadequacy of the physicalistic approach becomes gradually more clear, as the complexity of man–machine systems increases." (Hollnagel, 1983, p. 135)
- "We are still making the same seemingly contradictory statement: a human being is a poor monitor, but that is what he or she ought to be doing". (E. L. Wiener, 1985, p. 87)
- "The designer who tries to eliminate the operator still leaves the operator to do the tasks which the designer cannot think how to automate". (Bainbridge, 1987, p. 277)
- "Is it possible that our advanced command and control systems will require cognitive human performance that defies our ability to measure and predict? What none of the existing models is much good at is analysis of cognitive behavior". (Miles, 1993, p. 3).
- "As the machine is inanimate and cannot have sentience, the situation awareness of the overall system relies on the system operator's situation awareness as supported by situation assessment from the engineered portion of the system. Awareness like belief is program resistant (Gunderson, 1971, p. 144), (MacLeod, 1998b, p. 669).

Absent System Functions

What all the preceding quotations indicate is that there is a gulf between system functional design and the system-related performance requirements placed on the system operator. It is argued that this gulf is partly because of an absence of consideration of cognition as encapsulating functions pertinent to the system. Within a system a function can be considered as a system property or broadly defined as anything that a system might be required to do. We believe that there is an important distinction between system tasks and functions, a distinction that is often blurred. A *function* is a property of a system but is latent until it is evoked as required for system performance. In contrast, a *task* may utilize one or more system functions but always requires the application of effort.

System-related cognitive functions have traditionally been catered for by the system operator(s). With ever-changing technologies being applied to system design, accompanied by an associated increase in the automation of system functions, system designers are paying less and less attention to the work needs of the human operator of the system. Designers argue this lack of attention as being viable in their design approach, if they even are aware that there is a lack, because the engineered system is deemed to be reliably performing certain tasks hitherto involving operators. Further, the argument continues, the fewer operators required, the cheaper the system is to operate.

Nevertheless, the system operator is still considered to be in charge of the system and is required to direct its performance. However, the design trends in automated systems do not support this operator role, especially with relation to the quality of system feedback to the operator, frequently leaving the operator in a state divorced from the rest of the system. The operator is thus unable to provide the necessary cognitive functions that the system requires in support of its safe quality performance. Human cognitive functionality as applied to the system has been engineered out but not replaced by the engineered substitute of automation. The inevitable result is that some needed system cognitive functions are frequently absent from modern system designs and the subsequent operation of that system. Thus the nature of the design unknowingly and adversely changes the nature of operator work and its associated error forms. Adverse changes to work are especially exacerbated when system feedback to the operator is missing or inappropriate.

In this chapter, we contend that the SCF, inadvertently removed from the remit of the operator and absent from the system, is still required to maintain system integrity and quality performance. As the arguments of the chapter are developed, this issue is revisited.

Problems in the Appreciation of Unobservable Mental Life

The greater the complexity of a human–machine system, the greater the problems in efficiently integrating the human component into the system. As already argued, with a complex system, the human may have difficulties maintaining a concept of system performance and fitting that concept to the human role within the system. Such difficulties may not only exacerbate problems that the human finds in system control or supervision, but may also encourage the human to enter incorrect or inappropriate inputs into the system. Therefore, the human performance at the HMI of a complex system must be assisted in an attempt to ensure that situation and situational awareness are sustained. The maintenance of system support to the human requires that the human is aided in the obviation of human system-related errors, is helped to skillfully maintain a necessary defined role within the system, and is neither overworked nor bored. *Situational awareness* refers to an understanding of all the factors affecting mission performance, including the status of the aircraft and its mission system, and the tactical, spatial, and geographical environment external to the aircraft. Situational awareness is an integral part of human expertise and skilled performance (MacLeod, Taylor, & Davies, 1995). An older, higher level concept used by U.K. aviators was termed *airmanship*.

There are innumerable HF standards and guidelines on how to define and certify simple systems. These standards and guidelines are normally advisory and invariably stress the physical aspects of systems and the use of empirical evidence. They mimic the form of system specification in that components of a system are specified by their physical or manifest functions and the logical interrelation of these functions. Examples of these standards are discussed later. However, the human catalog of skills transcends the physical, especially when the human has to cope with complexity and uncertainty. As an example, Welford (1976) identified three types of human skill as:

> *Perceptual*—The skills that code and interpret incoming sensory information.
> *Motor*—The skills associated with skilled movement but controlled by perception and cognition.
> *Intellectual or cognitive*—skills considered by Welford to be the most important as they link perception and action through decision processes.

Cognition is hidden and may be either abstract or have manifestations in observable human activity. Thus, the processes of human judgment are also hidden. For example, the human assessment and judgment of the quality of equipment related information might be a continual background task with no directly associated manifest actions on the part of the equipment observer. However, if the observer's judgment leads to choice, and the choice requires a system control action, an observable human activity will result. Therefore, physicalistic system functions may or may not have equivalence within the cognitive functions of the human component of the system.

Moreover, physical systems are constructed by logical rules of engineering, whereas human logic is dependent on gleaned knowledge, human mental functions, and heuristics that are based on eons of human evolution. Training and experience can tune human abilities into skills with respect to the human role within a system. Training cannot mold a human into a metaphor of an engineered system component. At best, training can develop human work skills to better match the human appreciation of time with relation to a machine system-clocked time. Thus, the overall system must be built to consider the possible contributions of human and machine, to allow each to complement and appreciate the capabilities and system inputs of the other.

Traditional System Functionality

Traditional systems engineering stresses that the concept of design must be based on a detailed understanding of the functionality of the system. A function is stated here to be a system property that is latent until required, has an expected level of performance, and is appropriately evoked by engineered automation or the system operator through the application of effort (MacLeod & Scaife, 1997).

Thus, the transposition of engineering functions into required equipment performance is brought about by design based on a choice from a limited range of solutions. However, the manifestation of performance by the human depends on the individual human's innate mental abilities, developed physical and cognitive skills, the existing level of fatigue, and personal and organizational mores and ambitions, to name but a few influences. Underlying that human performance, human cognitive functions rely on human mental processes mediated by experience, and are related to human tool-assisted progress toward goals. A *goal* can be defined as the end result toward which a mental or physical effort is directed. A goal can also be described as an objective toward which the individual consciously or unconsciously strives (Adler, 1929). Functions directed toward goals may or may not be associated with the parallel performance of certain equipment functions within an engineered system as suggested by the traditional approach. They may however be part of designed or intended system processes that encompass both engineered and human system functions. Active system processes imply progress along a course of action as constrained by system functionality. Active system processes require the performance of system tasks.

System tasks include both human and machine tasks. Consideration of human tasks is approached by the discipline of ergonomics. One ergonomic based approach is based on task analysis. This analysis can be in the form of a predictive analysis used for the consideration of human tasks with relation to the operation of a system toward system-related goals. However, this predictive modeling and analysis is usually based on engineering-related functionality and is thus biased toward the observable aspects of human performance, ignoring cognition (two examples of such analyses are operational sequence diagrams and hierarchical task analysis).

Such task analysis is concerned with the analysis of human tasks. However, human tasks require the human to apply both cognitive and physical effort. This effort is needed to direct a system toward the achievement of preconceived goals both tactical and strategic (tactical and strategic goals are discussed in more detail later). The problem aired here is that not all human tasks can be considered and analyzed if only mechanistic-based or observable tasks are considered.

Moreover, system task performance involves directed effort by both human and machine and is a system's planned application of its functionality toward the satisfaction of explicit system goals. System tasks may involve one or more functions and solely reside within the engineered system, may be unique to the work performed by the system operator, or may involve the use of functions from both. Tasks can be physical in nature, cognitive in nature, or a combination of the two (MacLeod & Taylor, 1993). Further, the term *cognition* can apply to functions and tasks resident in either man or machine or both (Hollnagel & Woods, 1983). With the human operator, task-related activities and actions are necessary to control system task performance.

Mediating between operator system tasks and activities are the system operator's cognitive functions that include system pertinent SCFs. Thus:

Task >> Cognitive Function >> Activity >> System Feedback

As previously discussed, the SCFs can be absent within a system if not carefully considered by system design.

SCFs

With respect to system design, SCFs differ from the system operator's cognitive functions. SCFs are concerned with system performance issues and not the individual expertise and cognitive processes of an operator. Whereas operator cognitive functions, as already introduced, are tuned to the living needs of the individual, SCFs are functions that are required solely to allow the system to meet its designed performance.

Importantly, the specified cognitive functions should not try and represent a model of the human system that has far too many variables to make it of practical use (Chapanis, 1996). Rather, the SCFs should cover what the operator has to understand and activate with relation to the work situation and its associated operating procedures for control, direction, supervision, and management of the system.

Within a system, SCFs represent a complementary functionality to both engineering functionality and the operator's cognitive functionality. The trigger to evoke an SCF in a system task would normally be through a designed association with a system-engineered function, or associated with the need to evoke such a function.

Therefore, SCFs may be supported by operator cognitive functions, but they are concerned with purely system issues such as system direction, control, supervision, and management. As such, it is possible to engineer some SCFs. For example, some system management functions could be sensibly automated to complement the human tasks in the management and supervision of system operation.

However, designed SCF relocation should only be performed through knowledge of the system requirements for SCFs, the type of system, and an appreciation of the performance required from the system operator(s). This knowledge of SCF requirements should be derived at an early design stage—at the stage of logical specification of system functionality and performance.

Through knowledge of SCF requirements, any SCFs removed from the operator must nevertheless reside somewhere within the system. Thus, if it is a design decision to present a system operator with a diminished suite of SCFs, there must be engineered a complementary suite of SCFs to satisfy system requirements while still allowing the operator to be in charge of the system.

Traditionally, it can be argued, this problem area has been approached through the design of the HMI. However, an underlying argument of this chapter appears to refute such an approach. Traditional HMI practice can ably address the quality presentation of information (although not necessarily of advice where nuances associated with the intenstionality and extensionality of language [Hayakawa & Hayakawa, 1990] have to be considered [MacLeod, 1998b]). However, the HMI information is based solely on the systems-engineered functions, functions that do not explicitly include or acknowledge SCFs. HMI issues are briefly considered later in the chapter.

Consolidation of Cognition and Operation

To reiterate, the human may be considered a complementary part of an engineered system, but not as a piece of equipment that can be easily specified. The traditional partitioning of system functionality to allocate functions to humans and machine so that each performed the most appropriate one (machines are better at or humans are better at) assumed that physicalistic system functionality could be directly transposed to either human or machine. In defense of the traditional approach, developments on the theme considered the complementary nature of human and machine in a system and acknowledged that some functions could be performed by either with equal efficiency.

Indeed, the simple physicalistic transposition of functions to human or machine may have held true in situations where the human used machines as tools to be directly applied to work performed under immediate human attention. However, the subject transposition is much less likely to be true where humans work through complex systems toward mission goals. With complex systems the direct allocation of numerous system functions to the humans or the machine becomes more difficult to determine at anything but the highest level of consideration. Partly, this difficulty is caused because the nature of human work and system-related functionality is based more on human cognitive performance than on psychomotor performance.

There appears to be a gradual realization that the HF standards and guidelines produced in the early 1980s are set in the physicalistic engineering mores of the 1970s. In the 1970s systems were less complex. They were usually analogue (mechanical or electrical), and the human was generally closely involved in operating directly with systems to achieve goals (often in a one-to-one relationship as in the use of a computer based word processor application). However, under current design initiatives the human operates through systems to meet goals (as with an airborne mission system in which the tactical performance of the aircraft is directed by the operator through an HMI updated from advanced navigation, communication, and sensor avionics-based equipment).

As already suggested, one of the problems with the drive to automation with airborne systems is that the human has been forced by design, in many instances, out of the primary role of a system operator and into the primary role of system supervisor. This enforcement has been made without the development of tools to assist in the performance of the new role (or even a determination that a new role has been created). Many studies have shown that automation may have decreased the occurrence of certain error forms but has introduced new categories of man–machine system error that have yet to be fully understood (E. L.Wiener, 1985; Woods & Roth, 1988). As previously discussed, poor consideration of SCFs must be partly responsible for the poor adoption of new technologies and their associated levels of automation.

Advanced systems are being designed forgetting the underlying tenet of systems design—that the whole is greater than the sum of the parts. The human and machine components of a system must complement and assist each other within the system. It is not fruitful to enhance the speed at which a system can operate if the quality of system support to the human ownership and control of system processes is impaired.

Thus, the preceding supports an underlying contention of this chapter that the new forms of complex system operating problems and errors tend to be purely cognitive rather than psychomotor based. In addition, new forms of errors should be considered mainly dependent on the achieved efficacy of human–machine system design rather than as mainly resident with the human operator of the system.

Part of the blame for the lack of quality in the adoption of new technologies lies in an overrobust adherence to traditional system design practices. Exacerbating the poor quality consideration of new technologies by design is the poor consideration placed on technology and design life cycles by most HF standards.

PROBLEMS INHERENT IN HF SPECIFICATION WITH TRADITIONAL SYSTEMS DESIGNS

The problems inherent in HF specification with traditional systems design are illustrated using examples from two military HF standards.

Traditional Design by DEF STAN 00–25

Systems design and development rely on a traditional series of analyses through system planning and preliminary design to detailed design and development. The initial system requirement analysis is usually conducted by the customer and considers such needs as system purpose, sphere of operations, types of system components, and system reliability, to name but a few. This initial requirement is stated at a high level. The system requirements analysis is the basis of the specification that initiates the system process. conceive that: "The major system requirements are physical...but there are always explicit behavioral requirements" (p. 11).

Design by STANAG 3994

In STANAG 3994AI the problems inherent in straight physicalistic function allocation to the human are recognized in that a potential operator capability analysis is mooted alongside analyses of human decision, error, information requirements and control requirements. Indeed, the task analysis mooted explicitly covers human cognition as it states that the analyses:

> shall show the sequential and simultaneous manual and cognitive activities of the operators/maintainers, and include those aspects of their tasks which involve planning and maintaining situational awareness, as well as decision making and control activities. (p. 4)

In Fig. 14.1, it can be seen that it is presumed that an allocation of system functions can be performed in the traditional manner. Of interest, the particular U.K. concept of task synthesis entails "the design team, using their judgement and expertise, proposing a combination or sequence of tasks appropriate to the function" (DEF STAN 00–25, Part 12, p.14). Note that the previous forms of analyses on which the task synthesis is based are wholly physicalistic and conceive that " The major system requirements are physical...but there are always explicit behavioral requirements" (p.11).

Fig. 14.1 shows the U.K. DEF STAN 0–25 model of HF activities conducted during system design.

However, the task analysis is to be based on preceding analysis. The emphasis of the preceding analyses is still seen as placing an overreliance on physicalistic functionality. From the descriptions of the preceding analyses, there appears to be an underlying assumption that a form of mapping can be made from the systems functionality of advanced aircrew systems onto human functionality within the system and the associated human cognitive processes. To give an example of difficulty in such mapping, the STANAG example of Function (e.g. control air-vehicle) is decomposed under function analysis through "successive levels of detail to a point where individual functions can be unambiguously identified, prior to allocation to human, hardware, or software system components" (p. 3).

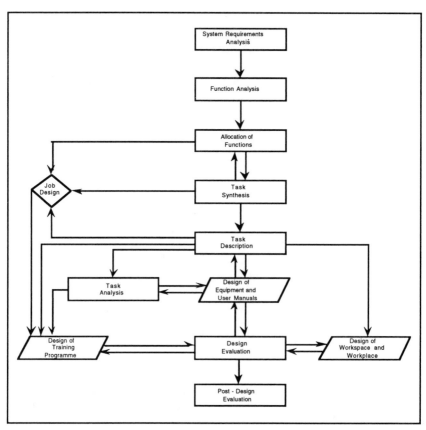

FIG. 14.1 DEF STAN 00–25—human factor activities for system design.

STANAG 3994 defines functions as a broad category of activity performed by a system usually expressed as a verb + noun phrases (e.g., control air-vehicle, update waypoint). Contrast this definition with the definition given earlier. It has already been argued here that function is an engineering concept within systems engineering and that human and engineering functions differ. Therefore, from that conceptual base it can be conceived that human mental facets such as human understanding, beliefs, judgment, and choice cannot be easily mapped onto a system function such as control air-vehicle. This is true especially as many complex system control functions must be hidden to the operator and many human cognitive processes must be governed by factors such as previous human training, experience, the effects of the flying environment, and immutable human heuristics that appear to be common to all humans (Tversky & Kahneman, 1974).

Nevertheless, although the STANAG concept of function allocation still mainly relies on the assumption that physicalistic functions can be mapped into human functions, it is strongly influenced by "the review of potential operator capabilities" (p. 3). It is suggested that, in reality, some of the human human–machine functions requiring consideration might emanate solely from the potential operator capability analysis area, especially through a review of operator tasks in similar systems. Importantly, it needs to be recognized that some essential human–machine system functions can be purely cognitive (e.g. SCFs), albeit open to influence from human understanding on the significance of information available from the pertinent human–machine system or other sources.

Figure 14.2 gives an indication of the initial analyses required by the STANAG 3994. It can be seen that both HF standards support analyses and design tasks that are predominantly sequential and have no reference to the life cycle model that they are suited for. In addition, the cited tasks' relevance to other engineering activities are not considered. Neither is the effectiveness of the contribution of the tasks' products to the quality of system design. This latter omission is symptomatic of the reasons most human engineering methods do not directly relate to other systems engineering activities. It also offers an explanation for why human engineering is frequently applied later in detailed design as a palliative to earlier ill-advised design decisions. This late application of HF is usually concentrated on the design of the system HMI.

Traditional Approach to HMI

The study of HMI, a term used to encompass human–computer interfaces, is also supposed to consider the amalgam of human and machine system components. However, what an HMI normally shows is an interface design tuned to foreseen needs for the human to equate to the engineered functionality of the equipment. Engineered system functions are the predominant driver of HMI content. Only recently have there been any signs of a consideration under certain applications for the cognitive needs and abilities of the human operator or maintainer (e.g., Macintosh™ and Windows™ WIMP in desktop metaphor). However, the concept still appears one way, that the human has to appreciate the machine.

There are many HMI paradigms in existence but they are not further considered here. As an aside, prototyping is meant to be an exercise where an HMI can be demonstrated and tuned to obtain the optimum interface between the machine and the human for a particular application. In reality, HMI prototyping with advanced aircrew systems is frequently used only as an exercise of demonstration and not as an analysis of the human–machine system performance capability allowed by the HMI. The design of the HMI should strive for required system performance through a balance between system automation requirements and human performance requirements.

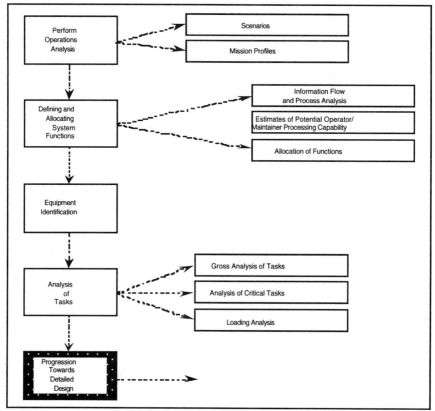

FIG. 14.2 STANAG 3994—Model for early human engineering program.

Thus the argument is that technology should not be adopted blindly as the predominant driver of design and the HMI.

Systems Engineering Models

It has been stated that what both these standards lack is true ties to any engineering model of design. One such model of design is shown in Fig. 14.3. This model presents design as a process including a logical implementation-free phase of requirements capture and a physical design phase of physical verification. Throughout the design life cycle there is control of the design processes. This control requires iteration of design, traceability of design to requirements and design decisions, validation of the design, design-related analyses, assessments, and trade-offs within the design process. Implicit in the model is the tenet that an invalid design can be reliable but that a design cannot be verifiable or valid without reliability.

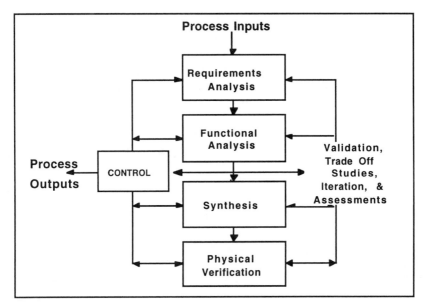

FIG. 14.3 Example of a generic process model of system design (after IEEE P 1220).

The given systems model is only one example of many design models. It is essential in engineering that the appropriate design life cycle model is adopted to suit the complexity of the system and the form of technology that is to be used for system building. It is vital that human engineering activities are fitted to the adopted design model. After all, one of the basic rules in ergonomics and HF is that the practitioner carefully consider the myriad of influences of the working environment on human work.

To reiterate, it needs to be recognized that some essential human–machine system functions can be purely cognitive. Thus, it is necessary that better consideration is given to the important functionality of human cognition within human–machine systems, this during the specification of advanced aircrew systems (MacLeod, 1998a). Before attempting to further advance such a consideration, it is first sensible to have a more detailed appraisal of some of the extant components of the system design process, and methods of their classification, starting with system functionality.

It has been argued to this point that the abstract concept of functionality used for systems design has a physicalistic, engineering, or empirical bias. This is historic in origin and is still the usual concept of functionality (Ackoff, 1972). However, functionality can be categorized in many alternative ways including material or informational, necessary or accessory (Price, 1985). It is argued here that for an advanced or complex system it can also be categorized as cognitive, equipment, or a combination of both.

SPECIFICATION OF SYSTEM CAPABILITY THROUGH ITS
REQUIRED FUNCTIONALITY

The specification of system functionality delineates the span of system capability and is one of the fundamentals of the traditional system approach to design. Functionality is based on a knowledge of intended system purpose, usage, the technology available, previous like systems (if any), and the level of human behavioral involvement with system operation. The engineering emphasis on specifying physicalistic functions rather than SCFs has already been discussed.

Engineered functional requirements normally have associated performance requirements. At the logical stage of system requirements and performance specification, the human is normally only considered through the definition of system constraints. The specification of constraints is essential but it should be noted that a system constraint has no explicit performance requirements (e.g,. aircraft cockpit dimensions are to cater for 5th to 95th percentile of the population; displays are to cater for a human viewing distance of between 500 mm and 630 mm).

Regardless of the constraints placed on system specification, there must also be a method of classifying the importance of the functions being specified. One method is to classify functions as either as necessary or accessory. *Necessary functions* are functions that are deemed to be essential to allow a system to successfully meet its goal(s). Absence or failure of a necessary function will result in a failure to meet the system goal(s). *Accessory functions* provide system redundancy, allow alternative paths to task completion, or add capabilities that enhance the system. The failure of an accessory function is not critical to the successful performance of a system. This classification method should be useful regardless of the nature of the functionality examined, although traditionally, systems design only considers tangible physical functions. It could be argued that if anything is tangible it can be considered under material functionality. However, this is not the case in the consideration of material functionality as it attends to things both tangible and physical with relation to a system. For example, fear is sometimes described by individuals as tangible, and it may be related to a system operation, but it is not physically part of a system.

Of interest, the necessary–accessory functional classification may appear to map conveniently onto the standard U.K. Ministry of Defense (MOD) specification of system features as essential or desirable. However, essential–desirable are indications of the MOD priority on requirements and may or may not be associated with considerations on the criticality of the feature with respect to the achievement of mission success.

It has already been argued that the physicalistic functionality required by a system cannot be simply or easily mapped across to the functionality pertinent to the operation of the human component of the system. For advanced and complex aircraft systems, it is important that a method be devised to classify functionality from several standpoints: the physicalistic or equipment standpoint, the standpoint of the human cognitive component, and the standpoint of amalgamated equipment and the cognitive system.

However, one improved approach to functionality classification verges on the recognition of the prime importance of human knowledge. In it Price (1985) mooted cognition in his classification of material and informational functionality.

Material functionality is seen as purely physicalistic and refers to traditional system-engineered or equipment functionality. Depending on the intent of the particular study this functionality can be generic or specific to adopted equipment. Informational functionality is closely associated with SCFs and will be discussed next

Informational Functionality concerns information that is associated with the system usage of physicalistic functionality. Thus, informational functionality is closely associated with material functionality. As an example, an informational function might be "provide plan to drop weapon" with a direct material equivalent of "provide means to drop weapon." Considering both suggests that a system function is a property of a system and covers aspects of what a system is intended to do.

Nevertheless, there may not necessarily be an obvious direct link between informational and material function. For example, the informational function of "consider weather effects on radar" has no direct material equivalent in the system, as humans cannot yet dictate the weather and are only considering it through the radar sensor. Moreover, the material radar functionality that could be related is diverse (e.g., adjust radar picture, adjust radar scanner tilt, switch off radar, inform crew of radar effects, etc.). Any departure from the basic verb + noun phrase defining informational functionality leads to difficulties in determining the equivalent material functionality. Therefore, it should be questioned as to whether this direct matching is of any use in determining the total functionality needed for an advanced or complex aircraft system.

The traditional answer to the preceding question would be affirmative as the high-level function "plan drop weapon" might be decomposed into several subfunctions such as "plan select weapon," "plan open bomb doors," "determine time to warn crew of attack," and so on. Again, however, the traditional approach would be to only consider the informational functionality directly associated with a material function or functions of "drop weapon."

Consideration of more complex functions highlights some of the application problems with the material–informational classification. More complex functions such as "choose best weapon and attack tactic" are obviously an amalgam of material, informational, and cognitive functions with the cognitive being paramount. Such functions can also suggest processes at work: the threads, information flows, and controls that tie functions together and give them meaning within system operation and performance.

However, a function such as "question evidence" is essentially cognitive and may be prompted by human knowledge and experience rather than by the physical evidence presented by a system. This latter function cannot be described as informational as there is no obvious association with a material function. Thus, the traditional informational approach must be questioned as only in the case of simple material functionality (e.g., "lower seat heigh") are there likely to be near direct associations of informational functionality.

Still considering the example of "drop weapon," in reality the associated human–machine related functionality is likely to be vast and could involve a cognitive association of information considering aircraft performance (height, speed, and attitude), an assessment of target performance (speed, aspect height, maneuver), a recollection of given rules of engagement, an awareness of positions of friendly forces that might be at risk, and an appreciation of aircraft stores remaining, to mention but a few possibilities. Therefore, it is argued that this attempt to break the physicalistic description of functionality into its material and associated informational components is still engineering associated and would require a redefinition to allow it to fully consider human and system cognition within a human–machine system.

Some attempts have been made through knowledge engineering to encapsulate human cognitive functions as knowledge encompassed by the materialistic functionality of advanced aircrew computer-based systems. It is beyond this chapter to consider whether knowledge engineering can successfully capture complex human cognitive-based expertise, and then usefully incorporate that expertise as system-related functionality within a dynamic and advanced aircrew system. Nevertheless, to be truly successful, any such attempts must explicitly recognize human expertise and the importance of human cognitive functions to the operation of a human–machine system.

System Performance

Performance is the manifest result of the work undertaken by a system. A system's required functionality is encapsulated by design to achieve a required quality or level of performance. In reality a system achieves a standard of performance that is seldom truly equivalent to that aimed for by the design. The designed amalgam of desired system functionality and, sometimes, planned capabilities of the human component within the system determine the predicted system standard of performance. In reality, the achieved standard of performance is mediated by the environment, the achievable performance of the designed system, system reliability and actual human standard of performance as allowed through the use of the system, and influences external to the system. The human ultimately directs advanced aircrew system performance. System performance cannot be fully addressed unless the functionality and its expected performance are considered alongside equipment performance during the early process of logical system requirement specification. As previously stated, the human is considered within logical system requirements as a system constraint with no associated performance requirements. The function of the human within system requirements is normally poorly considered. Therefore, any HF-related certification of a system is fraught with difficulties.

Summary of the Assessment of Functionality

It has already been argued that much of the physicalistic functionality associated with a complex system has no obviously associated human equivalent as the processes of cognition are hidden and may result in no visible human action or system input. Further, in aircraft systems there are important sustainers of human situational awareness that are not specific to system design but that are important to system performance. Such sustainers may be forgotten to the detriment of aircraft operation if an advanced aviation system design was to be purely based on the physicalistic or material approach. These include, but are not limited to, environmental or system associations such as arise from in ambient noise, vibration, or visual sightings. Indeed, sometimes the meaningful indicator is an absence of system derived information when the information has been determined by other means and should be present. Finally,

> Much is yet to be done, especially in analysing human cognitive requirements in working with automated machines and in putting a methodology into effect that will bring humans and machines systematically together to do those things that each can do best and that they can accomplish jointly to improve system performance. (Price, 1985).

Traditionally, the performance of complex airborne systems has been inferior to that predicted or required, a compromise has been accepted, and the system has only reached the desired standards after system enhancements introduced some time after the system becomes operational. It is suggested that such enforced compromises, some discussed later, will become less and less viable in the future, and are not supportive to safety and the HF certification of systems.

TRADITIONS OF ENFORCED COMPROMISE

Extrapolation of past practices suggests that the more complex the system, the longer it will take to bring an inferior system performance up to the level of expected or acceptable performance. In the military there are several advanced systems in which the latter point has been borne out.

The solutions applied to the introduction of systems that are inferior to the requirement appear to be selected from one or more of the following:

1. Compromise and accept the system as better than previous systems. (Do what you are told route).
2. Compromise and accept the system limitations until time and finances can be found to improve the system. (traditional route often involving an expected midlife update even before the initial system delivery).
3. Compromise and increase the number of personnel operating the system. (throw manpower at the problem—the serfdom route).
4. Compromise by expanding the training program and improving the quality and experience of personnel employed with the system (expensive, but can be blamed on the quality of the personnel available in the past, or past mistakes in recruitment or on the needed complexity of the system).

Methods Outside the Traditional Compromises

5. Reappraise the design method. Ensure that the next system is designed using the lessons learned (a form of reappraisal of design method should always happen after any system design).
6. Immediately cancel the production of the system (a final and shameful resort with blame attributed where and when possible and probably an obfuscation of any lessons to be learned).
7. Develop and apply suitable methods for specifying and certifying the design and building of advanced aircrew systems (this depends on a realization that current methodologies are inadequate and the existence of a will to improve).

Of the mooted seven solutions given, the first four compromises have been in existence for decades. The fifth solution is what is needed now and in the future to prevent the recurrence of expensive failures. The sixth has certainly been evoked recently (several programs in the United States. and the United Kingdom spring to mind). The seventh is obviously what must be effective and of which, it is argued, SCFs would have a part to play. Of course, no method will ever be perfect and certain necessary compromises or trade-offs will always be inherent in design and should not be decried.

Recapitulation on Design and Methods

Comprehensive and efficient design is the key to the achievement of required performance from complex systems but depends on the standard of system specification. Design is the bridge between system specification and the achieved system performance. In this chapter, we have argued that complex system designs have traditionally considered the human in the physicalistic and mechanicalistic sense.

Until recently, the problems associated with the parallel needs of promoting human understanding alongside system operation and direction were generally equated by the natural flexibility and adaptability of human skills. The human performance contribution to overall system performance was poorly considered by design and usually catered for postdesign by operator training.

However, the information rates and complexity of modern systems often place system processes beyond the supervisory or manipulative capabilities of the human, because human cognitive attributes and performance have not been properly considered, within the design of the system. Where human cognition has been considered it is normally only where the concept of human cognition has parallels into the physicalistic mold of determining material functionality. The concept of SCFs was introduced as a means of system-related consideration of the functional requirements of cognition within a system.

The next section considers an example of a design that partly approached some aspects of cognition within its logical specification. The actual specification produced for the Royal Navy (RN) Merlin helicopter is considered. Some of the lessons learned from that specification's consideration on system functionality are discussed.

FUNCTIONALITY SPECIFICATION FOR THE RN MERLIN HELICOPTER

The emphasis of this chapter is on the consideration of system functionality and its influence on the processes of HF certification. The overall specification process for the Merlin helicopter is presented in detail in Taylor and MacLeod (1994).

The definition of Merlin functionality constituted one of the three parts created for the Merlin specification exercise. The functionality document was termed the functional requirements definition (FRD).

FRD of Merlin Specification

The FRD considered system functionality for the Merlin, as "functionality is not solely derived from the definition of the requirements for the individual systems and their interaction, ...those systems interact with the crew and systems outside the remit of Merlin, and the operational environment in which they are to operate". (U. K. Directorate of EHIO, 1990)

Primary and secondary objectives were used to consider functionality. The primary objective was to define the minimum acceptable functionality for the Merlin. This involved specifying the functionality for existing and new systems. The functionality considered here was material functionality. Thus, considering the existing equipment "a set of "Major Functions" were identified. To allow the current documentation set to be as effective as possible, the Major Functions are chosen to be approximately equivalent to the systems fitted to the current helicopter [EH101] and are therefore not intended to be a 'pure' functional breakdown from operational requirements. (U. K. Directorate of EHIO, 1990)

The secondary objective was to define system management functions fundamental to the successful integration of all systems on board Merlin. System management is defined as "the usage of the Merlin's System through the tools devised from the amalgam of Human Engineering and other engineering approaches to the system design   (U. K. Directorate of EHIO, 1990). The management areas were considered as flight, tactical and maintenance. The three management functions and specific sensor functions define the parameters that are to be displayed to the crew and the controls necessary to influence the operation of the Merlin. However, the interdependency between the management areas was not addressed in any detail.

Flight and tactical management was split into the following subsets:

- *Mission management:* Those functions necessary to permit procedures and equipment to be employed that assist the crew in conducting the tasks required of Merlin.
- *Information management.* Those functions necessary for the collation and processing of data to determine a future flight path or for the collation and processing of a tactical picture.
- *HMI management.* Those functions necessary for display of the tactical picture and system and equipment status, and crew interaction with the tactical picture.
- *Sensor management.* Those functions necessary to control the operation of the various related sensor functions in a consistent manner.

It can be seen that under this form of functionality classification there was an effort to consider the functions that the crew would have to perform to manage the specified material functionality of the Merlin within, but not between, each of the three defined management areas. However, there is a potential pitfall with the focus on the human component of the system as a manager.   First, teamwork does not depend solely on good management but on a myriad of influences.   Further, by definition, managers supervise and administer the resources available to them.   A manager does not necessarily lead but directs resources within certain specified rules.   In contrast, a leader (or the Merlin observer in the role of tactical coordinator) does not necessarily need to be primarily a manager of systems, but guides others using foresight and tactics as well as resources. Indeed, the leader's used resources may be beyond those immediately available and may not necessarily be closely governed by rules.

In the Merlin case, the rules applied to the appreciation of system management failed to consider the system as a whole and thus restricted the scope of the given secondary objective. It is a matter of conjecture as to how efficiently the system might support leadership. However, the lack of a whole system appreciation of system management must raise a question about the efficacy of the aid that the resulting system will offer to the leader or system manager.

These points on the management of real-time complex systems may be summarized: A human is a poor supervisor. If the human is not involved in the operation of the system, the human's attention and reactions to system cues are liable to suffer. Moreover, if the human functionality within a system is not fully specified it will be difficult to properly manage within the designed remit of the system. At the worst, the human resource is then managed in a procrustean manner to fit into the machine design and may become involved in an incessant combat with the machine to achieve system goals. Traditionally, tools are devised to aid operators with their use of the functionality of the system (both seen as separate agents) rather than to assist the operators to control and direct system work through the system (system seen as one agent).

However, particularly with consideration of information management, there was some consideration of the human cognition requirements for the management of the Merlin systems. Examples were:

1. The symbology used to represent information displayed to the crew shall be developed in accordance with human engineering principles given to take account of the cabin and cockpit environment and human *understanding*.

2. Classification also depends on human *understanding* of presented data. Aspects of this process shall be evaluated as part of the Human Engineering Programme Plan.

Nevertheless, the main source of management functions did reside in the material functionality of the aircraft equipment. This was partly because the definition of the major functions was mainly determined by the existence of equipment already adopted for the helicopter during earlier development. It was also caused by a lack of in-depth consideration of extant maritime tactics or future possible tactics for the aircraft. The particular reasons for this are not discussed here. However, the result is that the requirements of possible tactical performance cannot be fully equated with the existing equipment functionality.

Thus the problems associated with necessary task-related judgments and direction through human cognition at the human–machine system interface could not be fully addressed. For this reason, much of the functionality that could have been ascribed to human mental properties was instead placed for consideration within the responsibility of the human engineering program plan

(HEPP), a plan that had to be constructed under the mandate of the already discussed STANAG 3994.

Merlin HEPP and Design Requirements

The application of human engineering to the Merlin system was governed by a mandated and agreed on HEPP. Human engineering placed emphasis on the human component of the system and introduced a planned approach to this aspect of the Merlin's design so that important HF facets were recognized and addressed.

Because of the already designed equipment, the HEPP and other design requirements were mandated too "late in the day" to be as effective as they might have been if they had been in existence from the onset of the system analysis and design process. Moreover, the Merlin specification was not created under the full remit of STANAG 3994, although the STANAG was mandated by the completed specification. Therefore, some of the STANAG's recommended system analyses were not fully considered in the specification including mission analysis and potential operator capability analysis.

However, for the new aircraft equipment, and hopefully for consideration of the integration of human performance within systems operations, the HEPP served as a valuable aid for indicating areas of the system where improvements might have to be made or new procedures devised. Of course, before the system can be accepted into RN service it must pass a formal operational performance and acceptance specification (OPAS) without a need for too much compromise. The HEPP is one of many ways of assessing the risk of successfully completing the OPAS.

Moreover, the HEPP was based on the FRD and design requirements. The problems with the method of the FRD specification of requirements, in that SCFs were not considered, have already been discussed. Therefore, with some system analyses already conducted, the biases of the analyses were reflected in the performance of the HEPP. As the HEPP was also concerned with HF acceptance of all forms of HF analysis, tests, and trials, the HF aspects of Merlin system certification had to bear in mind the initial problems of specification.

Discussion on Merlin Specification

The RN Merlin was respecified as a system many years after the onset of the initial design process for the helicopter. However, the respecification process allowed the remaining aircraft development process to be defined both with relation to the expected aircraft system performance and the possible risks associated with meeting the OPAS acceptance requirements. Throughout the HEPP, the HF input to design and the aircraft certification process has been stated with appropriate qualifications. This makes clear the process and requirements of HF certification.

The important HF lessons learned from the specification work were:

- HEPPs are currently devised to be applied during the physical phases of the design life cycle. They should be devised at the logical requirement specification stage of design and be applicable to the whole design life cycle.
- To be fully effective, an HEPP should be produced and started early in the systems analysis process.
- It was obvious that more care has to be taken in the consideration of human cognition with respect to the design of a man–machine system, especially where cognition may have specialized functions within the human–machine system.
- The omissions in the specification with relation to human cognition must represent a source of unspecified risk within the process of HF certification.

Through these arguments, it can be reiterated that appreciation of system cognition, particularly SCFs, is an important facet of complex human–machine specification and, ultimately, certification. A outline of a possible method for the consideration of cognition during system analysis and specification follows. In part, the method is a development of areas of STANAG 3994 and other sources.

THE INCORPORATION OF HUMAN TACTICS AND STRATEGIES

Responsibility for the direction and control of a manned aircraft ultimately rests with the aircrew. Responsibility for the safe flight of a remote piloted vehicle ultimately rests with the human controller on the ground. Responsibility for safe air traffic control usually resides both in the air and on the ground. All of these rely on the use of plans. Important SCFs in these plans concern the anticipation of future events and intentions influencing the system. The form of these plans can be considered under the titles of tactics and strategies.

Introduction to Tactics and Strategies

A *tactic* is defined as an arrangement or plan formed to achieve some short-term goal. The goal may be an end in itself or serve as a stage in the progress toward a later objective. A *strategy* governs the use of tactics for the fulfillment of an overall or long-term plan. Tactics can be formal written procedures or resident in human mental processes. Normally the human tactic selectively directs the formal system-related procedural tactic.

Human tactics and strategies are not only physicalistic, they are mental. Explanations of the human usage of a mental model of the world have been given by many researchers from various viewpoints and considering many possible constituent parts (cognitive maps, schemas, frames, scripts, goals, plans, and schemes). The use of *tactic* and *strategy* in this chapter is for the sake of explanation and is not intended to supplant what has gone before but to aid in the current exposition. The terms are used here as they conveniently afford a mirror of cognitive activity onto related aircrew operating procedures (which can be broken down to tactics and strategies in the militaristic sense). Reference is given to Adler (1929), Bartlett (1932), Craik (1943), Schank and Abelson (1975), Neisser (1976) and Card, Moran, and Newell. (1983)

A tactic is procedural, may be mainly skill based, and is flexible and adaptable to a changing environment. In a human–machine system, the performance of tactics and strategies is enhanced by system equipment designed to aid the human in interpreting information contained in the working environment and, also, in surviving in that environment. Strategies allow the human to be selective in the use of tactics, to choose the most effective or most expedient for the fulfillment of the foreseen plan.

The human perceives the world through information gleaned from the senses. This perception can be achieved through direct observation of the world or through a human–machine system's interface with the world. The perceived information is interpreted using knowledge, rule-, and skill-based mental and cognitive processes that may vary among humans (Rasmussen, 1986). Human tactics and strategies then govern the use of the interpreted information. These tactics and strategies are tuned through training and experience and are governed by the human role within a human–machine system and the human's interpretation of that role. The performance of human tactics and strategies is manifest in observable human activities. The quality of the results of human tactical and strategic performance is closely associated with human expertise and its accompanying situational awareness.

Tactics and strategies are continually mediated by both the information that the operator already possesses and information gleaned from the working environment. In the performance of work, the former information is purely human in origin and influences the latter, whereas the latter originates from human or machine assessments and can eventually contribute to human work experience. It is beyond the scope of this chapter to argue, but SCFs do encompass most of the area of tactics and strategies as described here. It is also possible that baseline SCFs for tactics and strategies could reside in the engineered system to complement operator cognition-based SCFs.

Difficulties in Plan Concept and Application

One of the main difficulties of concept with human tactics and strategies is how to translate the abstract into something concrete akin to a system function, and then in a form through which it can be applied to human–machine system analysis. The first stage is to make the abstract tangible with respect to stages of job performance.

STANAG 3994 mentions the use of a review of tasks in similar systems. To be effective, such a review requires an in-depth examination of tasks, possibly using subject matter experts (SMEs) and knowledge elicitation techniques. Knowledge elicitation techniques can be considered under two categories: direct and indirect methods. Direct methods are used where experts can be asked directly to indicate their knowledge. Indirect methods are used to infer the expert's knowledge from the expert's performance at other or similar tasks to those in question. Direct methods include interviews, protocol analysis, Kelly grid, and concept sorting. Indirect methods include 20 questions, concept recall, and listing. Further reading includes Kidd (1987).

The matters to be examined include:

- Common problem areas requiring cognitive effort; for example the interpretation of sensor data, the determination of rates of change of data, or the understanding of particular types of information.
- The concepts utilized by the SMEs in their performance of work.
- The association of the common problem areas with jobs (i.e., tactics and strategies) or forms of task.
- Any timing data that may be available; that is, with this form of problem and that form of task, why does the operator take a certain time to gather evidence and resolve the problem?
- Any evidence that can be collected on how to ameliorate the operator's cognitive effort, if deemed to be excessive.

The two areas of system-related cognition that might be investigated by the examination are:

- The application of cognitive processes to the performance of a task or subtask.
- The application of cognitive processes to progress tactics or strategies associated with a group of tasks or parts of a mission.

From an appreciation of these points, and the necessary and accessory system material functions, an appreciation of the necessary SCFs may be obtained.

AN AVENUE TOWARD A NEW METHOD OF SYSTEMS DESIGN

The efficacy of the commonly used forms of the Fitts list must be questioned. If it is accepted that some essential human–machine system functions can be purely cognitive, then it is necessary to develop a method to identify these SCFs and their implications within the system design.

The method of incorporation of human cognition into system design requires a detailed system functional and performance analysis process leading to a logical specification of system functionality and performance. This analysis and specification examine and combine the required and refined functionality of the new system, including SCFs, with that obtained from the examination of similar systems. Some considerations on an avenue to such a method are presented next.

Examination of Similar Systems

The initial action is to look at what systems have been used before for similar aircraft roles. This does not necessarily involve the examination of near identical jobs to those envisaged for the new system, although such an examination is preferable. It may be that there are no near identical jobs and that equivalent jobs will have to be used. For example, the operation of a collision avoidance radar screen might give indications of problems that may be encountered in the operation of weather avoidance radar.

From the initial action, a system (or systems) is chosen for further examination. Obviously, the greater the number of systems examined the better within the constraints of time and budget.

The next stage is to use knowledge elicitation techniques, as appropriate and with the cooperation of suitable SMEs, to determine where the SMEs assess the strongest cognitive constructs or greatest cognitive loads. To assist this process, the concepts of human tactics and strategies should be explained as well as the equipment and tasks being considered. The SMEs' use of tactics and strategies, both mental and in the material and militaristic sense, was approached by MacLeod and Taylor (1994) during the process of the predictive examination of workload for the RN Merlin. This study also included an examination of the effects of operator errors and decision processes on system tactical performance. It is important that a series of tasks is considered so that both tactics and strategies can be properly addressed. For example, it is contended that SCFs can only be determined through a consideration of the dynamic use of system functions. If possible, the questioning of an SME during the operation of an aircraft system or simulator is preferable. If actual equipment of some form is not available, some form of task-analytic simulation might suffice, provided the SME is well acquainted with the form of task representation used. It must be accepted that there is no current method of ensuring that all system-critical cognitive processes will be examined.

It is suggested that the elicitation of knowledge in the task-related area is easier than the elicitation of knowledge on tactics and strategies. For instance, a task is normally performed under operator-focused attention, whereas an operator's consideration of a tactic or strategy may not necessarily be continuous and, in the case of high expertise, may be resident in nondeclarative memory. Considering verbal protocols as an example of a task elicitation technique, concurrent and retrospective protocols may both be suitable to gain a fair indication of a task-related use of cognition. However, to elicitate information not in immediate memory (i.e., strategies) might require some form of prompting or the use of several knowledge elicitation methods and the triangulation of results.

The examination of the verbal protocols must be based on a model and some decided categories of cognition. The question of whether verbal reports are or are not epiphenomenal is a subject of continued debate that will not be addressed here. For detailed coverage of the debate see Ericsson and Simon (1984) and Nisbett and Wilson (1977). The concern here must be the benefit that any method brings to the final application compared to the benefits possible from other methods. Considering the inadequacies of the traditional approach, it is suggested that even a classification of human–machine system tasks as requiring associated system cognitive functions of a high, medium, or low nature is better than no consideration at all.

This discussion is currently focused on knowledge elicitation using old physical models. Once an early understanding is obtained on the operators' use of cognition or cognitive functionality, system task-related use of SCFs should be considered with relation to the chosen system's equipment functions embedded in that task. Where any association of cognitive and equipment functions is not possible, but the engineered function is understood, each form of function or functions should be specifically labeled as a *task=related engineered* function associated with a specific task or tasks (e.g., lower undercarriage). Where a task related function is deemed purely cognitive it should be labelled as a *task-related cognitive function* (e.g., assess visibility). Where system related equipment and cognitive functions must be associated, their association should be labelled as a specific *task related associated function* (e.g., determine position on glide-slope of airfield approach). Any cognitive functionality that cannot be related to a specific task, but to a series of tasks, a tactic, or a process, should also be considered and labelled as a *strategic system cognitive function* (e.g., consider tactics to be applied to surveillance of a maneuvering target).

Appreciation Within New System Design

If there is no preceding and similar system model on which to base the classification of function with expected task(s) or task forms, then a best estimate must be made that will be subsequently checked through iterative processes of design throughout the design life cycle. With many new technologies, such as Knowledge Based Systems (KBS), the simple waterfall model of the design life cycle is inappropriate and new models are required (e.g,. evolutionary, spiral).

Once knowledge is gleaned on SCFs, the next step should be to apply the data on cognitive processes obtained from the examination of similar systems as outlined earlier. The difficulties in determining and incorporating values of cognition into the system life-cycle design processes are considerable. To strive for the synergy necessary for a complex human–machine system, cognitive task analysis techniques may be applicable here, although the utility of many of the current methods for this task must be questioned.

The knowledge gleaned in this fashion could then be checked and further refined using mock-ups, different levels of prototyping, and simulators. Of course, the ideal scenario would be to continue the refinement process using data collected from actual aircraft equipment and flight prototype trials.

Whatever method is used, a careful consideration of cognition will require an iterative process in the early stages of the system analysis process. This iterative examination is seen as essential to consider and amalgamate the information gleaned from old systems (i.e., cognitive tactics and manifest operating procedures) with the detailed functionality and expected performance of new equipment.

Figure 14.4 represents a model of process control for a military aircraft system. Examination of the figure shows that the problem space of SCFs work would have to reside to a greater or lesser extent across three areas of the model, namely:

- Sensor control.
- Cognitive control.
- Equipment, aircraft, & weapons control.

With a human–machine system including SFCs, situational awareness must be shared through some form of dynamic apportionment within the team (MacLeod et al., 1995). Further, the model deals with uncertainty as might exist in a hostile military environment and indicates how that uncertainty may be caused and its affects ameliorated depending on the applied level of aircrew expertise. Thus both expertise and planning and tasking mediate the cognitive loop appraising the environment.

This loop is susceptible to artefacts of uncertainty that can evoke unplanned observations. These unplanned observations can be detrimental to human–machine system control if outside the scope of applied expertise.

Finally, the area of cognitive function in Fig. 14.4 represents the area suggested as being of prime concern for SCFs. SCFs are only capable of situation assessment in cooperation with other functions as, lacking sentience, they are incapable of situation awareness. Nevertheless, their situation assessment should help the operator focus on and maintain appropriate situational awareness.

Valid basics for the understanding of cognitive functionality can only be determined through practice in investigation and application. What is eventually required for the system designer is a set of rules through which the subject of human cognitive functionality can be effectively approached within the realms of system design as a whole.

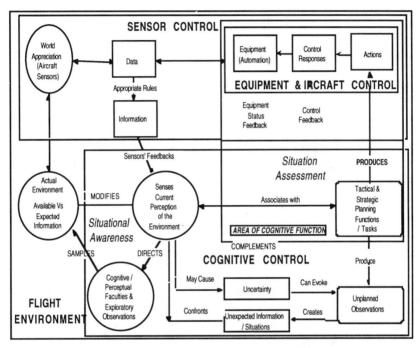

FIG. 14.4. A simple process model of an aircraft system (acknowledgements to Neisser, 1976, and Hollnagel, 1995).

It is easy to pay lip service to theory and say that equipment should be built to appreciate the human and the human trained to appreciate the machine. The basis of such mutual appreciation must be a better practical understanding of human cognition as applicable to advanced aircrew systems. However, such an understanding will involve a great deal of experimentation, preferably in the field rather than the laboratory, and an education of system designers to convince them that such an effort is necessary.

Finally, it is suggested that a careful appreciation of cognition within the specification of the functionality for a new system should allow improvements in the following:

- The initial assessment of the numbers and quality of personnel required to operate the system.
- The determination of necessary and accessory system functionality.

- The assessment during system design of the form of operator training required.
- The production and progression of a HEPP for the new system.
- The design of the system HMI.
- The efficiency of trade-off process during system design.
- The creation of formal operating or tactical procedures.
- The assessment of achievable system performance and the risk inherent in the design.
- Achieved system performance.
- Life cycle costs.

Certification Issues Revisited.. The difficulties involved with the certification of a system vary with the system application and the complexity of its constituent parts (e.g., its software, hardware, and liveware). The certification of software based avionics is based on well-established verification and validation principles and developing methods such as employed by software static code analysis.

Verification is the process of evaluating a system or components to determine whether the products of a given development phase satisfy the conditions imposed at the start of the phase (DEF STAN 00–55, 1991). In contrast, *validation* is the evaluation process that ensures that the developed product is compliant to the given product requirements.

Hardware principles are also well established. System acceptance principles are less well established but are based on specified system performance criteria and a regime of testing, evaluation, and acceptance. Aircrew certification is in the form of an endorsement of their reliability and expertise. This endorsement process includes an appraisal of the standard of aircrew ab initio and after continuation training, and includes frequent in-flight and simulator based evaluations of their level of competence with relation to the conduct of safe flight and associated aircrew tasks.

Current U.K. military certification of safety critical systems in aircraft involves satisfaction of many requirements, much of these contained in the U.K. software engineering Def Stan 00–55,1991; Def Stan 00–56, 1996; and Def Stan 00–58, 1992. Applied with less rigor, these standards can also be applied to nonsafety critical systems such as mission critical systems. The focus for the certification of a mission critical system is the criticality of that system performance to mission effectiveness. In contrast, in certifying a safety critical system, the criticality of that system to the safe performance of the mission should be the prime consideration. These standards do not directly address HF issues and, indeed, there is some divide between the application of safety engineering and HF.

Control theory has long been capable of specifying the dynamic control changes required to direct the necessary machine processes to adapt to changes caused by environmental influences or operator inputs to the machine processes. The seminal works by Wiener (1948) and Ashby (1956) on cybernetics introduced the idea of homeostasis in human–machine systems; adaptive control to maintain a stable and required human–machine system state. However, a system that includes a human, and possibly engineered SCFs, will involve a control loop the homeostasis of which relies on a sharing of cognitive aspects of machine and operator teaming, including the team application of pertinent knowledge, expertise, situation assessment, and situation awareness. Such a loop would have to reliably perform its required control functions to gain acceptance for its adoption or certification. In other words it must do what it is required to do in a variety of situations and in a sufficiently consistent manner to allow confidence in its acceptance. Certification of SCFs presents particular problems if imbued in a machine. The primary problem is related to how to verify and validate software that contains knowledge. The validation of the coding should not present any insurmountable problems. However, the verification and validation of the contents and application of the knowledge base will be harder to certify.

Certification of SCFs resident in the human are checked through selection, training, continuation training, periodic aircrew standardization, and performance checks. In the case of aircrew, certificates are awarded as ratings indicating that the aircrew has reached acceptable levels of competence. Possibly, future technologies (KBS knowledge bases being just one case in point) may be certified under some form of usage maturity model—a model that presents a higher certification, accompanied by an improved license expanding the use of the system the longer it has been in effective operation.

SCF-related certification should involve verification and validation examination of at least four separate processes as associated with:

- System architecture certification.
- Certification of the elicited inputs and knowledge bases for the SCFs (human- or machine-based).
- Certification of the SCFs build (software or human).
- Overall human–machine system certification.

The first process can arguably be catered for by logical generation of a system architecture model, including the anticipated tests for validity and for good result replication. The second process would probably require iterative examination of the knowledge and rule (possibly skill) inputs by several appointed and independent assessors, these assessors having roles similar to those of present-day flight crew assessors (MacLeod, 1997). The certification through such a process might result in an agreed on rating of the SCFs depending on their performance and maturity. The third process could be catered for by current certification methods: Does the system reliably and validly meet the requirements of the system specification and allow verification of its quality of build? The last process would encompass the results of the previous processes and might also include an assessment of the competence level of the system manufacturer as suggested by the current Software Engineering Institute (1996) Capability Maturity Model.

To be operationally effective an SCF would have to be continually developed (trained). Changes in any one of the first three of the preceding processes could affect the veracity of the whole machine and operator team, as might a change of operator. Therefore, the examination of all of these processes would need to be iterative and the findings applied as refinements to the system throughout the system life cycle. The amount of iteration and refinement allowed by the certification process would have to be carefully considered and administered (e.g., the problems of configuration control).

Further, it is to be expected eventually with SCF design, because an operator's cognitive control performance varies between and during a flight and also with changing flight conditions, that the SCFs could adapt to changes in the environment. The system would have to be aware of such changes and be able to adapt its teaming assistance accordingly. Thus, the concept of a future adaptive controller must cover a team interaction policy as well as a man–machine system control policy, and all this detail has to be unambiguously specified during concept and development before the system build and eventual certification.

In addition, the effects on the system control process of diverse, hostile, and uncertain environments must influence consideration on the requirements for the whole system–SCF–operator team (MacLeod & Wells, 1997). Further, it should be noted that the SCF consideration in this article refers to a one-to-one pairing between a single member of the aircrew and a system. The certified use of any SCF-associated technology within a multicrewed aircraft environment would present additional unique and complex teaming problems.

The current requirements for avionics and aircraft certification are well known. Starting from the baseline of the current requirements, we need to practically complement and supplement these requirements to meet the certification case for a system built using SCFs, this considering the SCF supporting technologies. It is important that a practical engineering emphasis is maintained, drawing from academic and research studies in the field, as the overall certification process will probably be very similar to those employed currently. Current certification processes require more than a method of conducting the certification: The process also requires real-time system operation, system evaluation, and the production of sufficient quality evidence to support the independent certification of the subject system.

SUMMARY AND CONCLUSION

System cognitive control relies on an understanding of what system functionality is necessary to support that level of system control required to allow system direction toward the fulfillment of system goals and the effective satisfaction of system purpose.

This chapter has examined some of the requirements of HF specification and certification within advanced or complex aircrew systems. It suggests reasons for current inadequacies in the use of HF in the design process, giving some examples in support, and suggests an avenue toward the improvement of the HF certification process. The importance of human cognition to the operation and performance of advanced aircrew systems has been stressed. Many of the shortfalls of advanced aircrew systems must be attributed to overautomated designs, designs driven by the introduction of new technologies and burgeoning system complexity. These new automated designs show little consideration for the cognitive functional requirements of the human–machine system. This lack of consideration forces the absence of system cognitive functions from design consideration. Moreover, the continued application of traditional HF methods and standards lacks pertinence to design using new technologies.

Traditional approaches to system design and HF certification are set within an overphysicalistic foundation. Further, traditional methods and HF standards are poor in their consideration of the introduction of new technologies. They also lack any appreciation of models of the system life cycle outside that of the traditional Waterfall model of design, code, test. Further, HF standards assume rather than make explicit the usefulness to design of their proposed design-related stages and tasks.

Fitts (1951) argued that physicalistic system functions could be attributed to either the human or the machine. However, in the past any problems associated with the parallel needs, those of promoting human understanding alongside system operation and direction, were generally equated in reality by the natural flexibility and adaptability of human skills. Thus, the consideration of the human component of a complex system was seen as being primarily based on manifestations of human behavior, this to the almost total exclusion of any appreciation of unobservable human mental and cognitive processes. The argument we presented in this chapter is that the considered functionality of any complex human–machine system must contain functions that are purely cognitive.

In this chapter we discussed one example of an attempt to devise an improved method of specification and certification with relation to the advanced aircrew system, that of the RN Merlin helicopter. The method was realized to have limitations in practice, these mainly associated with the late production of the system specification in relation to the system development process.

The need for a careful appreciation of the capabilities and support needs of human cognition within the design process of a complex man–machine system has been argued, especially with relation to the concept of system functionality. A new classification of system functionality is proposed, namely:

- *Engineered:* System engineering related.
- *Cognitive:*   System cognition related.
- *Associated:* Necessary combination of system related engineering and cognitive functions.

We did not define a new method for a fuller consideration of cognition within systems design, but we suggested the need for such a method and indicated an avenue toward its development through the appreciation, and eventual logical specification, of system cognitive functions. Importantly, the HF certification of advanced aircrew systems is argued as only being possible in a limited sense until the important system cognitive functions are considered within the system design process.

REFERENCES

Ackoff, R. L. (1972). Science in the systems age: Beyond IE, OR, And MS. Address to the Operations Research Society of America, Atlantic City, NJ.

Adler, A. (1929). Individual Psychology. London: Allen& Unwin.

Ashby, W. R., (1956). An introduction to cybernetics. London: Methuen.

Bainbridge, L. (1987). Ironies of automation . In J. Rasmussen, K. Duncan, & J.Leplat (Eds.),New technology and human error(pp. 276-283) Wiley.

Barlett, F.C. (1932). Remembering: A study in experimental and social psychology.Cambridge, UK: Cambridge University Press.

Card, S. K., Moran, T.P., Newell, A.(1983). The psychology of human computer interaction. Hillsdale, NJ: Lawrence Erlbaum Associates.

Chapanis, A. (1996). Human factors in systems engineering. New York: Wiley.

Craik, K. J. W. (1943). The nature of explanation. Cambridge,UK: Cambridge University Press.

Department of Defense. (1991). Defense acquisition management policies and procedures (Department of Defense Instruction 5000.2). Washington, DC: Office of the Secretary of Defense.

Ericsson, K. A., & Simon, H. A. (1984). Protocol analysis: Verbal reports as data. Cambridge, MA: MIT Press.

Fitts, P.M. (Ed.). (1951). Human engineering for an effective air navigation and traffic control system. Washington, DC: National Research Council, Committee on Aviation Psychology.

Gunderson, K (1971), Mentality and Machines. New York: Doubleday . as quoted in J. Heil., (1983). Perception and Cognition, (p. 144) London: University of California Press.

Hollnagel, E., & Woods, D.D. (1983). Cognitive systems engineering: New wine in new bottles. *International Journal on Man–Machine Studies, 18,* 583–600.

Hollnagel, E. (1983). What we do not know about man–machine systems. *International Journal of Man–Machine Studies, 18,* pps 135–143.

Hollnagel, E. (1995, August). *Lecture notes.* Industrial Summer School on Human-Centered Automation, Saint-Lary, Pyrenees, France.

MacLeod, I. S. (1996). Cognitive quality in advanced crew system concepts: The training of the aircrew-machine team. In S.A. Robertson (Ed.)*Contemporary Ergonomics* (pp. 41-48). London: Taylor & Francis.

MacLeod, I. S., (1997, September), *Certification of the EC / aircrew team—A cognitive control loop,* Paper presented at the 4th Human Electronic Crew Conference, Kreuth, Germany.

MacLeod, I. S. (1998a), A case for the consideration of system related cognitive functions. In E. E. Basker, A. Morison., & K. Tothe. (Eds.)*Proceedings of the 8th International Symposium of the International Council on Systems*

MacLeod, I. S. (1998b). Information and advice: Considerations on their forms and differences in future aircraft systems, In P. A. Scott, R.S. Bridget, & J. Charteris (Eds.)Proceedings of the Global Ergonomics Conference, Cape Town (pp.667-662).. Oxford: Elsevier Science, Amesterdam Engineering, (pp. 57-65), Boeing Company.

MacLeod, I. S. & Scaife, R (1997). What is functionality to be allocated? In E.F. Fallon ,L. Bannon, & J. McCarthy (Eds.). *Proceedings of ALLFN'97* Galway, Ireland (pp. 125-134), IEA Press, Louiseville, USA.

MacLeod, I. S. and Taylor R. M. (1994), Does human cognition allow human factors (HF) certification of advanced aircrew systems? In: J. A. Wise, D. J. Garland, & V. D. Hopkin (Eds.). *Human factors certification of advanced technology.* (pp.163-186) Daytona Beach, FL: Embry-Riddle Aeronautical University Press.

MacLeod, I. S., Taylor, R. M,. & Davies, C. L. (1995). Perspectives on the appreciation of team situational awareness. In D. J. Garland & M. R. Endsley (Eds.).*Experimental Analysis and Measurement of Situational Awareness* (pp. 305-312). Daytona Beach, FL: Embry Riddle Aeronautical University Press.

MacLeod, I. S.& Wells, L. (1997, September), Process control in uncertainty, Paper presented at the 6th European Conference on Cognitive Science Approaches to Process Control, Baveno, Italy.

Meister, D. (1985). Behavioral analysis and measurement methods. New York :Wiley.

Miles, J. L. (1993, July), TASK analysis - Foundation for modern technology._Paper presented at MRC Workshop on Task Analysis, University of Warwick, U. K.

NATO Standardization Agreement (STANAG) – STANAG 3994AI First Draft.(1993, June) *The Application of Human Engineering to Advanced Aircrew Systems.*

Neisser, U. (1976).*Cognition and reality: Principles and implications of cognitive psychology.* San Francisco: Freeman.

Price, H. E. (1985). The allocation of functions in systems. *Human Factors, 27*(1), 33–45.

Rasmussen, J. (1986). *Information processing and human machine interaction* New York: Elsevier.

Schank, R. C., & Abelson, R. P. (1975) Scripts, plans and knowledge.In
    *Proceedings of the fourth International joint conference on artificial
    intelligence*, Tbilisi.: as reproduced in (1980) Thinking: Reading in Cognitive
    Sciences, Johnson-Laird, J. & Wason, P.C. (Eds.). The Pitman Press, Bath U.K.
    421–432

Software Engineering Institute. (1996)*The capability maturity model*, Guidelines for
    Improving the software process, (SEI–CMM V1.1) Pittsburgh, PA: Carnegie
    Mellon University.

Taylor, R. M. & MacLeod, I. S. (1994), Quality assurance and risk management.
    Perspectives on human factors certification of advanced aviation systems.   In J.
    A. Wise,., D. J. Garland, & V. D. Hopkin (Eds.).   *Human factors certification of
    advanced technology* (pp. 97-118). Daytona Beach, FL: Embry-Riddle
    Aeronautical University Press.

UK DEF STAN 00–55. (1991). *Requirements for Safety Related Software in Defence
    Equipment*. The Ministry of Defense, Directorate of Standardization. Glasgow,
    UK.

UK DEF STAN 00–56. (1996).*Safety management requirements for defence
    systems.*The Ministry of Defense, Directorate of Standardization. Glasgow, UK.

UK DEF STAN 00–58. (1992).HAZOP *Studies on systems containing programmable
    avionics*. The Ministry of Defense, Directorate of Standardization. Glasgow,
    UK.

UK DEF STAN 00–25. (1989), *Human Factors for Designers of Equipment*. (Part 12:
    Systems). The Ministry of Defense, Directorate of Standardization. Glasgow,
    UK.

Wiener, E .L. (1985). Beyond the sterile cockpit. *Human Factors, 27*(1), 87.

Woods, D .D. & Roth, E. M. (1988). Cognitive systems engineering, In M. Helander
    (Ed.), *Handbook of human-computer interaction*. (p.16) Amsterdam: Elsevier,
    North-Holland.

V

PARALLEL VIEWS AND TOPICS

15

Practical Guidelines for Workload Assessment

Andrew J. Tattersall
Liverpool John Moores University, UK

There are numerous factors that need to be taken into account in a comprehensive human factors certification process. One major factor of human factors concern when humans are required to interact with complex systems is workload. If the process of human factors certification of systems is to succeed, then workload assessment must be incorporated into the evaluation and certification process. Proper and effective evaluation will ensure that the workload experienced by users of any system is taken account of in system design and development.

There is now a vast amount of literature on workload assessment. This interest has been stimulated primarily because of the need to design complex task environments that do not place undue demands and requirements on the human operator. The principal applications have typically been in process control and aviation settings such as air traffic control and aircraft cockpit design. This chapter addresses a number of practical issues that need to be taken into account when any evaluation of existing systems or future systems is carried out.

There is little dispute that workload is a multidimensional concept (Damos, 1991; Gopher & Donchin, 1986), but one distinction that is not very often made explicit is between acute and chronic dimensions of the impact of workload. If workload is defined in terms of the costs that operators incur in performing tasks (Kramer, 1991), then the distinction is even more apposite. An understanding of this distinction should aid evaluations of the nature of workload in work settings.

First, if one is concerned with the acute effects of workload, then the main focus of the research will be on the interference between tasks in dual- or multiple-task situations. A principal question to be asked is whether the tasks are too demanding in terms of the human information processing requirements, in which case performance on one or more of the primary work tasks may be degraded. The logic of many of the approaches in this area is based on multiple-resource theory (e.g., Wickens, 1991), in which it is proposed that there is a variety of processing resources that are limited in their capacity. The extent to which tasks will interfere with each other when carried out concurrently will depend on the extent to which they compete for common resources. Evaluation of workload in this case may include primary and secondary task performance measures, subjective measures, and certain psychophysiological measures. The ultimate intention of such approaches is to predict performance in multiple-task situations.

Second, if one is interested in chronic symptoms of heavy workload then the main concerns are with the effects of managing the demands of work over a day, a week, or a more prolonged period of time. The aftereffects of work are an important consideration. In other words, to what extent is the physiological and emotional state of an individual affected over a period of time, and does subsequent performance at work show any decrement due to these changes in state? For example, individuals may become increasingly fatigued because of the sustained demands in the job that need to be managed effectively. This may result in breakdown of skills over the longer term, and current models of workload are unable to predict these outcomes with any degree of certainty. Assessment techniques might again involve the use of performance measures, subjective measures, and physiological measures of workload. In addition, subjective and physiological measures of individual state will be useful as there may be implications of the longer term effects of workload for emotional and physiological well-being, health, performance, and safety at work. Studies of occupational stress and health provide a useful way to gain background information about possible sources of stress and the coping strategies and behavioral styles that are effective in moderating the effects of work demands (e.g., Tattersall & Farmer, 1995). The development of a model of stress and workload based on observations of individual patterns of response to various demands will lead to the more accurate prediction of states or situations in which a breakdown in skills might occur.

QUESTIONS THAT SHOULD BE ASKED

Before any evaluation is carried out, a number of questions need to be addressed that will enable the appropriate measures to be taken for the particular problem that is to be tackled. An approach that has been used to great effect in assessing the usability of human–computer systems (Ravden & Johnson, 1989) is to provide users or designers of systems with a checklist of items to consider. The following questions are based on part of that checklist, and they may be useful for workload assessors to identify the key areas of concern about workload in the initial stages of an evaluation.

1. What is the general question? Is the concern with the overall system or a specific piece of equipment, and is the focus primarily on performance or on the health and safety of the workforce? In other words, what is the aim of measuring workload?

- Is the primary concern with task design (the scheduling or allocation of tasks within or between individual operators' jobs)?
- Is the primary concern with equipment design (e.g., to evaluate the effects of the introduction and integration of a new item of technology, or to evaluate the difficulties of working with one item of equipment)?
- Is the primary concern with health and safety (the outcomes of working within an existing or proposed system, perhaps as a result of sustained task demands, underload or boredom, etc.)?

2. Is there a specific problem? That is, do operators complain about specific tasks or functions?

- What are the best aspects of the system?
- What are the worst aspects of the system?
- What parts of the system give the most difficulty when carrying out the task?
- Are there parts of the system that are confusing or difficult to understand?
- What are the most common mistakes made in using the system?
- What performance problems exist with the system?
- What changes do operators suggest might be made to the system to make it more effective and usable?

These questions will help to define the problem and the goals to be set for the workload assessment exercise. By focusing on these issues, the task of choosing a set of workload assessment techniques and specifying the research environment and design should be more straightforward and the results of studies easier to interpret and act up.

FACTORS TO CONSIDER IN CHOOSING A PARTICULAR TOPIC

Many different measures of workload have been developed, but their effective use in particular situations will depend on various factors, including their sensitivity to changes in demand, their ability to distinguish different kinds of demand, and their suitability or relevance to that situation. Extensive reviews of these techniques have been produced before, many of which discuss criteria for application (Damos, 1991; O'Donnell & Eggemeier, 1986; Hancock & Meshkati, 1988). The main factors to consider are given in the following sections.

Validity

Moray (1988) argued that because no clear, precise definition of workload exists, it is difficult to establish the validity of different techniques. He suggested that the reliability of measures has to be sufficient for practical purposes until such a definition is agreed on.

Reliability

Measures should be accurate and provide similar values from different operators doing the same task. There should be a good correlation between the values produced by different techniques if they are used to assess the same dimensions of workload. They should also have test–retest reliability, although as yet few reliability studies have been carried out.

Sensitivity

The concern here is with the effectiveness of the technique to discriminate between different levels of primary task load. In other words, does the technique actually measure changes in task demands and identify conditions of extreme workload?

Diagnosticity

Diagnosticity refers to the extent to which the technique is able to distinguish between different types of task demands and to identify the particular components within complex tasks that result in difficulty. Some techniques provide a general measure of resource allocation or effort, whereas others, such as secondary-task methodology, may be more sensitive to variations in different domains of processing. They may, for example, distinguish between verbal processing requirements and spatial processing requirements of tasks. Primary task measures of performance provide a global measure and are not really suitable for this purpose. Subjective measures, unless used to assess different task components, systems, or functions, for example, are generally not very diagnostic. The need for diagnosticity really depends on the aims of the study. If the aim is to assess the introduction of a new piece of equipment or change in working procedure then diagnosticity may not be critically important. If, on the other hand, there is a need to assess the demands of different control actions (e.g., manual or spoken) in relation to different modes of information presentation (e.g., visual or auditory), then the diagnosticity of the technique will be an important factor.

Intrusiveness

This refers to the extent to which the workload assessment technique disrupts the performance of the primary work task. The disruption could result from the use of obtrusive equipment or the application of a technique, or, in the case of secondary-task methodology in particular, simply the requirement to perform a concurrent task. If safety is a major concern (e.g. in air traffic control), then clearly workload assessment techniques that may degrade performance should not be used. Simulation exercises may provide useful data should techniques that may be intrusive also provide other useful attributes.

Generality, Acceptability, and Applicability

On a pragmatic level, it may be useful to choose a technique that can be used in different situations, perhaps so that comparisons can be made between different conditions. Ease of use and special requirements that may restrict the application of a technique, such as the need for special, expensive equipment, can be important considerations, as can the extent to which operators are accepting of the particular technique.

WORKLOAD ASSESSMENT TECHNIQUES

The concept of mental workload has proven to be more difficult to define and measure than physical workload, which causes some concern when the focus is on the tasks of air traffic controllers and pilots, as these tasks primarily involve cognitive processes rather than place great physical demands on personnel. It is not easy to estimate the demands of these tasks and therefore to predict the consequences of different levels of demand. We need not only valid, reliable and sensitive measures of workload, but also good methods based in sound theory to evaluate the cognitive activity that is required to perform these types of complex tasks.

A further important factor to consider is that different types of measurement technique may relate to different dimensions of workload and therefore may provide a different perspective of the particular demands of the task. Indeed, some subjective rating techniques are designed to assess a number of different dimensions, such as time pressure, frustration or anxiety, and mental effort. Certain physiological measures may be most sensitive to one particular dimension; for example, heart rate variability may be more likely to be associated with changes in effort, whereas mean heart rate may reflect changes in anxiety or physical effort.

Furthermore, there needs to be consistency in the way that terminology is applied, and finally, the operational procedures should be standardized as much as possible.

SELF-REPORT MEASURES

Subjective measures are relatively easy to employ, and asking workers to rate the levels of demand they experience and their state of well-being and health at least has face validity. It is intuitively attractive simply to ask workers about the levels of demands they are experiencing and the impact of work demands at different levels. A number of subjective workload assessment techniques have been developed. Among the validated scales that are widely used in aviation settings are the NASA Task Load Index (TLX; Hart & Staveland, 1988), and the Subjective Workload Assessment Technique (SWAT; Reid & Nygren, 1988). They both assess perceived workload on a number of dimensions, usually after the task has been performed. Nygren (1991) suggested that they are both useful measures of workload and that SWAT is sensitive at both individual and group levels. Hill,et al. (1992) compared four subjective workload rating scales for sensitivity, operator acceptance, response requirements, and any special procedures they require. All were found acceptable and sensitive to different levels of workload. Nygren (1991) pointed out, however, that the psychometric properties of these scales need to be fully understood, in addition to their implications for task performance, before they are applied extensively.

A recent development in this area has been the attempt to design subjective techniques that provide ratings of workload during primary task performance rather than after carrying out tasks. One such example is the Instantaneous Self-Assessment (ISA), technique which was initially designed for use with air traffic control tasks. Few evaluation studies have been carried out, but it appears to be a relatively sensitive measure of workload (Tattersall & Foord, 1996). A similar technique was used by Rehmann, Stein, and Rosenberg (1983), who suggested that gaining concurrent workload evaluations was more accurate than posttask ratings. In complex tasks that involve multiple elements or phases, the ratings may be more clearly related to changing task demands than retrospective ratings. Tattersall and Foord (1996) found, however, that ISA responses, although correlated with other subjective workload measures, interfered to a certain extent with the primary tracking task, whether responses were made by speech or manually.

One problem with subjective measures has been their diagnosticity and to a certain extent their reliability, whereas their validity and sensitivity have been fairly well established. A further worry is that they are not always found to correlate with measures of performance. Tasks that are performed better are sometimes found to have higher ratings of workload (e.g., Yeh & Wickens, 1988).

Self-report scales of a different type have been developed to assess mood and longer term health (e.g,. Warr, 1989). Importantly, significant relationship have been found between subjective responses and specific physiological responses, such as that between cortisol and subjective distress (Frankenhaeuser & Johansson, 1986), and effort and heart rate variability (Aasman, Mulder, &

Mulder, 1987). If one is interested in the relation between workload, performance, and individual state, then repeated measurement of mood and other variables is likely to be necessary.

PHYSIOLOGICAL MEASURES

These techniques include measures of cardiac function, brain function, and other physiological processes (Kramer, 1991). In terms of electrocardiogram measures, a number of studies now suggest that the power in the midfrequency band of the heart rate variability spectrum (0.07–0.14 Hz) is related to the level of mental effort invested in a task by an individual (Aasman et al., 1987; Mulder, 1980; Tattersall & Hockey, 1995). Such variability has been found to decrease as a function of task difficulty in a number of laboratory tasks and field studies (Jorna, 1992; Mulder, 1980). Mean heart rate may offer a more sensitive measure of response load, and is certainly influenced by physical activity and perhaps anxiety, which may limit its usefulness in relation to workload. The assessment of pupil dilation and electroencephalogram measures, including evoked potentials, have also been used effectively, particularly in laboratory situations, but the advantage of measures of cardiac function is that they can be continuously and independently applied without intrusion to the primary task.

Other techniques, such as analyses of urine and blood, provide measures of changes in physiological state through assessment of cortisol, adrenaline, and noradrenaline excretion. Urine analysis allows measures of longer term changes in state through assessment of cortisol and catecholamine concentration. Sustained stress states tend to show increased levels of these hormones. Blood or saliva samples may provide shorter term measures of fluctuations in state.

There are potential problems with the diagnosticity of physiological measures, and a further problem to be aware of is that physiological processes are sensitive to the effects of physical activity and to emotional factors that may have an effect on physiological functions.

PERFORMANCE

Two major approaches to performance assessment are primary task and secondary task techniques. Primary task measures are normally only useful for giving an indication of the impact of gross demands. These measures may be easy to obtain in some situations, but in others, such as air traffic control, it is difficult to generate a simple measure of a controller's level of performance that would meet the criteria outlined earlier. If one examined the safety record in air traffic control, for example, one might conclude that workload is only a minor problem. However, such a measure may not be immediately sensitive to the effects of changes in task load or working procedures and will only give a crude indication of the cumulative effect of sustained and high task demands over a long period. Task strategies may differ between skilled and inexperienced operators, and the health and state of the operator may determine the perceived difficulty of a task. The effects of these factors may only be detected by primary task measures once performance suffers or errors are made. From a safety perspective this may be already too late a stage at which to investigate the adequacy of human factors aspects of the system.

Secondary task measures may be more sensitive to changes in demand or working procedures but unless the allocation of information processing resources to the two tasks is controlled, it can be difficult to interpret the observed secondary task performance decrements. Norman and Bobrow (1975) introduced the important concepts of resource-limited processes, which are limited by the effort invested in a task and the priority placed on task performance, and data-limited processes that are constrained by the quality of information rather than by increases in effort. In work situations, operators may compensate for any increase in task demands by increasing the amount of effort invested in the task. Therefore observed performance levels may remain constant but the operator experiences increased workload. Conversely, a reduction in the level of performance may result either because operators cannot maintain the level of effort expenditure required, or because they lower their criteria for adequate performance. Therefore task performance in resource-limited tasks may be limited by the effort put into the task (related to the priority an individual places on performance), as well as the difficulty of the task. Secondary tasks have to be chosen carefully to not introduce structural interference with the primary task. However, secondary task performance measures can provide a more systematic technique for analyzing interference in multiple-task situations than many of the other workload measures.

WORKLOAD IN APPLIED SETTINGS

An important point to be made is that there is variation in the way that people do tasks and therefore in the effects and consequences of what is termed workload. Prolonged active management of the resources required to meet task demands may lead ultimately to a deterioration in performance, but there may also be implications for short-term well-being and longer-term health. The experience of workload is thus unlikely to depend simply on task load, but rather on the interaction of task demands, how these demands are dealt with by an operator, and the level of performance achieved. Task demands are important but are mediated by effort and the priority placed on the particular tasks.

The level of control that operators are able to exert in complex systems is an important factor in the relation between task demands, performance and well-being. Studies typically show an advantage for active control over passive control (Hockey, 1997), whereby open-loop strategies, involving a greater degree of planning and broader understanding of the system as a whole, are seen to be more skilled and efficient than the closed-loop mode. However, Umbers (1979) found that even experienced operators resort to the closed-loop mode when under high levels of workload or when unfamiliar situations or problems occur. This could be a cause for concern as remedial action may be applied once critical events have occurred rather than the more desirable situation in which impending catastrophic events are predicted at a time when something can be done to prevent them from occurring. Sperandio (1978) found that air traffic controllers also vary their strategies according to task demand, taking fewer variables into account as the traffic load increases. Similarly, in a study of process control, Bainbridge (1974) found that individuals under pressure used quicker, less accurate methods of finding data values.

Although discretion to use open-loop control may normally be preferred by operators and lead to enhanced performance and safety, this discretion could be seen as a demand in itself imposed by the structure of the task or job. Performance may be maintained at a desirable level (determined by personal and perceived organizational goals) but the effort required to deal with the demands and exert control over situations is observed as costs in other psychophysiological systems (e.g., Frankenhaeuser, 1986; Mulder, 1986). Frankenhaeuser demonstrated various changes in catecholamine and cortisol excretion with increased work demands. The patterning of these changes reflects active management of work and opportunity for control over work. Generally it has been found that increased catecholamine excretion and lower levels of both cortisol excretion and anxiety are associated with active processing strategies linked with increased control and effort investment. Distress and both increased catecholamines and cortisol levels are associated with passive conditions or strategies (Frankenhaeuser, 1986). Using different physiological measures, Tattersall and Hockey (1995) identified different activities in a simulated flight engineer task that resulted in different cardiovascular costs and subjective ratings of effort and concern. Heart rate appeared to be associated with concern, particularly during activities such as landing and takeoff, but suppressed heart rate variability and increased subjective ratings of effort were associated with the requirement for problem-solving activity in different activities.

Thus, in order to understand the relation between demanding situations and changes in performance, well-being, and health, it is necessary to investigate changes in different domains. This involves the short-term and long-term assessment of individual state, both physiological and affective state (in terms of mood and well-being), as well as cognitive activity (as implicated in performance). An example of such a study was carried out to investigate the impact of naturally varying workload in air traffic control on a range of measures including performance and physiological and affective state (Farmer, Belyavin, Tattersall, Berry and Hockey, 1991; Tattersall and Farmer, 1995). In comparing two different working shifts, operationally defined as busy or quiet on the basis of traffic load, a number of interesting differences were found, indicating the negative consequences of dealing with sustained demands. Subjective ratings of workload were higher during the busy shift. Dimensions of mood were affected in different ways by increased workload. Anxiety showed a significantly greater increase during high workload days, but preshift levels were not affected by workload. In contrast, levels of depression and fatigue were both higher at the start of the day under high workload conditions and were also elevated during the high-workload shift. The sustained demands of the busy summer months in air traffic control appear to result in chronic aftereffects of fatigue and depression, whereas anxiety was affected more transiently. Salivary cortisol concentration was greater during high-workload than low-workload shifts, declining during the day but less so in the second half of the high-workload shift. Noradrenaline excretion decreased over the low-workload shift but increased in the second half of the high-workload shift. These findings perhaps reflect active coping with the quantity of demands during high-workload shifts. The pattern of hormone excretion during high workload is consistent with the findings of Frankenhaeuser (1986), for example, in that cortisol and noradrenaline excretions are greater under conditions associated with lowered control and increased distress. This pattern may have long-term consequences for the health and well-being of controllers if sustained over long periods. Finally, the performance of air traffic controllers tended to show an improvement over the day with the important exception of visual vigilance sensitivity, the ability to detect signals in noise. This measure showed an improvement on low-workload days but not on high-workload days, suggesting that heavy work demands in air traffic control may have a detrimental effect on monitoring performance.

It is argued that multilevel measurement techniques can provide a broad assessment of the impact of different work demands. Further studies, both controlled laboratory-based studies and field-based studies, are necessary to refine the techniques, but a model of stress and workload management based on findings from such studies should allow the more accurate prediction of states or situations in which a breakdown in skills might occur. Such a breakdown is referred to by air traffic controllers as "losing the picture", when they experience difficulties in attending to, and remembering accurately, relevant information about aircraft under their control. It is precisely this kind of situation that should be avoided in work in which safety is critically dependent on performance.

CONCLUSIONS

The practical problems that might be encountered in carrying out workload evaluations in work settings have been outlined. Different approaches have been distinguished that may determine the type of research design used and provide assistance in the difficult choice between workload assessment techniques. One approach to workload assessment is to examine the short-term consequences of combining various tasks. Theoretical models of attention allocation will underpin specific studies of interference and the consequences of task demand and task conflict for performance. A further approach with a different temporal orientation may lead us to a better understanding of the relations between work demands and strain through the analysis of individual differences in cognitive control processes. The application of these processes may depend on individual differences in long-term styles and short-term strategies, but may be used to prevent decrements in work performance under difficult conditions. However, control may attract costs as well as benefits in terms of changes in affective state and physiological activity. Thus, strain associated with work demands may only be measurable in the form of trade-offs between performance and other domains of individual activity. The methodological implications are to identify patterns of adjustment to workload variations using repeated measures and longitudinal sampling of performance as well as subjective and physiological measures.

Possible enhancements to workplace design must take into account these human factors considerations of workload to avoid potential decrements in individual performance and associated organizational problems.

REFERENCES

Aasman, J., Mulder, G., & Mulder, L. J. M. (1987). Operator effort and the measurement of heart-rate variability. *Human Factors, 29,* 161–170.

Bainbridge, L. (1974). Analysis of verbal protocols from a process control task. In E. Edwards & F. P. Lees (Eds.), *The human operator in process control* (pp. 146–158). London: Taylor & Francis.

Damos, D. L. (1991). *Multiple-task performance.* London: Taylor and Francis.

Farmer, E. W., Belyavin, A. J., Tattersall, A. J., Berry, A., & Hockey, G. R. J. (1991). *Stress in air traffic control II: Effects of increased workload.* (RAF Institute of Aviation Medicine Rep. No. 701). Farnborough, Hants, England: Royal Airforce Institute of Aviation Medicine.

Frankenhaeuser, M. (1986). A psychobiological framework for research on human stress and coping. In M. H. Appley & R. Trumbull (Eds.), *Dynamics of stress* (pp. 101-116). New York: Plenum.

Frankenhaeuser, M., & Johansson, G. (1986). Stress at work: Psychobiological and psychosocial aspects. *International Review of Applied Psychology, 35,* 287–299.

Gopher, D., & Donchin, E. (1986). Workload—An examination of the concept. In K. R. Boff, L. Kaufman, & J. P. Thomas (Eds.), *Handbook of perception and human performance: Vol. II. Cognitive processes and performance*(pp. 41/1–41/49) New York: Wiley.

Hancock, P. A., & Meshkati, N. (1988). *Human mental workload*. Amsterdam: Elsevier.

Hart, S. G., & Staveland, L. E. (1988). Development of a NASA TLX (Task Load Index): Results of empirical and theoretical research. In P. Hancock & N. Meshkati (Eds.), *Human mental workload* (pp. 139-181). Amsterdam: Elsevier.

Hill, S. G., Iavecchia, H. P., Byers, J. C., Bittner, A. C., Zaklad, A. L., & Christ, R. E. (1992). Comparison of four subjective workload rating scales. *Human Factors, 34*, 429-439.

Hockey, G. R. J. (1997). Compensatory control in the regulation of human performance under stress and high workload: A cognitive-energetical framework. *Biological Psychology, 45*, 73-93.

Jorna, P. G. A. M. (1992). Spectral analysis of heart rate and psychological state: A review of its validity as a workload index. *Biological Psychology, 34*, 237-258.

Kramer, A. F. (1991). Physiological metrics of mental workload: A review of recent progress. In D. L. Damos (Ed.), *Multiple-task performance* (pp.279-328). London: Taylor & Francis.

Moray, N. (1988). Mental workload since 1979. *International Reviews of Ergonomics, 2*, 123-150.

Mulder, G. (1980). *The heart of mental effort*. Gronigen, The Netherlands: University of Groningen.

Mulder, G. (1986). The concept and measurement of mental effort. In G. R. J. Hockey, A. W. K. Gaillard, & M. H. G. Coles (Eds.), *Energetics and human information processing*(pp. 175-198). Dordrecht, The Netherlands: Nijhoff.

Norman, D. A., & Bobrow, D. G. (1975). On data-limited and resource-limited processes. *Cognitive Psychology, 7*, 44-64.

Nygren, T. E. (1991). Psychometric properties of subjective workload measurement techniques: Implications for their use in the assessment of perceived mental workload. *Human Factors, 33*, 17-33.

O'Donnell, R. D., & Eggemeier, F. T. (1986). Workload assessment methodology. In K. R. Boff, L. Kaufman & J. P. Thomas (Eds.), *Handbook of perception and human performance: Vol. II: Cognitive processes and performance* (pp. 4/21-42/29). New York: Wiley.

Ravden, S., & Johnson, G. (1989). *Evaluating usability of human–computer interfaces: A practical method*. Chichester UK: Ellis Horwood.

Rehmann, J. T., Stein, E. S., & Rosenberg, B. L. (1983). Subjective pilot workload assessment. *Human Factors, 25*, 297-307.

Reid, G. B., & Nygren, T. E. (1988). The subjective workload assessment technique: A scaling procedure for measuring mental workload. In P. A. Hancock & N. Meshkati (Eds.), *Human mental workload*(pp. 185-218) Amsterdam: North-Holland.

Sperandio, J. (1978). The regulation of working methods as a function of workload among air traffic controllers. *Ergonomics, 21*, 195-202.

Tattersall, A. J., & Farmer, E. W. (1995). The regulation of work demands and strain. In S. L. Sauter & L. R. Murphy (Eds.), *Organizational risk factors for job stress* (pp. 139-156). Washington, DC: American Psychological Association.

Tattersall, A. J., & Foord, P. S. (1996). An experimental evaluation of instantaneous self-assessment as a measure of workload. *Ergonomics, 39*, 740-748.

Tattersall, A. J., & Hockey, G. R. J. (1995). Level of operator control and changes in heart rate variability during simulated flight maintenance. *Human Factors, 37,* 682-698.

Umbers, I. G. (1979). Models of the process operator. *International Journal of Man–Machine Studies, 11,* 263–284.

Warr, P. B. (1989). The measurement of well-being and other aspects of mental health. *Journal of Occupational Psychology, 63,* 193–210.

Wickens, C. D. (1991). Processing resources and attention. In D. L. Damos (Ed.), *Multiple-task performance*(pp. 3–34). London: Taylor & Francis.

Yeh, Y. Y., & Wickens, C. D. (1988). Dissociation of performance and subjective measures of workload. *Human Factors, 30,* 111–120.

Is There A Role For A "Test Controller" In The Development Of New ATC Equipment?

Ron Westrum
Eastern Michigan University, USA

In the aviation field, test pilots have long performed a valuable function in the evaluation and improvement of new aircraft. Through their special experience and training, test pilots are able to provide expert feedback for the development process. Although often glamorized by films and books, the role of the test pilot is basically that of a member of the engineering team. The test pilot checks out the plane in the air, explores the performance envelope of the various aircraft systems, notes "bugs" and other infelicities of the equipment, and makes suggestions for improvements. Test pilots have unusual piloting skills, but more important. They have training in systematic check-out and a high sensitivity to performance quirks that others might miss (Hallion, 1981).

Testing new air traffic control (ATC) equipment necessarily involves similar skills to test piloting. Check-out is expected to take place according to systematic protocols, and problems in operation are expected to be spotted and removed. Who is qualified to do this? What impact will the involvement of controllers at various stages of the development process have on the effectiveness of equipment finally released? Dujardin (1993) suggested that early involvement of controllers in the research and development (R & D) process may discourage important advances, as controllers will feel comfortable only with equipment that seems familiar. There is broad agreement, in fact, that early involvement of working controllers is likely to lead to compromises or kludge designs. The regular controller is unlikely to want to push the envelope. Many observers have remarked that equipment used by the Federal Aviation Administration (FAA), both currently and in the near future, reflects this conservative attitude.

A test controller, however, would not share the same bias against new equipment. Note that this is a different role from the evaluation of finished systems. The test controller would be used to seeing equipment in raw form, just as a test pilot would be. If we follow the analogy with airplane development, a test controller would have to be recognized as a top practitioner, with the respect of other controllers. Such a person's certification of the equipment to the working controller, then, would be one guarantee that the equipment, if not trouble-free, would be at least safe and efficient to use.However,use of highly skilled controllers mat "pass" equipment hard for low-skilled users to operate (Cardosi, 1999).

New hardware and software now face stiff resistance if they originate from someone other than facility automation specialists. For instance, FAA-produced software for airport ATC systems has many credibility problems. It is not perfect, and in any case needs to be customized to handle site-specific problems. Currently, having capable controllers is the best guarantee against software's inadequacies. Most controllers use "so-so software developed by someone" as Schmidt (1995) of Martin Marietta put it. Controllers must carry on from where the software leaves off, bridging between it and the operating situation. Potentially, certification by a test controller that new equipment satisfies human factors requirements might give controllers the confidence they need to master complex new equipment and procedures.

Another important feature of the test controller is checking out the far corners of the envelope. In the R & D process, early efforts are focused on getting the system to work. However, test controllers need to try to make it fail, to exercise it, as it were, beyond ordinary limits, and to eliminate the hidden bugs. Ideally, of course, a better process would be developed for getting error-free software. In real life, however, automated systems are likely to possess "glitches" that are difficult to eliminate. For instance, a *Wall Street Journal* article reported that the Honeywell autopilot installed in Boeing 747s behaved in mysterious ways. The FAA noted about 30 incidents, including a recent near-crash over Thunder Bay, Canada, involving malfunctioning autopilots. Experts have been unable to isolate the fault (Carley, 1993). Thus, test controllers need to check the system out using impolite actions. This is very similar to what Popper (1961) recommended in his book *The Logic of Scientific Discovery:* Propound bold hypotheses, and then give them severe tests.

Before we go further, however, we need to consider the innovation process itself, because a test controller will have to fit into it. Innovation in the U.S. FAA is a very problematic process. We need to examine it in a bit more detail.

INNOVATION: A LONG HAUL

Charles Franklin "Boss" Kettering once said that "getting a new idea into a factory is the greatest durability contest in the world" (Schiaro, 1997, pp.141-176) He might have said this about the FAA's approach to ATC. The current process by which new ATC equipment is introduced is slow and inefficient in many ways.

1. There are excessive delays, on the order of a decade or more, from the time new equipment is developed until it is actually used. Thus, by the time the equipment is installed, it has usually been obsolete for years. It may nonetheless represent a real advance over what was used before. Many U.S. airports function with equipment that controllers think properly belongs in museums. This is demoralizing both to innovators and operators.

2. An incremental approach is used. This approach forces new equipment to be compatible with current equipment, often leading in the end to an inelegant, kludge design, rather than an optimal redesign from the ground up (Heppenheimer, 1997). Although each new piece of equipment may function well on its own, nothing guarantees either its compatibility or lack of redundancy with current equipment.

3. Use of political fiat is sometimes used to impose quick fixes that need better testing before implementation takes place. In many cases these programs fail to work as planned and thus increase barriers to further innovation. Problems include failure to introduce new equipment effectively, failure to take learning curve considerations into account, inappropriate check-out by test controllers, and ill-conceived instructional methods.

4. Very high stakes are involved in securing government contracts, leading to intense struggles on the part of private firms to get their product accepted. Because competing parties often resort to legal action to block or reverse decisions already made, delay is common. As with defense contracting, the long haul involved and the "winner-take-all" outcomes often result in selection of contractors who are good at lasting through the many rounds necessary to win a contract; these are not necessarily the contractors with the best systems.

The current system is designed to include three parties: private firms that do the actual hardware and software development; FAA higher officials, who make decisions about which devices to install; and controllers, who will actually use the devices once they are officially accepted. In principle, controllers develop needs, these needs get expressed as FAA requirements, and private industry responds to the requirements by hardware or software innovation. However, there is a built-in paradox: Controllers do not know what they can ask for until they know what can be developed. Vendors, on the other hand, often do not understand what controllers need. FAA higher authority, trying to bridge the gap between needs and products, is hemmed in on one side by political and legal constraints and on the others by vendors jockeying for contracts and FAA facilities fearful of clumsy automation. The situation is made worse by the FAA's ineptitude in project management (Schivaro, 1997, p. 146)

Improving the technology for ATC will now be more difficult with the failure of the Advanced Automation System (AAS), the budget for which reached $7.5 billion, but turned out to be a colossal flop (Wald, 1996). The FAA, after contracting with IBM to produce the hardware and software for this next-generation system, did not manage the acquisition well, and in the end had to abandon it. This complex system had a stack of specifications 3.5 ft.t high and would require about 1.6 million lines of code. Among the many faults in project management committed by the FAA in trying to bring the system online were unrealistic goals, failure to conduct hands-on management, poor communications, and changing specifications. IBM, in turn, allowed software engineering to take place without systematic integration between various projects, and apparently ignored contract provisions. In many respects the fiasco resembled a typical big-ticket military acquisition contract, with the usual challenges one finds in such R & D efforts, but without the military's expertise in managing big projects

Failures of this kind cost more than money and time. They lose credibility for the entire system of acquisition. One can be sure that the resistance will be stiffer for the next attempt to "revolutionize" the system.

Since the FAA acquisition system has long delays, attempts to get around the normal channels are common. Lee Paul wrote in 1979 that "The number of years required by an orderly development process results in irresistible pressures to bypass the system." Seemingly, very little has changed. Frustration often leads to ill-considered moves, including "designs by fiat" that not only fail but also prejudice future attempts at innovation. The formal system also largely ignores the automation specialists (see below) and other members of the system who often have excellent ideas, but who are not considered partners in the innovation process. On the other hand, there has been long-term involvement of controllers on work teams.

These complex dynamics do not bode well for getting the right control equipment to the right people in the right time frame. A top priority for the FAA might well be to examine its own innovation process,

A COMPARISON WITH CHINA LAKE

I have just completed a study of the China Lake Naval Weapons Center and its methods of developing new military equipment (Westrum, 1999; Westrum & Wilcox, 1989). China Lake's Technical Director from 1954 to 1967 was William Burdette McLean, inventor of the Sidewinder missile, and one of the most highly regarded scientists in the Navy. McLean, in working with the R & D process, was convinced that the usual mode of developing new equipment was badly flawed. The usual linear process starts with a specification that is worked into a concept, which becomes a prototype, and eventually (we hope) becomes a production model and an operational system. McLean's critique of the system was that its linear nature made the specification too difficult to change, even as information emerged that the specification was flawed or could be improved. Often, he found, specifications asked for the impossible (as apparently did the FAA's AAS) or failed to capitalize on emerging capabilities.

Instead of a linear process, McLean wanted the development to take place through a dialogue between designers and users. After a careful study of users' needs, the designers would use the best technical capabilities at hand to create something that would work for the users. At China Lake, constant user input was ensured because test pilots on the base were rotated off aircraft carriers, but were specially chosen for their role as test pilots (few in the 1950s had had any formal test pilot training). At the end of their China Lake tour, the pilots were sent back to the fleet, where they would have to defend their recommendations. Using such test pilots assured that input was relevant, informed, and reflected the judgment of top performers. The pilots themselves often became instrumental in promoting the new equipment.

However, the key feature of the innovation process under this style of development was its flexibility. McLean noted that with a firm specification for development, many projects were continued until doomsday although they needed radical change or termination. At China Lake projects could change in midstream if technical capabilities allowed a different direction. Only when the concept was finalized and ready for production was there to be a firm specification (Westrum, 1999). China Lake also had a variety of mechanisms for bridging the gap between the laboratory and the fleet, both to assure that designs were working as planned and that users understood the equipment's capability and how to use it.

McLean's concept of a creative dialogue in the R & D process would do much to assist the FAA in restructuring its innovation system. One of the features of the development of Sidewinder was the close connection between all parts of the design process. The test pilots especially understood what the team's mission was and worked hard to support it. The program made sense to those working on it, and the central vision (and firm hand) of McLean kept the team working as a connected group. McLean also was very good at understanding the kind of features that would make the new missile appealing. The two most important features were that its simplicity made it easy to use and also made it likely to work.

It would appear that in the AAS project both the FAA and IBM lost sight of their customers' needs. But what of the controllers working with the developers, the test pilots?" Apparently the controllers had shifting patterns of preferences with shifting groups of controllers. Cardosi (1999) suggest that better human factors involvement would have integrated the controller's preference effectively. But a real lack of compelling vision also created, I believe, a limited boy-in. Because neither the FAA nor IBM seems to have had a compelling vision of what each wanted, the relation between daily activity and final goal may have been elusive. The Sidewinder team, by contrast, had high morale. The test pilot could see the evident value in the system they were developing, and they were motivated to make it work. People developing a system need to believe in the system they are developing.

The Sidewinder team had one other advantage over the AAS group. The motivation for Sidewinder was clearly to help the pilot. The reason for developing the AAS, however, was to allow the FAA to operate with fewer controllers. Yet every controller knew in his or her bones that the controller is what keeps the airplanes safe when the radar or the automation breaks down. The idea of a system with fewer controllers did not make sense. No doubt the key motive for many test controller in AAS was to keep the FAA from doing something stupid. If this is correct, it is hard to see anyone being inspired by such an intrinsically negative task.

PATCHING IT UP

Ironically, controllers often seem to do better themselves through informal networking when it comes to customizing ATC software. Where as some software used to control aircraft can be originated locally, hardware, on the other hand, necessarily is produced off-site and centrally tested and introduced. Still, because each site has slightly different requirements, software can be generic only up to a certain point. Beyond this point, software must be customized for the specific site. This is done through *patches*: software oriented to site-specific problems, and written by local automation specialists. For instance, at Detroit Metropolitan Airport the FAA-produced A–305 ARTS–III software (introduced in 1993) was voluminous. Documentation for the software was four large volumes, each a folio volume the size of a desk encyclopedia. The cross-reference book alone weighed 20 pounds. Yet the software required 190 patches to adapt it to local conditions. The automation specialists on the staff estimated that development of this supplemental software required several months to complete, exclusive of already existing site-specific software, which also had to be changed. In principle, software received from the FAA is supposed to be implemented as is. The reality is that this cannot be.

At Chicago O'Hare Airport, for instance, there was at one time a rule that strictly limited the number of local patches. This rule, however, did not make sense, and so was constantly bypassed. Each time local controllers asked for a specific change, the patch would be added to an existing patch, which clearly flouted the spirit of the rules, although superficially legal. This bypassing of the system was never formally acknowledged. Nonetheless, the local programmers thought FAA officials must have been aware of it.

Although sites are often different, many sites share the same problems. The obvious thing, then, is to make sure that a site has available to it any patch in the system that may aid in solving its problems. Although all patches used anywhere are included on a list sent to all automation specialists, this list is seldom seen by controllers, and it is hard to interpret in any case.

There are about 200 automation specialists in the United States. They are former controllers now responsible for software management at the control centers. They are expected to act as the local interface between the needs of controllers on one hand and the provision of new automation through borrowing patches or getting local technicians to do the programming. However, they usually have their hands full with programming responsive to demands by the local controllers for various kinds of minor fixes. One major airport had a list of 30 such patches waiting to be programmed; this is fairly typical. These demands are often either made by top management or presented through union channels, which makes them hard to ignore. Useful patches, however, may often get lost in the system's complexities because they are not available in a user-friendly way.

The automation specialists have more credibility because they are former controllers. Knowing the job that the controller must do in some detail makes their products far more user-friendly than they might otherwise be. However few have college degrees in computer science. Some do not have college degrees at all. They get considerable in-service training from the FAA, but both they and others believe that their programming would be superior if they had more computer training.

No one knows how much of the innovation in the system is actually due to the local automation specialists. For instance, a program called Cenrap allows the local facilities to get radar screen pictures even if their antenna goes out, by getting information on plane positions from more powerful Regional Center radars. This program was reportedly suggested in about 1985 by either an automation specialist or a technician who realized that capabilities already in use, with a little extra work, could provide backup radar pictures for facilities that lost their radar but not their system (ARTS–III) software. It has since become an important safeguard.

LOCAL CONTENT

To compensate for the system's inadequacies, informal networking often provides the primary channel for patches to travel from one center to another. In the FAA southern region, for instance, an informal computer bulletin board provides information about patches. Other information comes about through individuals who move from one site to another, or who through union duties or curiosity circulate through the system. Actually watching a patch in operation may be more valuable than reading an abstract about it.

Yet informal networking clearly is a second-best to a user-friendly system for spreading information about patches. Why is there not a dedicated patch specialist who knows what is available and who travels through the various sites?

Similarly, site development of patches is often done partly *sub rosa*. Legally, patches must be run through Washington, DC for regulatory approval before they are used. The formal approval process takes about 1 year. After approval, the patch is sent back to the site for online testing. However, local automation specialists do not want to send a patch through the system until they know it works. And how do they know it works? They try it out. To try it out, they need a computer. Often the only computer available for the purpose is the center's main computer. It would be best to try the patch out on a mock-up computer offline, but mock-up computers can cost as much as the main computer. So the patch is run on the main computer at a lull time, such as 2 a.m. Controllers will almost never try to control aircraft with experimental software. Locking up the system would both be dangerous and would jeopardize their jobs. But without a realistic (i.e., live) test, they do not want to release the software. So while planes are controlled through some other method, the patch is tried out. Once it is known to work, it is sent through the formal process.

I was unable to gain any information regarding site-generated patches for the regional (en route) centers. Ostensibly, all patches for regional centers must originate from the FAA. Technical Center in Atlantic City, New Jersey. To proceed otherwise risks severe legal sanctions.

Higher echelons of the FAA must know that this kind of covert experimental activity goes on, although they cannot publicly either acknowledge or condone it. However, although obviously better than a paper check-out of the software, this "skunk works" approach has some dangers. One of the problems is that fewer programs result from it than would if it were openly acknowledged. Controllers and automation specialists would both get in serious trouble if they were caught operating with an illegal patch. It would be better if the test were carried out openly. The best would be creation of what Paul (1979) called a more forgiving environment, where experiments with patches could be run offline, using a full-scale simulation facility that could be customized temporarily to run a center's software.

Controllers' experience with innovation has largely been negative. Good ideas by those lower down in the system often seem to get stonewalled or put on the back burner. Ideas that come down from the top are often half-baked or flawed. However, the strongest message about innovation is the equipment that controllers in the United States. are forced to use. State-of-the-art aircraft are controlled by ATC equipment that is often two or three generations out of date. Whether the explanation for this state of affairs is politics, bureaucracy, or sheer conservatism, the message it sends is one of stagnation and indifference. "Good enough for government work" seems to be the limit the controller can expect. Controllers have to be good; their equipment is not.

TEST CONTROLLERS

The innovation process for new hardware and software occurs along a time line that can largely be considered in three phases: R & D, preliminary testing, and full-scale deployment. There is a role for controllers in testing new hardware and software in each of these phases.

R & D

A role in R & D means that the controller would act in the same role as a test pilot. He or she would encounter new equipment in its formative stages and would be able to help suggest improvements that would move the system from the prototype to the operational stage. The FAA used controllers on its work teams for the new consoles that IBM was developing in Rockville. Some of these controllers were on the teams for 10 years. However, unlike the system for test pilots, there was, no way, to record the experience of these controllers as test controllers other than in the design of the equipment. They were simply sent back to their centers after they finished their tasks. Note also that at many ATC facilities, local software will be tested by individuals who serve the role of test controller for that facility, even though the term as such may not be used. Donald Pate at the Standards Development Branch of the FAA in Oklahoma City similarly used journeymen controllers for his experimentation and standard-setting.

One anonymous observer pointed to a problem with the use of ordinary controllers in the R & D process. Early involvement of routine users tends to lower the team's sights, and thus may lead to incremental changes rather than redesign from the ground up. An example was the FAA's use of a sector suite validation team involving ordinary controllers early on in the process. Ultimately, the console produced by the team was a kludge design, according to this individual. Thus early involvement can lead to dangerous compromises. A test controller, in principle at least, would have enough experience with the innovation process to be less bothered by radical innovation.

A second problem to which union representative Larry Barbour called attention is the attrition of skill among controllers who are promoted to supervisor. Several individuals noted that supervisors could no longer be considered proficient in acting as controllers, once promoted. During the PATCO strike, many of the supervisors actually had to do some controlling.

This was a frightening experience for some of them who had lost their skills. "I remember watching some of these guys with sweat pouring down their backs," said one observer. Yet several of the IBM-design work teams contained a majority of supervisors by the time the project was finished. One work team had 1 controller and 11 supervisors! This, however, would be a problem for test controllers, too. Some method for alternating actual control experience and innovation activities would be necessary.

Preliminary Testing.

In this phase, the overall design of the software or equipment is fixed, and the purpose of testing is to eliminate any remaining bugs. The role of controllers here is to act as intelligent customers rather than test pilots per se. It is often during this phase also that software can be customized for a particular site. Bergeron and Heinrichs (1993) reported their experiences in using controller "cadres" first to test software, and second to act as trainers, both at Denver and at Dallas/Fort Worth. These experiments seem to have been very successful, although the system was not completely ready for them.

Bergeron and Heinrichs' (1993) cadres might be seen as somewhat analogous to the New Equipment Introduction Details (NEID.) used by the U.S. Signal Corps in World War II.   The Signal Corps, finding that newly developed equipment typically was not accepted in the field, developed a kind of special detachment under the leadership of a Lt. Col. Jensen.The detachment included no one without a uniform, and no one ranking above a major or below a sergeant. It always accompanied the equipment from the point of origin (the I factory) to the field with no hand-offs. The Signal Corps discovered that this scheme was so successful that it could not get equipment into the field without an NEID.   Unfortunately, the value of this device was not recognized after the war, and so no one studied it. Its only mention in the official history of the Signal Corps in World War II is a tiny footnote.

Full-Scale Deployment

During this phase, controllers are still important as intelligent users. Cadres who have been used in the second phase to work out bugs can act as brokers between laboratory and ATC facilities to transmit information backward and forward between designers and users. As equipment and software is given a full-scale test, limitations and bugs will become apparent. Often this may mean moving the novel hardware or software to a new location, with new demands. IBM, using the Seattle Regional Center as a test site, is rumored to have eliminated phone jacks from the controllers' positions, which a "D man," used extensively in the busier centers to work computers and assemble flight strips, could use. Protests by the Cleveland Center (and others) quickly got the jacks back. Developing effective channels for user feedback is thus very important.

DISCUSSION AND CONCLUSION

Wiener (1995) pointed out that human factors problems fixed during the R & D stage are paid for once. When they are not fixed during R & D, they are then paid for every day. How users are involved in the R & D process to assist in developing equipment is a critical issue. Effective involvement can produce real improvements. Ineffective involvement can produce inefficient kludges or systems that are actually dangerous.

The underlying problem is the management of information and ideas. To develop a really generative system (see Westrum, 1993) a great deal would have to change in the way the FAA innovates. Use of test controllers would solve only some of the problems. For instance, we have cockpit resource management now for pilots; we are getting it for controllers. However, the management of ideas in the innovation process also needs intellectual resource management. Simply involving users is not enough. Brought in at the wrong point in the development process, users can block or compromise innovation. Failure to inspire and empower the controllers acting as test pilots for the system during development will rob them of the sense that the work they do is valuable. A test controller may be a solution to problems with using ordinary controllers, but if the project does not make sense, the best test controllers in the world will not be able to fix it. Furthermore. involvement by itself is not enough. Controllers need integration into an intelligent human factors program Finally, it might be necessary to have several kinds of test controllers (en route vs. TRACON, for instance). No doubt further problems would surface in getting test controllers into operation.

I would recommend that the FAA engage in a series of case studies of controller involvement in the innovation process. A systematic comparison of effective and ineffective cases would do much to clarify what we ought to do in the future (further, see Cardosi, 1999). One study of the introduction of automation at London Air Traffic Control Center showed that British controllers could be just as resistant and suspicious as their American counterparts. Yet Gras, Moricot, Poirot-Delpech, and Scardigli (1994) described a far more positive interaction between engineers (designers), controllers, and official bodies in France during the introduction of a new system called Phidias. Perhaps because the introduction was designed to be evolutionary rather than revolutionary, and because the control community is smaller in France and less bureaucratic, the introduction of this new system seemed to take place with less conflict than in the United States. More of a dialogue seems to take place as the new system takes shape.

Yet in the United States controllers work for the FAA, analogous to France, but few report the warm, friendly feelings toward the FAA Technical Center in Atlantic City expressed by the French controllers toward CENA, the air navigation design center. And the situation in France with the controllers is in violent contrast with the French pilot community. In the same Gras et al.

(1994) study the pilots reported feeling very estranged from the designers (e.g., of the Airbus 320). Gras et al. explained this difference between pilots and controllers thus: The ATC designers belong to the same organization as the controllers. The pilots, however, work for individual airlines, and the planes are designed by a totally separate company (in this case, Airbus). But American pilots, whose planes are designed by several distinct companies, seem to feel that Boeing, for example, understands their needs more than Airbus is said to do by the French pilots. Clearly the factors that lead to a user population's identification with the designers are not simple. However, these factors producing identification with the designers need further investigation. If a user community cannot identify with those who design their equipment, they are likely to reject the equipment.

My basic concept is that test controllers might well play the role played by the NEIDs in World War II, and act as the brokers and bridgers for the new technology. This role has been played by test pilots for guided missiles, as shown at China Lake. Clearly something of this nature is taking place with the French "expert" controllers discussed in passing by Gras et al. (1994). The lack of such personnel may account for the suspicion that British ATC controllers feel toward new equipment.

If we do not study the process, and get a better handle on why ATC equipment introductions fail, I am afraid that the same sorry state of affairs documented in many studies so far will persist for many more years.

ACKNOWLEDGMENTS

This chapter is based on interviews and fieldwork carried out in 1993, with additions and corrections in 1999. Many individuals took considerable time and effort to assist the author in writing this chapter. These persons include Larry Barbour, Alex Becker, John Dill, Dick Edsall, Ron Leoni, Don Pate, Lee Paul, Willis Richardson, Jim Schmidt, Steve Walter, Marty White, and above all, Dud Tenney. Although these persons contributed, the opinions expressed in this chapter are solely my own and do not necessarily reflect either the views of the meeting's sponsors or of any individual who assisted in the chapter's preparation.

REFERENCES

Bergeron, H. & Heinrichs,H. (1993). *A training approach for highly automated ATC systems.* Paper presented at the Seventh International Symposium on Aviation Psychology, Columbus, OH.
Burgess, J. (1993, March 8). Out of control contract. *Washington Post*, Washington Business Section, pp. 1, 22, 23.

Cardosi, K. (1999). Lessons learned in the design and implementation of air traffic control systems. *The Controller* pp. 11–15.

Carley, W. M. (1993, April 26). Jet's near-crash shows 747's may be at risk of autopilot failure." *Wall Street Journal,* pp. A1, A6.

Dujardin, P. (1993, April). The inclusion of future users in the design and evaluation process. *Le Transpondeur,* pp. 36–39.

Gras, A., Moricot, C., Poirot-Delpech, S. L, and Scardigli, V. (1994). *Face a l'automate: le pilote, le controleur, et l'ingenieur.* Paris: Sorbonne.

Hallion, R. P. (1981). *Test pilots: The frontiersmen of flight.* Garden City, NY: Doubleday.

Heppenheimer, T. A. (1997, Summer) The antique ,machines your life depends on. *Invention and Technology 13* #1   pp.42–51.

Paul, L. (1979). How Can We Learn From Our Mistakes If We Never Admit We Make Any? *Proceedings, 29th Air Traffic Control Association Conference,* Fall, pp. 194-198.

Popper, K. (1961). *The logic of scientific discovery.* New York: Science Editions.

Schiaro, M. (1997). *Flying blind, flying safe.*New York: Avon Books.

Wald, M. L. (1996, January 29). Ambitious update of air navigation becomes a fiasco. *New York Time;* pp. A1, A11.

Westrum, R. (1993). Cultures with requisite imagination. In J. A. Wise, V. D. Hopkin, & P. Stager, (Eds.). *Verification and validation of complex systems: Human factors issues*(pp. 401–416) New York: Springer-Verlag.

Westrum, R. (1999). *Sidewinder: Bill McLean and the creative culture of China Lake.* Annapolis, MD: Naval Institute Press

Westrum, R., & Wilcox, H. A. (1989, Fall). Sidewinder. *American Heritage of Invention and Technology.* Fall, pp. 56–63.
****

This paper is based on interviews and fieldwork carried out in 1993, with additions and corrections in 1998. Many individuals took considerable time and effort to assist the author in writing this paper. These persons include Larry Barbour, Alex Becker, John Dill, Dick Edsall, Ron Leoni, Don Pate, Lee Paul, Willis Richardson, Jim Schmidt, Steve Walter, Marty White, and above all, Dud Tenney. While these persons contributed, the opinions expressed in this paper are solely those of the author and do not necessarily reflect either the views of the meeting's sponsors or of any individual who assisted in the paper's preparation..

17

Towards a Framework of Human Factors Certification of Complex Man-Machine Systems

Birgit Bukasa
Austrian Road Safety Board

The recognition of the importance of human factors to system safety, especially in aviation, is constantly increasing. Foushee (1993), in his keynote address at the International Civil Aviation Organization Flight Safety and Human Factors Symposium, even talked about the human factors revolution, emphasizing how quickly human factors thinking has infiltrated the world of aviation and high technology in some parts of the world. Consequently, the Federal Aviation Administration (FAA), in its 1990 National Plan for Aviation Human Factors, amongst others placed stronger emphasis on human factors as part of the aircraft and avionics certification requirements (Foushee, 1993).

This claim for stronger consideration of human factors principles in the designs of complex and integrated human–machine systems is the reaction to changes and foreseen further future changes in the aeronautical world. These changes have and will have a great impact not only on the aviation community, especially the operators such as pilots or air traffic controllers, but also on society as a whole.

Advanced automation in aviation, including its implications for the users as well as for safety concerns, is the main point at issue. Together with the extensive application of the information and communication technology summarized under the keyword *glass cockpit*, the safety-adverse effects of the human's role as a system supervisor through extensive monitoring who nevertheless has to manage emergencies, had been underestimated by the technical disciplines.

Taking the example of the work management in transformation plants, Brehmer (1996) confirmed that since the introduction of computer-controlled technical processes in the 1970s, the registration of as many as possible signals in the shortest time was the maxim principle. The exact analysis and elimination of disturbances within a tenth of a millisecond or shorter was the dream. The human with his or her restricted physiological possibilities was far less considered and the recognition that the best (technical) system with a poor user interface will be only poor or at least average was neglected, for far too long.

Molnar (1992) outlined major differences between old and new technologies (i.e., automated data processing systems with all their hardware and software technological realizations vs. the former kind of tools and machines) that have a strong impact on the organizational culture and the individuals working with it (see Table 17.1).

TABLE 17.1.
Differences Between Old and New Technologies
(According to Molnar, 1992)

	Old Technologies	New Technologies
Tools/machines		
Purpose	Physical output	Intellectual output
Range of application	One-purpose machines	Multi-purpose machines
Scope	Local, individual	Global, structural
Product		
Object of work	Concrete objects	Abstract processes
Results	Material	Immaterial
	Three dimensional	Two dimensional
Work		
Qualification	Specific	Generalistic
Usability	Static, analogous	Dynamic, digital
Locus of control	More on the user's side	More on the tool's/
Kind of work	Sensoric-motoric	machine's side
Demands	More physical	Mental-cognitive
		More psychic

Consequently, behavioral disciplines, like software design and software ergonomics have become increasingly important during recent years, focusing on the optimization of the human–computer interfaces as well as on the sociodynamic implications of new technological solutions, like design and allocation of functions between the human and machine, and the design of future procedures and structures based on the socio-organizational realities.

Besides, accident and incident analysis in aviation, as well as in other nonaviation environments (e.g., ferry, tanker, train, or car accidents), still identifies human performance or behavior problems, the so-called human error, as a major contributing factor. Taking the expected future air traffic growth into consideration, Foushee (1993) predicted one major aviation accident every week despite considerable improvements in technology in the next couple of decades

Undoubtedly, this very broad categorization called human error caused many controversial discussions. It is still the remaining category in accident analysis (i.e., if the accident could not be explained by other technical reasons), Berninger (1991), for example, tried to clarify the role of human error in aircraft accidents by stating that the conclusion of human error only proves that the human could have prevented the accident but not that the human (pilot) caused it. Instead, he argued that the system characteristics working against human performance cause the human to fail.

According to Foushee (1993), there is enough evidence that the automation philosophy—automation being an easy way to remove human error from the system—has to be critically examined and that new, more, human-centered approaches to automation have to be considered. The interaction of human and machine in advanced automation systems obviously brought about conflicts and new kinds of errors that did not exist in times of lower levels of automation. According to Scardigli, Moricot & Gras (1991), the mixture of human and "intelligent" machines leads to an ambivalent situation because neither the automated system nor the pilot controls the flight. Berninger (1991) emphasized that systems that are compatible with humans seem to be a promising approach to further system safety improvements.

The project of human factors certification has to be seen in this context. One approach to reaching the goal of generally more human-centered automation or technology is by putting pressure on system designers and producers to incorporate human factors considerations into the design process.

DEFINITION OF TERMS

According to the law of accreditation (e.g., Austrian Standards Institute, 1992), certification is defined as the formal certificate of conformity carried out by accredited independent representative impartial third persons or bodies. Certification can be a forced legal act or a voluntary measure based on documented and accepted rules, procedures, and processes.

Certification is, above all, a measure of quality assurance, often connected with safety goals. Only those products or systems that have proven their conformity to defined standards are allowed into operation. According to the subject of certification, the time of validity might be limited or unlimited. In the latter case, certification is the temporary end of a process that can last over the life span of a system.

Human factors certification specifies the subject that has to be certified. It is meant in the sense of certification of human factors. Following this understanding, human factors certification means to certify human factors issues as part of general system certification. Therefore, the previously mentioned general definitions and principles of certification are valid too.

PROBLEMS OF HUMAN FACTORS CERTIFICATION

When tackling human factors certification, the specialty and complexity of the subject in question has to be considered. Human behavior depends on many conditions from inside the individual and the outside world, from past experience and future plans and expectations, from man's interactions with others and from social, cultural, political, economical, ecological, and technological conditions and developments.

Humans are in a permanent process of adapting themselves to the external world as well as adapting the external world to their needs. Yet, the speed of adaptation is different depending on whether the intellectual, behavioral, or emotional side is considered. Schulz von Thun (1994) pointed out that where as intellectual progress advances in seven-league boots, human behavior and feelings still follow the old trot running behind at a snail's pace, millimeter by millimeter only. This can be verified when comparing the contribution of technical and behavioral sciences to scientific progress. The natural sciences or technical disciplines progress in a more and more accelerating way, plunging into areas like artificial intelligence, virtual reality, nonlinear self-organizing systems, or genetic technology, where as the behavioral sciences in general have no substantial impact on research programs determining man's presence or future.

What Makes Human Factors Certification So Difficult?

One of the particular problems of behavioral sciences are the so-called soft data. Referring to the previously mentioned characterization, human behavior is not deterministic but rather probabilistic by nature. Therefore, measuring methods, results, and predictions is not that exact in behavioral sciences compared to natural sciences. A higher amount of uncertainty or failure rate has to be accepted when dealing with human factors data.

This leads to basic questions about human factors certification: which degree of certainty is certain enough, how to define human factors standards, how to consider sociocultural differences and social changes, how to evaluate not only short-term but also long-term effects, where to set cut-offs, if it is enough to distinguish between optimal and still acceptable solutions and what the price for it is (e.g., the loss of safety).

Besides enhanced efforts in recent years to establish quality control or quality management systems in areas others than technical or industrial ones—for example, psychology, psychotherapy, or health services (Kaba, 1997; Kroj, 1995; Laireiter, 1997; Pietsch-Breitfeld, Krumpaszky, Schelp, & Selbmann, 1994)—show that new behavioral-related concepts of quality management have to be developed for these domains.

Certified quality management systems according to the international norms DIN ISO 9000ff claim to be applicable to all branches. In fact they are strongly based on structures, terms, and procedures of industrial production and industrial quality management (Eversheim & Jaschinski, 1995; Schmetzstorff, 1995). Consequently, they show a lot of weak points and deficits when applied to the "soft" domains. For example, the customer (client) and the purchaser are often not the same in psychological fields, like research, selection and training. A clear relation between customer and supplier does not exist. Thus, if several customers or recipients have to be considered the simple, one-dimensional measurement of customer satisfaction is no adequate way to measure the quality of a psychological product. This underlines the necessity to develop systems or models that fit the subject in question.

APPROACH TO HUMAN FACTORS CERTIFICATION

Referring to the previously mentioned problems, the approach to human factors certification must be a behavioral one. This means that human factors evaluation should be based on models, concepts, and methods of the behavioral disciplines only. Consequently, the experts involved should be exclusively from the behavioral disciplines.

Model of Human-Centered System Issues

In the context of what should be certified, a general model of human-centered system issues that can be applied to different man–machine systems can help to specify the relevant dimensions on a scientific basis. Harwood (1993), for example, introduced an approach distinguishing three broad categories of human-centered system issues that all have to be considered. These categories, which have to be specified according to the context of domain (e.g., air traffic control, flight deck, power plant), are technical usability, domain suitability, and user acceptance.

Technical usability refers to perceptual and physical aspects of the human–computer interface. *Domain suitability* refers to the content of information and display representation for domain tasks as well as functionality and decision-aiding algorithms. Finally, *user acceptability* refers to the ease of use and suitability of the system for supporting cognitive task requirements as well as aspects of job satisfaction (Harwood, 1993).

The model of Koch, Reiterer, and Tjos (1991) is even more comprehensive. It covers general criteria for human adequate labor and is applicable to different domains, too. According to this model, the potential for individual learning and development is the ultimate criterion besides the protection of personality. To reach this criterion, aspects like individual autonomy, getting feedback, transparency of tasks, completeness of tasks, dealing with different tasks, fostering cooperation, and communication have to be guaranteed at the work place. This leads to software ergonomical requirements like potential for individualization, controllability, ability of self-description, error tolerance, clearness of tasks, and conformity of technological procedures with expectations. Besides, task suitability and trainability have to be considered, too.

Tools for Human Factors Certification

Common tools for certification purposes are handbooks that are available for human factors. Yet, they are not sufficient.

Even if they cover a very broad area of knowledge, human factors handbooks are of limited value to evaluate new systems from a human-centered point of view. General human factors principles have to be adapted, modified, and evaluated in regard to the specific application. This means that any test more than a decade old cannot do justice to the current state of human factors (Kantowitz, 1993). Especially in the era of globalization, systems will be implemented worldwide and operated by personnel with different sociocultural backgrounds. The strong influence of the social environment on man's behavior was demonstrated by the work of Mischel (1971, 1973).

The effects of substantial changes in the operator's working place, his role, functions and working conditions in new human-machine systems may neither be identified by other existing tools, like accident or incident reporting systems. Therefore, human factors certification must strongly focus on empirical data, revealed by testing and evaluation, especially on validation studies.

According to the outcomes revealed at the NATO ASI on verification and validation of complex systems (Wise, Hopkin, & Stager, 1993) validation starts in the early stages of system development and lasts over the life span of the system. Following this definition, the validation process includes evaluation in different phases of the system. Linking certification to the validation process would mean that certification is not the final point but rather a culminating process (Endsley, 1994). Thus, human adverse effects might be identified earlier, saving costs of redesign in later phases. In addition, not only short-term but also long-term effects on human factors as well as any significant changes in human behavior can be identified.

CONCLUSIONS

More than ever before, advanced technology requires the collaboration of technical and behavioral disciplines. The combination of technical and social components in human–machine systems demands not only the contribution from engineers but at least to an equal extent from human factors experts. This has been neglected for too long. The psychological, social, and cultural aspects of technological innovations were almost ignored.

Yet, along with expected safety improvements, human factors gain importance. The institutionalization of human factors through certification would give an answer to Foushees (1993) question: The recognition of human factors represents a meaningful change and not a temporary infatuation.

REFERENCES

Austrian Standards Institute (1992). *Law of accreditation.* Vienna: Author.

Berninger, D. J. (1991). Understanding the role of human error in aircraft accidents. *Transportation Research Record, 1298* 33–42.

Brehmer, R. (1996). Visualization of technical processes. *Human Ware, 4,* 16–18.

Endsley, M. R. (1994). Aviation system certification: Challenges and opportunities. In J. A. Wise, V. D. Hopkin & D. J. Garland (Eds.), *Human factors certification of advanced aviation technologies (pp._9-12).* Daytona Beach. Embry-Riddle Aeronautical University Press,.

Eversheim, W., Jaschinski, C.M. (1995). Quality management for non-profit organizations. *VOP, 2* 101-106.

Foushee, C. (1993). The human factors revolution: Meaningful change or temporary infatuation? In ICAO (Ed.), *Human Factors Digest No.9. Proceedings of the Second ICAO Flight Safety and Human Factors Global Symposium* (pp.11-30). ICAO Circular 243-AN/146, 11-30.

Harwood, K. (1993). Defining human-centered system issues for verifying and validating air traffic control systems. In J.A. Wise, V.D. Hopkin & P. Stager (Eds.)._Verification and Validation of Human-Machine Systems. NATO ASI Series (pp. 115-129).* Berlin: Springer.

Kaba, A. (1997). Quality control in traffic psychology. In F. Baumgärtel, F.-W. Wilker & U.Winterfeld (Eds.), *Innovation and Experience. Analyses, Plans and Practical Reports of Psychological Fields of Activity* (pp. 63-71). Bonn: German Psychologist's Press.

Kantowitz, B.H. (1993). Selecting Measures for Human Factors Research. *Human Factors, 34*, 387-398.

Koch , M., Reiterer, H. & Tjos, A.M. (1991). *Software-Ergonomy*. Vienna: Springer.

Kroj, G. (Ed.). (1995). *Psychological Expertise Fitness for Driving*. Bonn: German Psychologist's Press.

Laireiter, A.-R. (1997). Quality control in psychotherapy. *Psychology in Austria, 5*, 178-183.

Mischel, W. (1971). *Introduction to personality*. New York: Holt, Rinehart & Winston.

Mischel, W. (1973). Toward a cognitive social learning reconceptualization of personality. *Psychological Review, 80*, 252-283.

Molnar, M. (1992). Software-Design and Ergonomy. *Informatik Forum, 2*, 54-63.

Pietsch-Breitfeld, B., Krumpaszky, H.G., Schelp, B. & Selbmann, H.K. (1994). Description of existing quality control measures in health services. In Ministry of Health (Ed., *Measures of the medical quality control in Germany - state-of-the-art (pp 6-163)*. Baden-Baden: Nomos Press.

Scardigli, V. (1991). Automation in Aeronautics: A French Research Program. In M.C. Dentan, P.Lardennois (Eds.), Human Factors for Pilots. Report on the XIX Conference on the Western European Association for Aviation Psychology. Paris: Air France.

Schmetzstorff. M. (1995). Building stones of a quality management system for services. *Office Management, 7-8*, 20-23.

Schulz von Thun, F. (1994). *Talking Together*. Reinbek: Rowohlt.

Wise, J.A., V.D. Hopkin, P. Stager (Eds.). (1993). *Verification and Validation of Human-Machine Systems. NATO ASI Series*. Berlin: Springer.

VI

REFLECTIONS OF CERTIFICATION IN AVIATION

18

Successful Management of Programs for Human Factors Certification of Advanced Aviation Technologies

Rod Baldwin

Baldwin International Services (Aviation Management Training), Luxemburg

In recent years there have been immense pressures to enact changes in the air traffic control (ATC) organizations of most states. In addition, many of these states are or have been subject to great political, sociological, and economic changes. Consequently, any new schemes must be considered within the context of national or even international changes.

Europe has its own special problems, and many of these are particularly pertinent when considering human factors (HF) certification programs. Although these problems must also be considered in the wider context of change, it is usually very difficult to identify which forces are pressing in support of HF aspects and which forces are resisting change.

There are a large number of aspects that must be taken into account if HF certification programs are to be successfully implemented. Certification programs would be new ventures, and like many new ventures it will be essential to ensure that managers have the skills, commitment, and experience to manage the programs effectively. However, they must always be aware of the content, and the degree of certainty to which the HF principles can be applied, as Debons and Horne (1993) carefully described.

It will be essential to avoid the well-known pitfalls that occur in the implementation of performance appraisal schemes. Although, many of these schemes are usually extremely well thought out, they often do not produce good results because they are not implemented properly and staff therefore do not have faith in them. If the manager does not have the commitment and interest in his or her staff as human beings, then the schemes will not be effective.

Thus, one aspect of considering HF certification schemes is within the context of a managed organization. This chapter outlines some of the management factors that need to be considered for the ATC services.

Management and organizational issues certainly need to be included in any frame of reference by those who may be involved in developing certification programs.

DEFINITION OF HF

The concept of HF issues is broad and still somewhat vague as the subject tends to include any aspect of human behavior. However, experience in the design, operation, and maintenance of large and advanced systems has shown that there is a human element that needs to be more carefully considered if the system is to perform as required. Unfortunately there are still too many glaring examples of poor HF aspects, and many managers could describe numerous examples.

A formal definition of HF is difficult, but many groups, such as The European Study Group for Human Factors in Air Traffic Control (1991), accepted the MANPRINT (U. K. Ministry of Defense Army Department, 1991) definitions of HF as a useful guide:

- Manpower.
- Personnel.
- Training.
- HF engineering.
- System safety.
- Health hazard assessment.

However, the structure is based on functionality rather than applicability, which is a line manager's requirement if HF is to be properly used. A general approach (ICAO, 1989) addressed this need, although it was oriented to HF in the cockpit, called the SHELL model (software, hardware, environment and liveware). Subsequently ATC was addressed (ICAO, 1993) but a specific approach for ATC (Eurocontrol, 1996b) was built on the SHELL concept with the particular categories of application of:

- Communication/teamwork/customer focus;
- Rostering and working practices;
- HF techniques/career and personnel development.

Unfortunately, these two categorizations are approaching HF from the different requirements of the air and ground and thus possibly helping to reenforce the ground air paradigm gap, which, as Shaffer and Baldwin (1996) showed, needs to be overcome. They suggested (Shaffer & Baldwin, 1997) that one possible way is to combine the ICAO and Eurocontrol categorization in a way that might help the aviation manager understand the broader applications. The format is shown in Table 18.1. This is, admittedly a broad approach and one of the problems for performing certification will be deciding who, and what type of person, should perform the HF certification. It has been suggested (Endsley, 1994) that for the bottom line it is the job of the systems customer—but who is the customer? This is definitely a major problem and is dealt with elsewhere in this volume. Fortunately, there are many areas, such as accident investigation (Zotov, 1997), where HF issues are included, and those responsible for safety aspects of ATC are including more consideration of HF (Eurocontrol, 1996a) but much more still needs to be done in integrating the expertise of the safety and the HF experts.

TABLE 18.1
A Possible Combined Categorization for Aviation Human Factors

HF Category	Meets ICAO Characteristics	Meets Eurocontrol Characteristics
Physical requirements	Physical size and shape Physical needs	
Working practices	Output characteristics	Rostering and working practices
Information processing	Input characteristics Information processing	Communication/teamwork/c ustomer focus
Organizational, regulation, management, and training aspects	Environmental tolerances	Human factors techniques (management) Career and personnel development

ATTITUDES

There is no doubt that most management and certainly most employees in ATC have negative attitudes to the concepts of HF and human-management techniques (Baldwin, 1994). Most instructors of management courses, especially those for technical and operational personnel, know the problems that arise when they suggest that there are theories for dealing with people. There is an initial suspicion that the instructor is suggesting that the theories will provide answers for dealing with life, the universe, and everything (Adams, 1982). It takes time to persuade them that the theories serve as a framework for putting the human problems into context and some theories can provide guidelines for dealing with the problems. When the term *psychological theory* is introduced, unease appears due to the trainees not appreciating the difference between psychologists and psychiatrists. The former implies that the other person is the problem, but that is okay, whereas the latter implies that I am the problem, and that is certainly not okay!

Such attitudes are varied between types of industry and of staff employed for certain kinds of work. Again, this varies from country to country and it is interesting to see the different responses to situations, such as those already described, by controllers and technicians from the various European countries in comparison to those from other countries.

It would be interesting to study the more highly educated leaders of the air traffic services and see how receptive to HF certification ideas they are, then compare them with those of more modest academic levels of achievement. Any such assessment would, of course, run into the problems of cultural issues. This aspect has been addressed with respect to flight crews (Helmreich, Merritt, &Sheraman, 1966) and so it seems logical to perform similar studies on aviation managers. Obviously there is no doubt that much more needs to be done in this area with respect to ATC if European integration is to be successfully achieved.

The issues will certainly need to be addressed if such schemes are to have support from the top levels of the ATC organizations. With the decline of the Renaissance man there are now too many managers who confuse the issue of HF with the old themes of the humanists and their suggestion that achievers are not sufficiently interested in the human being. An encouraging development is to note that a database (Ministry of Defence Evaluation and Research Agency, 1998) to improve contacts amongst those involved in HF includes categories for managing director and other senior staff. Interest in HF from senior staff is welcome, as a regular comment from many lower level staff is that HF experts have difficulty in trying to communicate with top management.

In practice, much of the misunderstanding and confusion about HF is a result of fear that manifests itself in outright antagonism to acceptance of HF issues. But what are they afraid of? Is it that the HF experts are seen in the parent role? Perhaps they are identified with those other parent figures, the teacher, instructor, lecturer, professor, who knew them so well in their formative years and perhaps knew that they were not up to the job.

As usual, the writers of fiction are ahead of the managers of reality, as novelists can express their views without having to substantiate their comments or deal with the practical issues. Nevertheless, their comments can be valid and pertinent as is shown in the following quotation from Mann's (1924/1960) *The Magic Mountain* in which Settembrini says:

> We humanists have all of us a pedagogical itch. Humanism and schoolmasters, there is a historical connection between them, and it rests upon psychological fact: the office of schoolmaster should not , cannot, be taken from the humanist, for the tradition of the beauty and dignity of man rests in his hands. The priest, who in troubled and inhuman times arrogated to himself the office of guide to youth, has been dismissed; since when, my dear sirs, no special type of teacher has arisen. The humanistic grammar-school, you may call me reactionary, Engineer, but in abstracto, generally speaking you understand, I remain an adherent . (p.64)-

Perhaps an even more pointed comment comes from the famous psychological detective Porfiry Petrovich (Dostoevsky, 1866/1920) when he told the murderer Rodion Raskolnikov—"that's the whole catch, that this cursed psychology is double-ended!" (p.453).

So, will HF experts have to overcome feelings of guilt in managers and is antagonism based on a feeling that if the HF expert was good enough he or she would be in a line managerial job?

If these ideas are correct, then the HF experts will have to:

- Mount a massive public relations exercise to convince the line management staff that HF issues are important.
- Give concrete evidence to line management staff that HF schemes will increase performance, quality, capacity, and so on.
- Be prepared to give firm advice and say if HF aspects are not satisfactory.

ORGANIZATIONAL TASKS

Strictly speaking, the whole organization should be analyzed as a total system of interacting parts. However this does require great effort and it is questionable whether such a large picture is indeed meaningful.

This question was intended to be answered by the Federal Aviation Administration's (FAA) National Plan for Aviation HF (FAA, 1990). The program was certainly comprehensive and thought to be realistically aimed by putting implementation responsibility into each line unit of the organization. Although not implemented as envisaged, the plan seems to have resulted in a thorough study of the HF needs in the design of ATC systems (U.S. Department of Transportation, 1995).

Large-scale plans have their place, but sometimes it is more appropriate to select a few key tasks of an organization, starting with the end user of the system, and concentrate on the most significant interactions.

For instance, what do air traffic controllers actually do and how do they do it? When these aspects are established, we can provide better management for the activities in general and ensure that they receive all aspects of information support they need, as opposed to what they think they need.

A number of programs have addressed these matters (e.g., Day, Hook, Warren, & Kelly, 1993) and Eurocontrol has supported its studies in this area with work on how an ATC team works together (Eurocontrol, 1998).

However, the definition of the tasks must be related to the nature of the organization, the type of management, and the leadership styles. The effect of the latter on the other aspects has been extensively studied (Handy, 1985) with particular attention given to directives against consultative styles of management.

It is interesting that most staff seem to believe that a more consultative style of management will produce the best fits for achieving task requirements in conjunction with the needs of the organization, the team, and the individuals. That this is not necessarily so was indicated by Fiedler (1967), who showed that both styles can be effective if used appropriately. He also stated that an individual's style cannot be changed through training, as style is a stable characteristic of the individual. However, this view was refuted by , who believed that a leadership development program could enable a person to widen his or her range of management styles and so be more appropriate to a particular situation. Indeed, there are serious programs in the field of crew resource management that are devoted to training in this area (Hill, 1998).

There is a need for a study of the contingency approach to leadership and management in the air traffic services to establish more understanding about the relations between the different tasks. A very subjective attempt to place these factors onto a bar chart (Fig. 18.1) shows the wide misfit that can occur, and possibly explains why there are so many problems in the ATC services. These occur despite the evidence that controllers actually enjoy the detailed workings of their job, and that their monetary rewards are generally toward the higher end of the remuneration scale. However, it has been suggested (Baldwin, 1995a) that better controller–management relationships will not be established until many of old attitudes are jettisoned by both sides. A particular point of concern (Baldwin, 1993, 1995b) is shown by the continuous antagonism of many controllers to the ATC privatization programs "fewer, better, simpler," (1998). This feature certainly makes an impact on all of the bars shown in Fig. 18.1!

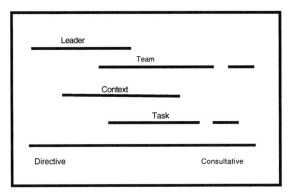

FIG. 18.1.   Management styles in ATC

ORGANIZATIONAL STRUCTURE

The type of organization will play a significant role in how a certification program is carried out. Unfortunately, many organizations have not fully clarified their objectives, how they are to be achieved, and how the various elements of the organization relate to each other. Thus, misunderstandings and dissatisfaction will occur when an organization attempts to introduce new schemes without having a clear understanding of the context into which the new scheme must fit.

Organizations can be classified in many different ways depending on performance or other criteria, that need to be analyzed. Burnsand Stalker (1968) distinguished three forms of organization as *mechanistic* (i.e., bureaucratic), *organistic* (i.e., flexible), and *pathological*. Westrum (1993), however used the more colloquial terms of *normal, healthy,* and *sick*!

Modern practice is to consider that individuals and teams can operate more effectively and with more motivation in organic style organizations as opposed to the more mechanistic forms. Thus, organizational development should start with a careful definition of the task or tasks to be achieved, and then consideration should be given to the individual and team requirements for maximum efficiency in achieving the task.

Most ATC organizations, however, have traditionally had a mechanistic structure, even, to some extent, with a slightly militaristic attitude. Now, many of them are attempting (or have attempted) to change (with varying degrees of success) into organic forms to meet modern task and personnel requirements.

One of the dangers in attempting to change an organization is the problem of becoming stuck in the pathological form. This can often come about by upper management trying to move the organization from the mechanistic to the organic, without wishing to lose the benefits of the former. Of course, *benefits* can imply material benefits or intellectual ones that provide a comfort zone based on previous experience.

This situation is particularly evident in the case of Air France (*Flight International*, 1998), where there was tension between government control and realistic commercial operation.

The present interest in privatization—or more correctly commercialization—of the ATC organizations has provided additional pressure for a move to more organistic organizational forms. This in turn, however, has necessitated a change in attitude among many staff who were formally civil servants. Unfortunately, in some cases a new problem has occurred through staff sectorization because some of the changes have been more beneficial to certain personnel than others. Consequently, many of these organizations may be actually moving into the pathological state, although offering a window dressing of go-getting private attitudes!

THE TEAM

In any organization, there are many teams operating at different levels. These are mostly interdependent, but effort has to be expended by management to ensure that each understands its role and its relationship to the other teams.

Definition of the organizational task should start with the end user team, but this should not necessarily imply any superior position. In fact, each team should see itself as an end user of another team's product and a supplier of a product (perhaps as in-house consultants) to another team or teams.

However, superior teams will arise if certification programs are introduced for some personnel but not for others. Initially there may be acceptance of the situation, but whereas equipment has an accepted life span and generally keeps to that if the original specifications have been adhered to, humans have a habit of changing according to circumstances. Thus, certified but poorly performing individuals introduce accentuated problems into the teams.

Hence, certification programs must give close attention to team composition and the shelflife of the certified personnel! This therefore introduces the need for currency checking and the consequent costs of maintaining such programs. In addition, consideration must be given to the degree to which the certification process can slow down an operation and whether the process is counterproductive to ensuring that an organization is adaptable to changing requirements as Hancock (1994) strongly urged.

The role of HF auditing programs needs much more consideration as a way to overcoming many of the disadvantages of certification as this process would also leave line managers in control of their operation. However, it is always difficult to identify appropriate criteria for assessing cross-functional activities and consequently HF criteria for certification will also be difficult.

REGULATION

In Europe, there are a wide variety of certification standards and training schemes, so mobility of staff is either difficult or impossible in many cases. There are two possible solutions if common acceptance is to be achieved "Europe Without Borders", (1993):

- Each state could recognize the qualifications of personnel from other states; how the training was performed is the concern of each state. This, of course, requires a careful definition of what each qualification allows the holder to do and the context in which the task can be performed.
- The qualifications and training programs are integrated throughout the European states. This requires very careful definition of the knowledge and skills required for the qualification, establishing other conditions that must be applied, and how standards are specified, set, and checked throughout the large number of states concerned.

At the moment, there appears to be pressure in Europe for the latter solution and a feeling that the problems can be overcome. The problem of maintaining common standards for training air traffic controllers was addressed by Baldwin (1988) some years ago, when he suggested the establishment of external examiners, where *external* means from another state. Initially, the external examiner would only act as an observer, offering comments on how a similar situation would be addressed and dealt with in his or her own organization. Naturally, there would be problems because each ATC system has grown up in its own way and comparisons are not very easy. However, as Europe moves toward an integrated system, a common certification program becomes feasible.

A major requirement for these external examiners will be that they have a strong sense of tact and a serious desire to study the other country's system, appreciate how it is structured, how it operates, and the context of the operating system. Success has been achieved in similar schemes, for the certification of aircraft, run by the Joint Aviation Authority (JAA), which is an associated body of the European Civil Aviation Conference (ECAC) through its maintenance and standardization (MAST) program. Several expert groups were formed from staff of the member states but given a briefing from the technical director of the JAA. These groups, in fact, act as integration groups for the maintenance of standards and are therefore very welcome by each member state.

The MAST program is appropriate as aircraft are common to the various airlines of the different countries, and thus common requirements are relatively easy to apply. The same conditions do not apply to the ATC until there is an integrated system. However, a start has been made in Europe and thus there is no reason why some form of central bureau should not be established.

In fact, it is an interesting exercise to compare possible requirements for the certification of advanced air traffic systems with those presently applied to the certification processes of developing a new aircraft. Much more thought needs to be given to which stages require the involvement of HF specialists, to what extent they should be involved, and who the specialists should be. In short, there appears to be a need for a progressive change from HF certification to checking, and then auditing as an overview.

Of course, in the European context, a major problem is that present organizations such as Eurocontrol, ECAC, the European Commission, and the JAA need to redefine their terms of reference with respect to what they can best achieve, what they are good at, what authority they can wield, and how they relate to each other.

THE LEADER AND TECHNOLOGICAL CHANGE

At one time, the leader could develop his or her technical skills, expand into leadership skills, and then apply them until retirement. However, with the present rate of technological and organizational change, this method of working is no longer possible and, in fact, can cause top-down obstruction to new schemes.

Although this phenomenon is not particularly new, it is so in terms of the rate of change together with the need to place more emphasis on the HF elements. However, although many managers are expert in dealing with the component parts of systems, they might not be expert in dealing with the special aspects of the whole system—especially when it is technologically advanced and large. Therefore, certification of the system can become particularly difficult and we will need to reconsider who should be carrying out the certification process, the methodology used, and who endorses the certificates.

Toffler (1970) made an impact on society with his book *Future Shock*. In one section, entitled "Taming Technology" he made this comment:

> Given that a majority of men still figuratively live in the twelfth century, who are we even to contemplate throwing away the key to economic advance? Those who prate anti-technological nonsense in the name of some vague "human values" need to be asked "which humans"? To deliberately turn back the clock would be to condemn billions to enforced and permanent misery at precisely the moment in history when their liberation is becoming possible. We clearly need not less but more technology. (p. 387)

After more than a quarter of a century many of these terms need to be changed, depending on the topic of discussion and the environment under consideration. However, although we now approach the new millennium, there is no doubt that the views still echo much of the thinking in the air traffic services. Certainly, there are many views expressing resistance to the introduction of advanced automation. At times, there appears to be a mind-set on the role of humans and whether they should adapt to automation. The main problem is the mind-set. Each of us has to ask if we are locked into set opinions and possibly politically correct thoughts. If the answers are 'yes' then we have to question how we can break out of this straightjacket.

In many cases the main constraint for consideration is the degree to which HF experts can commit themselves to certification. Do they have a problem with the question concerning validity, or do they just not want to commit themselves in the way that technical and operational staff have to? To some extent these problems are regularly addressed in discussion groups, but there is often a suspicion amongst the administrative, operational, and technical managers than HF experts find discussion and advising is easier that giving firm recommendations!

Westrum (1991) took the theme further by analyzing the intellectual resistance to innovation in terms of failures of imagination and failures of nerve. The examples given from military operations make useful parallels for air traffic where safety is the priority factor.

However, for some years now (Helms, 1993) there has been a strong feeling that we now have enough technology to meet whatever technical task is required. Future requirements should concentrate on the individual and team ability to use the technology. Baldwin (1991) pointed out that if the human element is ignored in the present massive European expenditure on new ATC systems and equipment, then the result might be no ATC capacity increase at all!

CONCLUSIONS

The corner has been turned, but the commitment to HF from senior staff is still not as strong as it should be (Baldwin, 1995b). Progress will be made if the HF experts conduct their work and present their results in forms that are readily understandable to the leaders, managers, engineers, and operators. That is, the HF experts must now study themselves and how they relate to their clients. This message has been made several times in this chapter and it will continue to be made until the subject of HF is regarded as an essential division of an aviation organization.

REFERENCES

Adams, D. (1982). *Life, the universe and everything.* New York: Simon & Schuster.

Baldwin, R. (1988). *ATC 2000.* Luxembourg: Eurocontrol Institute.

Baldwin, R. (1991, August 28-September 3). *Humans are the limit!*, *Flight International*, Vol. 140, letters.

Baldwin, R. (1993, October).*De-nationalization of ATC. Are we serious?* Paper presented at Convex International '93, Bournemouth, England.

Baldwin, R. (1994). The successful management of programs for human factors certification of advanced aviation technologies. In J. A.Wise, V. D.Hopkin and D. J. Garland (Eds.), *Human factors certification of advanced aviation technologies,* (pp. 238-239). Daytona Beach FL: Embry-Riddle Aeronautical University Press.

Baldwin, R (1995a, October). *Yes —it is all change, but for whom?* Convex International '95, Bournemouth, England.

Baldwin, R. (1995b, November–December), "Money talks - who owns air traffic control?" *Air Traffic Management*(pp. 18-21).

Burns, T., & Stalker, G. H. (1968). *The management of innovation.* London: Tavistock.

Day, P. O., Hook, M. K., Warren, C. & Kelly, C. J. (1993, April). *The modelling of air traffic controller workload.* Paper presented at Workload Assessment and Aviation Safety, Royal Aeronautical Society Conference, London.

Debons, A. & Horne, E. H., (1994) Information system certification: purview, perspective and projections, In J. A.Wise, V. D.Hopkin and D. J. Garland (Eds.), *Human factors certification of advanced aviation technologies,* (pp. 355-371) Embry-Riddle Aeronautical University Press.

Dostoevsky, F. (1992) *Crime and punishment* R. Pevear & L. Volokhonsky, (Trans.) London:Vintage. (Orginal work published 1886)

Endsley, M. R. (1994). Aviation System Certification: Challenges and Opportunities. In J. A.Wise, V. D.Hopkin & D. J. Garland (Eds.), *Human factors certification of advanced aviation technologies,* (pp. 9-12) Embry-Riddle Aeronautical University Press.

Eurocontrol. (1996a, September). Safety management awareness workshop, Eurocontrol Publications, 96 rue de la Fusee, B-1130, Brussels, Belgium.

Eurocontrol. (1996b). [Series of Human Factors publications as part of the EATCHIP program]. Available from DGS Logistics and Support Department, 96 rue de la Fusee, B-1130, Brussels, Belgium: Eurocontrol.

Eurocontrol. (1998, January). Teamwork in air traffic services: *Proceedings of the second Eurocontrol Human Factors Workshop.* Eurocontrol Publications, 96 rue de la Fusee, B-1130, Brussels, Belgium: Eurocontrol.

Europe without borders. (1993, June 28–July 11),.*Commercial Aviation News,* (June 28-July 11).

European Study Group for Human Factors in Air Traffic Control (1991), *Terms of reference.*

Federal Aviation Administration, (1990). *The National plan for aviation human factor.* Washington, DC: U..S. Department of Transportation.

Fewer, better, simpler: Cutting red tape—Deregulation —A sensible ATC policy? (1998). *Official Publication of the UK Guild of Air Traffic Controllers* (GATCO CAF). Surrey, UK: U. K. Guild of Air Traffic Controllers.

Fiedler, F. E. (1967). *A theory of leadership effectiveness.* New York McGraw–Hill.

Flight International, (1988, March). Number 4615, *153*, 4.

Hancock, P. A. (1994) Certifying life, In J. A.Wise, V. D.Hopkin and D. J. Garland (Eds.) *Human factors certification of advanced aviation technologies,* (p. 38) Embry-Riddle Aeronautical University Press.

Handy, C. (1985). *Understanding organizations.* London, U.K: Penguin .

Helmrich, R. L , .Merritt, A. C. & Sherman, P. J. (1996). Research project evaluates the effect of national cultures on flight crew behaviour" *ICAO, Journal, .51*(8), (p. 14).

Helms, J. L. (1993, June). *Air Traffic Control Association Conference,* Geneva, Switzerland.

Hill, D. (1998). Creating a culture of safety. In Baldwin (Ed.), *Developing the future aviation system,* (pp. 72–84). Aldershot,. England, UK: Ashgate.

ICAO. (1989). *Human factors digest No.1: Fundamental human factors concepts,* (Circular No. 241–AN/145) Montreal, Canada: Author.

ICAO. (1993). *Human factors digest No.8: Human factors in air traffic control,* (Circular 241-AN/145) Montreal, Canada: Author.

Mann, T. (1960). *The Magic Mountain.* London: Penguin (Orginal work published 1924).

Ministry of Defence Evaluation and Research Agency. (1998). *Human factors integration, point of contact directory,* (Issue 2). London : Author.

Shaffer, M. T. & Baldwin, R. (1996, March). *Automation and human performance: Implications for cockpit and air traffic control interactions.* Paper presented at the Second Automation Technology and Human Performance Conference, University of Central Florida, Orlando, FL.

Shaffer, M. T. & Baldwin, R. (1997). *A systematic approach for applying multimedia techniques to aviation:* Advances in multimedia and simulation: Human–machine-interface implications, In K. P. Holzhaisen (Ed.) *Proceedings of the European Chapter of the Human Factors and Ergonomic Society Annual Conference,* (pp. 93-102) Bochum, Germany, November 1997. Ed. Klaus-Peter Holzhausen.

Toffler, A. (1970). *Future shock.* London: Pan Books,

U.S. Department of Transportation. (1995). *Human factors in the design and evaluation of air traffic control systems.* (Final Report, 1995). Washington, DC: Author.

U.K. Ministry of Defence Army Department,. (1991) *.MANPRINT.* London: Her Majesty's Stationary Office.

Vroom, V. H & Yetton, P. (1973). *Leadership and decision-making.* Pittsburgh, PA: University of Pittsburgh.

Westrum, R. (1991). Originators and managers of technology. In *Technologies and society,* (pp. 151–156). Belmont, CA: Wadsworth.

Westrum, R. (1993). Cultures with Requisite Imagination. In J. A. Wise, V. D. Hopkin, & P. Stager (Eds.), *Verification and validation of complex systems:Human factors issues* (pp.401-416). Berlin: Springer-Verlag.

Zotov, D. V. (1997). Training accident investigators for the human factors investigation. In G. .J. F Hunt (Ed.), *Designing instruction for human factors training in aviation*, (pp.257–271). Aldershot, UK: Avebury Aviation.

19

Evaluation in Context: ATC Automation in the Field

Kelly Harwood
Beverly Sanford
Sterling Software (U.S.), INC., Moffett Field, CA

INTRODUCTION

Certification is defined as attesting as certain (Flexner & Hauck, 1983). "Certainty" however may be a rare commodity when the introduction of new technology into an existing system can "...destroy the blanket of established know-how" (Rasmussen and Goodstein, 1988; p. 179). It is impossible to foresee all emergent properties and interactions between system components and their implications. A complete set of requirements and criteria for safe and efficient system functioning is difficult, if not impossible, to define in advance of system implementation. Once the system is in an operational environment, requirements may need to be rejuvenated due to our imperfect foresight and lack of understanding. Christensen (1958) has referred to this dilemma as the "omnipresent criterion problem".

One way to tackle this dilemma is to incorporate field testing early in the system development cycle. This paper describes the field assessment process that has been applied to the development of an advanced ATC automation system, the Center/TRACON Automation System (CTAS). Field testing provides insight into the true characteristics of the system; that is, how it actually operates and any emergent properties as a function of being integrated into the operational environment. Such insight provides guidance for capturing and refining meaningful requirements for system verification and certification. By delaying field testing until late stages of development, solutions to design problems are likely to be technology driven with validation, verification, and certification relying on context-free guidelines for human-computer interaction.

275

Field testing conducted early during the development and demonstration phase of system development affords exploration of the users' experiences with the system in the context of their work domain. It provides the opportunity to understand the implications for system design of the interdependencies between the physical environment (lighting, workplace layout), task domain (goals/functions of the domain) and work activities (social aspects of team coordination; sources of motivation and job satisfaction). The richness and complexity of these context-based factors and the relationships between them is not accessible through design guidelines or standards. Guidelines and standards cannot provide insight into effective design solutions when system performance is highly contingent on context (Meister, 1985; Gould, 1988). Early field testing promotes the development and validation of a tool as a problem-solving instrument (Woods, Roth & Bennett, 1990), thereby increasing the likelihood of a match between the system's capabilities and its context of operation (Rasmussen & Goodstein, 1988; Bentley et al., 1992).

The FAA TATCA Program recognizes the importance of early field testing for the development and validation of advanced ATC automation. It is presently using rapid prototyping and early field exposure as part of the development of CTAS, using on-site system evaluations with active controllers and representative traffic flows and conditions. Iterative field testing is regarded as integral to the development process, with the objective of achieving a match between the system and context for its use. This approach deviates from traditional approaches to ATC system development and will expedite a possible national deployment of CTAS. Embracing the context of the ATC domain is particularly important because of our limited knowledge of the impact of advanced information technology on controller/team job performance and the stringent requirements for maintaining ATC system safety and continuity during system transition (Harwood, 1993).

The first section of the chapter provides a brief description of CTAS, followed by an overview of the field development and assessment process in the second section. In the third section, particular attention is paid to the *structured* assessments of CTAS. These assessments take a principle-driven approach, drawing on principles, perspectives, and methods from human factors engineering, cognitive engineering, and usability engineering. Activities are described that include the identification of human-centered system issues to help guide the collection and interpretation of data, method selection and tailoring, data collection, data analysis, and interpretation. Examples are provided of the types of findings that are a consequence of this development and assessment process. The fourth section discusses requirements definition and rejuvenation. This paper is not a comprehensive review of all possible methods that could be used, but rather a description of those that have been applied in tailoring a process to bring CTAS functions to a level of stability and usefulness. Emphasis is on the mechanics of executing the process, with mention made of the nuances of conducting development and assessment at an operational field site.

CTAS

CTAS is an integrated set of automation tools, designed to provide decision-making assistance to both Terminal Radar Control (TRACON) and Center controllers via planning functions and clearance advisories. CTAS consists of three sets of tools: the Traffic Management Advisor (TMA), the Descent Advisor (DA) and the Final Approach Spacing Tool (FAST). CTAS development has involved thousands of hours of laboratory simulation with controllers to refine and extend algorithms and to enhance the user interface. In order to bring system functions to a level of operational stability and to provide information to Air Traffic and System Development Organizations on a possible national deployment decision, further development and validation is being conducted at FAA ATC field sites. TMA is the first CTAS component to undergo the field development and assessment process and will be the focus of discussion for this paper. (For further information on CTAS see Erzberger and Nedell, 1989; Tobias, Volcker, and Erzberger, 1989; Davis, Erzberger, and Green, 1991; ATC Field Systems Office, 1992; Erzberger, 1993.)

TMA has been developed for use by the traffic manager at traffic management units within Air Route Traffic Control Centers and TRACON facilities. Unlike controllers, traffic managers do not control traffic directly. Instead, they monitor the demand of arrival traffic into the Center and Terminal areas, coordinating with TRACON, Center, and Tower personnel, making decisions to balance the flow of traffic so that traffic demand does not exceed airport and airspace capacity. Traffic managers use information about the arrival flow to decide whether the traffic should be delayed or metered, to distribute the load from one area to another, and to assign departure times for aircraft departing airports within the Center's airspace that will enter the arrival flow for the metered airport. Information about the traffic situation is accessed from multiple sources, such as flight strips, weather displays, operational personnel, and aircraft situation displays. Often, there is no steady state in the traffic flow; the location of a single *heavy* aircraft can disrupt the scheduled flow of traffic, as can poor weather, equipment outages, and emergencies. Given the extent of coordination required, the variety of sources of information accessed, and the dynamic and often variable state of the traffic flow, context through early field testing is crucial to ensure the robustness of TMA and its effective integration into the traffic management unit.

Representations of traffic flow are conveyed on the TMA by configurable moving timelines. Aircraft data tags move down the timelines and are color coded to portray landing schedule and sequence status information. The traffic manager can override TMA's automatically generated schedule at any time by resequencing aircraft, inserting slots for additional aircraft, or changing the airport acceptance rates. A traffic load display provides a graphical representation of various traffic load characteristics, and several configuration panels are available for modifying timeline displays and setting schedule parameters. The workstation consists of a SUN4 SPARC workstation with keyboard and mouse input devices. TMA presents the traffic management coordinator with new capabilities that are a significant departure from the current traffic management system. The next section describes the process that has been applied for developing and assessing TMA at an operational field site.

FIELD DEVELOPMENT AND ASSESSMENT PROCESS- OVERVIEW

Development and assessment of CTAS is currently underway at two FAA ATC field sites. This paper focuses on the development and assessment of TMA at the Center and TRACON of one of the field sites. TMA is accessible in the traffic management units at the Center and TRACON. A one-way interface with the current HOST system is available so that TMA can reflect the current traffic situation. Traffic managers reference TMA during traffic rush periods and in off times to explore its capabilities and to understand how the tool can be used to solve traffic management problems.

The field development and assessment process is geared for system refinement and enhancing our understanding of the potential impact of TMA on traffic management problem-solving and inter-facility coordination. The process also provides insight for various program objectives, such as operational procedures and requirements definition. To do this expediently, two mechanisms are in place to allow the timely transfer of information from the field site back to the primary development site at NASA-Ames. These are "unstructured" and "structured" assessments. Both are described briefly here, and key aspects of the structured assessments are elaborated further in the third section.

Unstructured assessments are performed by traffic management personnel on a daily basis, during traffic rush periods and during off-times. Here traffic managers access TMA representations, comparing data between TMA and the existing system at the Center, or with decisions made from compiling many separate sources of information together at the TRACON. Human factors engineers and development personnel may observe TMA-use for the purpose of understanding the system, but goal-directed data collection does not occur at this time. Unstructured assessments are for traffic managers to experience the system and provide feedback without an audience or intrusion. Having the system available on a continuous basis provides exposure to a variety of traffic flow and weather situations, allowing the users to "shape" TMA-use to fit the problem-solving demands of their environment. This process is instrumental for engendering trust in the system.

Structured assessments are conducted to systematically investigate tool use and to capture the user's experience with TMA. How the traffic manager uses the tool in response to various problem-solving demands is an important gauge of the match between TMA features and functions and the context for their use. It has been argued that a major cause of system failure is a mismatch between the system capabilities and the demands and constraints of the operational environment (Bentley et al., 1992). Calibrating the match is thus a key activity during development for ensuring system success. Structured assessments are conducted by human factors engineers and development personnel and provide feedback for further development and program milestones. Methods and approaches for structured assessments are described further in the third section.

Quick transfer of information from the users to the development site and back to the users again is critical for continuity at the field site. Timely feedback to the traffic managers on their questions and suggestions during unstructured assessment is essential for maintaining their interest and involvement as well as for streamlining the development and assessment process. An electronic-mail system connecting the traffic managers at the Center and TRACON to NASA-Ames and field-site development personnel has facilitated timely information transfer. Questions are addressed immediately, and design issues from structured and unstructured assessments are entered into a data-base managed at NASA-Ames. Resolution and augmentation of TMA features and functions are decided by committee, with representation from program management, developers, human factors engineers, and testing personnel. Issues are categorized and prioritized by the committee according to their pragmatic and technical implications; for example, implications for system usability and operational suitability, availability of development resources, the need for further analysis, and objectives of an upcoming structured assessment. Refined software is shipped back out to the field sites on a near-monthly basis.

STRUCTURED ASSESSMENTS – METHODS AND APPROACHES

Structured assessments of TMA take a principle-driven approach, drawing on principles, perspectives, and methods from human factors engineering, cognitive engineering, and usability engineering. These fields provide a knowledge base from which methods and approaches for validating system designs may be derived. Structured assessments focus on specific aspects of the users' experience with TMA and consist of several activities:

- issue identification
- method selection
- data collection
- data interpretation
- analysis, inferences and implications.

These activities are described next. Each activity relies heavily upon the operational context of the traffic management unit, focusing on exploration and discovery as well as assessment.

Issue Identification

Operational requirements for TMA are currently being defined and thus are not yet available for verifying the system design. This lack of guidance is compounded by the generality of ultimate criteria for ATC –'namely, safe, orderly, and expeditious flow of traffic, and the general lack of knowledge regarding performance of individual and controller teams in current and future ATC environments. In the absence of requirements and criteria, there is a risk of collecting data that may be expedient but inappropriate (Parsons, 1972; Hopkin, 1980). To compensate for this knowledge gap and to systematically guide the collection and interpretation of data to support the refinement and validation of TMA, we focused on three broad categories of human-centered system issues:

- technical usability
- domain suitability
- user acceptance.

These three categories are also of interest to the FAA for its specification of operational requirements and for formal operational test and evaluation. Others have distinguished previously between two or three of these categories (e.g., Hopkin, 1980; Gould, 1988; FAA, 1989; Rasmussen & Goodstein, 1988). Categories and approaches for defining issues are described briefly. Further details can be found in Harwood, 1993.

Technical Usability. Technical usability refers to perceptual and physical aspects of the human-computer interface such as display formatting, graphics, and human-computer dialog as well as anthropometric characteristics of the workstation. Issues in this category address the general question: Can the users extract and access the data needed to do their job? A tremendous amount of research in human factors engineering and human-computer interaction has contributed to the development of principles and guidelines for designing and evaluating human-system interfaces (See for example, Department of Defense, 1989; Shneiderman, 1987; Smith & Mosier, 1986; Van Cott & Kincade, 1974). These principles constitute the basis for defining technical usability issues.

Domain Suitability. Simply addressing issues of interface usability does not necessarily provide insight into the suitability of the automation tool for the domain. Here it becomes necessary to address domain suitability, which refers to the content of information and representations for domain tasks as well as functions and decision-aiding algorithms. Issues in this category address the general question: Does the representation support the problem-solving requirements of the domain? In contrast to technical usability, which is driven by issues of technology utilization, domain suitability requires an understanding of the "cognitive problems to be solved and challenges to be met" (Hollnagel and Woods, 1987, p.257; see also Rasmussen, 1986; Rasmussen & Goodstein, 1988).

The fundamental basis for understanding the types of cognitive demands that can arise is a description of the domain in terms of the domain goals to be achieved, the relationships between these goals, and means for achieving goals (Rasmussen, 1985, 1986; Woods & Hollnagel, 1987; Rasmussen & Goodstein, 1988). This sort of system description, in terms of a goal-means decomposition, is particularly useful for system evaluation: it guides the description of the cognitive situation that the design must support and it guards against narrowly focusing on problem-solving demands in only one aspect of the work domain.

User Acceptance. User acceptance is obviously enhanced by the ease of use and suitability of the system for supporting cognitive task requirements. Yet user acceptance also depends upon job satisfaction, professional esteem and opportunities for demonstrating individual merit. Hopkin (1980;1992) has argued that such issues are usually overlooked in the context of technology upgrades, but may have serious consequences for ultimate system safety and efficiency. Attention must thus be devoted to disclosing issues associated with the impact of new technology on ATC job satisfaction.

Context is critical for understanding the impact of new system upgrades on sources of job satisfaction and professional merit. What is satisfying and motivating about a job is as much a factor of the individual as it is the nature of the tasks and work domain. Ethnographic techniques for understanding the work environment are thus instructive for capturing valid descriptions of sources of job satisfaction. Such techniques are geared to the study of complex social settings to understand what aspects of activities are important and relevant to individuals. In general, ethnographic techniques have been recognized as essential to understanding, designing, and evaluating complex systems (Bentley, et al. 1992; Hughes, Randall, & Shapiro, 1992; Hutchins, 1992; Suchman, 1987; Whiteside, Bennett, & Holtzblatt, 1988; Suchman & Triggs, 1991).

Issue identification for TMA has been based upon hundreds of hours of observing traffic management activities, reading operational documents on traffic management, and interacting with traffic managers. Approaches for identifying issues are contextually based, that is, based upon an understanding of the physical characteristics of the environment, causal relationships between goals and functions in the task domain, and characteristics of the work activities. Focusing on only one or two of these factors risks collecting data that will not provide insight into sources of design deficiencies or provide a basis for defining meaningful human factors system requirements.

Method Selection

The operational field-site is important for gaining insight into the match between an automation tool design and its context for use. The complexity of the operational environment, with its inherent task demands and the access to operational personnel, allows discovery of unexpected feature use and assessment of the extent to which the tool will support its users. However, while testing at an ATC field site offers a unique perspective on system effectiveness, it also presents a number of constraints that preclude typical laboratory practices and techniques (see for example, Johnson & Baker, 1974).

The availability of controllers and scheduling and resource constraints can severely restrict the extent to which different conditions or system configurations can be investigated. In addition, sample sizes may be small, with the number of replications limited to a single trial. The physical environment is natural and intrusive factors are uncontrolled. Variables are driven by the system, not the experimenter, and the units for measurement are macro-units in the order of minutes. Measures are more often qualitative rather than quantitative.

Given these constraints, our expectations of field assessment must be adjusted appropriately. Field assessments provide an opportunity for capturing the users' ongoing experience with the tool, discovering how new functions will be used and where mismatches occur between the capabilities of the technology and the user's needs. Field assessment provides insight into the integration of a new automation device into an existing environment, indicating issues for transition training and operational procedures. However, field testing is only one level of system evaluation, often augmenting simulation and laboratory testing. Field testing is not a panacea, but it provides an important and necessary perspective for achieving system success.

To accommodate the constraints associated with field assessment and to maximize the opportunity of accessing the operational site, methods must be tailored accordingly. Several criteria guided the selection of methods for assessing TMA:

- Methods must capture the user's ongoing response to the system
- Methods must be sensitive to design deficiencies
- Methods must provide opportunities for discovering new strategies and system functions
- Methods must not disrupt traffic management operations.

Context-sensitive data collection techniques, that is, techniques based on observation and interpretation in the context of the user's work environment, meet these criteria (Whiteside, Bennett, & Holtzblatt, 1988). Such methods include observation and contextual interviews with active involvement of the users in the interpretation of the observations.

Field assessments of TMA, to date, have focused on capturing the traffic managers' experience with the TMA. Whiteside and his colleagues have argued for the importance of the users' experience as valuable information for engineers about user's needs. The appropriateness of features and functions "...exists in the experience of the user, and experience is driven by the context in which it occurs" (Whiteside, Bennett & Holtzblatt, 1988; p. 809). Capturing the users' experience with the tool is especially important for complex, ATC automation systems, where the implications of the interactions between system components and emergent properties are largely unknown prior to implementation. When validation of system designs rests on reconciling technological possibilities with work needs, the users' experience assumes an important role.

Data Collection

Assessments are conducted in the traffic management areas at the Center and TRACON. This location serves both technical and pragmatic interests. Traffic management involves extensive coordination with other traffic managers and area supervisors, communications with other facilities, and accessing and integrating information from a variety of different sources, such as weather displays, aircraft situation displays, and flight strips. Accessing TMA-use in the context of these operational activities is essential for addressing domain suitability and user acceptance. In addition, access to operational lighting conditions is desirable for validating such technical usability issues as color discrimination and readability. Lighting in the operational area is complex, with overhead lighting located in high ceilings and local lighting on work surfaces.

From a more pragmatic perspective, the location of the test area accommodates resource constraints and works well with the culture of the unit. To date, it has not been possible to schedule participants prior to the assessments. Instead, the supervisors on duty release traffic managers when staffing and the traffic demand allows. Having the supervisors control access to the traffic managers minimizes the impact on the unit, thereby increasing acceptance of the assessment process. Supervisors release and summon traffic managers as the conditions permit. Modular organization of data-collection materials, and non-intrusive observation are also flexible to accommodate this scheduling constraint.

Several different methods are used to collect data for assessing TMA. Scenario-driven surveys using prerecorded traffic data are used to assess technical usability. Shadowing of traffic management operations is used to assess technical usability, domain suitability, and user acceptance. These methods are described next, with particular attention given to the mechanics of their execution.

Scenario-Driven Surveys.   Scenarios systematically guide the traffic managers through the display and interactive features of TMA and instruct them to view or manipulate different features. Pre-recorded traffic data are used to ensure that everyone views the same traffic conditions during the exercise. Associated with each scenario are validation statements that focus on specific technical usability issues, such as color discriminability, symbol detectability, and ease of interacting with the input devices. Traffic managers indicate whether they agree or disagree with the validation statement, and space on the survey is provided for comments and suggestions. A human factors engineer sits with the traffic managers as they complete the survey, answering any questions, and observing TMA use. Scenarios generally progress from being easy and simple to more difficult and complex. This is arranged to gauge the level of understanding of basic TMA features in the implicit check of TMA proficiency If a participant is deficient in any area, instruction is provided, and the session is treated as training instead of as TMA assessment.

Technical usability issues are assessed for all TMA modifications, new features and new functions. The initial survey of all TMA features and functions lasted 2.5-3 hours per session, and subsequent assessments have lasted 45 minutes to an hour. The modular organization of the survey allows traffic managers to resume operational traffic management duties when necessary.

Shadowing Live Operations and Contextual Interviews.   Shadowing involves a traffic manager using TMA to make traffic management decisions, mirroring the operational traffic management position. The shadowing traffic manager has access to all other sources of information in the unit except for the operational traffic management system. One observer observes and queries the shadowing traffic manager, and the traffic manager's ongoing commentary is tape recorded for later analysis. Another observer watches the operational traffic manager. Here, traffic management activities and decisions are observed in a more passive mode to avoid disrupting operations. Understanding and interpreting TMA use, at both the Center and TRACON, depend upon an understanding of the operational context. The second observer is critical in this regard.

Shadow-mode operations are effective for discovering unexpected and serendipitous tool-uses and for assessing issues of technical usability, domain suitability, and user acceptance. Methods for data collection are similar at the TRACON and Center but tailored for their unique constraints. Efforts are focused on capturing the traffic managers' ongoing experience with the system using context-based interviews (cf Whiteside, Bennett and Holtzblatt, 1988). This technique involves observing and questioning the users about the tool as they are using it for various planning and problem-solving activities. A critical aspect of contextual interviews is involving the users in the interpretation of their experience with the system. This aspect is discussed further in the next section on data interpretation.

An important aspect of data collection in the field is the period of acclimatization that precedes actual data collection. Prior to conducting structured assessments, we spent several weeks in the traffic management units at the Center and TRACON, simply observing operations and answering questions on the purpose of our presence and the TMA assessment process. This acclimatization period allowed the traffic managers to become comfortable with us, making our observations less intrusive. It also allowed us to work out methodology issues, (e.g., optimum observation positions, and an effective observation checklist) and allowed us to gain a deeper understanding of traffic management operations.

Subjective Ratings. Subjective ratings of a system's usefulness provide another avenue for capturing the users' experience with an automation tool. Following a traffic rush period, traffic managers rate the usefulness of various TMA features, on a scale ranging from 1 to 5, for different traffic management tasks. Ratings capture the users' cumulative experience with the tool, in contrast to the momentary experience captured by a comment made during a specific activity. As a consequence, discrepancies between ratings and comments are possible and are a cue to dig deeper: while a feature may be useful in one situation it may be perceived to be insufficient in another   Such discrepancies underscore the importance of conducting observations at different times of the day, over several days, and preferably during different seasons to capture changes in traffic flow and weather disturbances. TMA has been at the field site for over a year and assessments to date have been conducted during the summer and winter months, each lasting about a month.

Data Interpretation

Data interpretation occurs on and off the field site. Observation alone is not sufficient for exploring and assessing tool use. The observer's interpretations of the observations must be shared with the user to verify their truthfulness (Whiteside, Bennett, and Holtzblatt, 1988). Mutual understanding of the traffic managers' experience with TMA is achieved during the traffic rush and immediately following the rush. The traffic managers are questioned in a debriefing interview on feature use for various problems, their experiences with TMA and their impressions of the traffic rush. In turn, the observers' interpretations of TMA use and the traffic managers responses to questions are also verified. Specific questions and observations, during and immediately following the traffic rush, are guided by a set of general questions:

- What is/was the the traffic situation?
- What decisions and planning activities are occurring/occurred?
- What information is/was accessed from TMA and non-TMA sources?
- How is/was TMA used to support various traffic management decisions?

- What information is/was lacking or hindered decisions?
- What improvements are necessary?

These questions provide a framework for systematically exploring and understanding TMA use in the context of traffic management operations. They also provide a basis for deeper probing of technical usability, domain suitability, and user acceptance issues; for example: Is the number of steps for a particular feature excessive given operational time constraints? Is sufficient information provided for determining whether the airport acceptance rate should be changed? Is the *right* information provided to support equitable decisions? All phases of the interview are tape recorded and conducted at the TMA, in the operational area, to provide a reference for discussing and interpreting the system. The merits of video, for this purpose, have been broadly extolled. Unfortunately, we were precluded from videoing activities in the control room.

Something that proved helpful for data interpretation was for the observers to spend time each day, off the field-site, reviewing and discussing the observations. Dovetailing the different observational perspectives was useful for identifying knowledge gaps and for recalibrating the focus to further explore unexpected discoveries of tool use and possible emerging strategies. Any outstanding questions or new interpretations were taken up with the traffic managers the next day.

Analysis, Inferences and Implications

Surveys, observations, context-based interviews, and subjective ratings provide multiple windows on the traffic managers' experience with TMA. These methods and data provide a qualitative assessment of the match between TMA features and functions and the operational context for their use. The challenge lies in elucidating a tractable set of inferences from this large amount of data. To date, the focus of the TMA development and assessment process has been on identifying design deficiencies, discovering unexpected feature uses, understanding how the tool is used for various problem-solving activities, and defining operational requirements. Analyses have been geared accordingly. Frequency counts of negative responses on surveys provide insight into deficiencies and discrepancies. Content analyses of observations and interviews, coupled with subjective ratings, also provide insight into design deficiencies and discrepancies and enhance the understanding of tool use. Analyses, inferences and implications are described next in some detail as guidance for requirements definition evolves from these insights.

Identifying Design Deficiencies Surveys. Scenario-driven surveys directly assess technical usability issues. Analysis is straightforward, focusing on negative responses to survey validation statements, which indicate difficulties in extracting, reading, discriminating, and accessing data from the TMA.

Comments made by the users during the survey suggest resolutions to these deficiencies.

Survey data can provide diagnostic insight into users' possible difficulties when using the tool to make traffic management decisions. For example, a survey finding helped account for what appeared to be less efficient decision-making during shadow-mode exercises. The survey had revealed that a particular configuration caused crowding of data. Later, during the shadow mode exercises, this finding helped pin-point why particular traffic management decisions were being altered to what appeared to be less efficient decisions: Data congestion was causing traffic managers to overlook data, leading them to alter their decisions as the traffic situation progressed. This interpretation was confirmed by the traffic managers, and the problem was remedied in re-design.

Survey data provide only a partial window on system usability. Display clutter, color coding, and data entry may be assessed differently when the users are actively engaged, using the tool to solve traffic management problems. For example, a particular feature that required manual setting received a positive response on the survey, but negative comments during actual use. During shadow-mode operations, the traffic managers were too busy with other traffic management activities to manually re-set a feature to reflect changes in the traffic flow. Too time constrained, they had to extrapolate the actual setting, making the feature effortful to use. While survey data provides useful information about system usability, this example illustrates the importance of accessing the users' experience with the system from different perspectives.

Observations and Interviews. Design deficiencies are also accessible from observations and interview data collected during shadow-mode exercises. Analysis of these data is time consuming, but the richness of the findings would not be available otherwise and outweighs the cost associated with the time spent. Observations and interview data from the two observers are merged into a single chronological description of each traffic rush. Such a description is useful for capturing the context of TMA use and provides a basis for various content analyses. Content analyses are performed in order to make qualitative inferences about TMA as a potential traffic management tool.

It is important to select categories for the content analysis that reflect the objectives of the assessment. To date, traffic managers' decisions and actions have been categorized according to design deficiencies, feature use for various traffic management activities, and unexpected discoveries. (For a concise description of content analysis, see Weber, 1990.) Data interpretation with the traffic managers during the interviews greatly facilitates the categorizing of observation and interview data. Some examples of the kinds of design deficiencies that can be inferred from content analysis of interviews and observations are presented next. Feature use and unexpected discoveries are described in the following section.

Technical usability deficiencies are defined as observed or reported difficulties in accessing, interacting with, or reading data. Examples of findings include ineffective coding of information for data search, the need for labeling to reflect operations, and too many steps required to implement various functions. In some instances, usability issues revealed here support findings from the surveys; in other instances new issues are raised.

Domain suitability problems are defined as occasions in which the traffic manager needs certain information that is not available, and where extracting information interferes with or hinders problem-solving or decision-making. Examples of findings include the need for organization of information on panels to reflect operational constraints, the need for display parameter settings to reflect current airport configurations, and the need for representation of specific categories of information to reflect the characteristics of the traffic flow.

User acceptance problems are not as easily accessible or apparent as usability and suitability issues because they tend to be incidental consequences of the information technology (cf. Hopkin, 1980; 1992). Understanding the operational context is thus essential for identifying user acceptance problems. System upgrades can affect job satisfaction and opportunities for recognizing professional merit by either affecting what was satisfying about the job in the current system or by causing new situations to emerge that disrupt job satisfaction. Findings from the TMA assessments have provided insight into both of these possibilities. Earlier observations, prior to data collection, had revealed that an important source of job satisfaction is in making decisions and plans that strike an equitable balance of restrictions across facilities and aircraft. Findings from the assessments to date have revealed that a key source of information for ensuring equitable decisions was quite difficult to extract from TMA. This difficulty reduced the use and preference for the representation, and pointed the way for system refinement. In contrast, TMA representations have also created new situations that appear to enhance job satisfaction. One situation is the elimination of a time-consuming counting task, and another is the pulling together of previously disparate sources of information into a single representation. Comments from the traffic managers indicate that such features provide them more time for important planning and allow them to keep up with the dynamic traffic flow situation.

In addition to helping disclose design discrepancies, content analysis of observation and interview data provides insight into feature use. This insight is important for understanding the users' needs and the extent to which they are supported by the automation's capabilities. Feature use and discovery are discussed next.

Discoveries and Description of Feature Use

The introduction of advanced technology and innovative display and interactive features, like that embodied in TMA, alters the way work is done and how problems are tackled. Understanding these changes and how the features are incorporated into the flow of work is as important for assessing the match between the user's needs and system capabilities as is the identification of design deficiencies.

Patterns of feature use elucidated from the analysis of observations and interviews led to the discovery of two different strategies for making a particular traffic management decision. One strategy involved feature use that solved the problem in a similar way to current practices; namely by accessing information that managed traffic demand at the level of individual aircraft. The other strategy solved the problem in a new and different way, by relying on representations that provided information about the aggregate traffic demand. This second strategy was an unexpected discovery. Decisions made with both strategies were equally efficient, relative to those made by the operational traffic manager using the current traffic management system. At one level, the finding of different strategies of feature use suggests that the TMA representations are flexible enough to support different traffic management styles and preferences. At another level, the finding has broader implications for operational requirements because the information needed to support different problem-solving strategies must be identified.

One of the biggest changes to traffic management as a consequence of TMA involves the level of coordination between the center and TRACON. With TMA, both facilities now have access to the same information about the traffic demand. Observations and interviews with the traffic managers indicate that this has enabled the TRACON to coordinate proactively with the Center on decisions regarding the distribution of the traffic flow. Such coordination between the two facilities is essential to avoid overloading the TRACON and to maximize airspace capacity. At a more subtle level, TMA elevates the role of the TRACON traffic managers, allowing them to be more active players in traffic management. This elevated status has obvious implications for job satisfaction.

Another change to current practices is the impact of TMA on the exchange of information between facilities. Analysis of the traffic management communications between the Center and TRACON indicates that well over 50% of the transmissions between the facilities involves the transmission of information that is accessible from TMA. This finding suggests that certain verbal transmissions between facilities could be eliminated, augmented, or reduced by electronic sharing of information via TMA. Changes in the level of coordination and exchange of information between the Center and TRACON alters the way traffic management is performed and have implications for operational procedures and requirements.

REQUIREMENTS DEFINITION AND REJUVENATION

Requirements are the services to be provided by the system and the constraints under which it must operate. Complex domains, like ATC, with their myriad interdependencies and interconnections, are difficult to understand and thus a complete set of requirements is not likely to be available prior to development. Instead, definition of requirements is likely to evolve with development, and modifications and refinements will be necessary as an understanding of the user's needs improves. Field testing conducted early in development catalyzes the requirement definition process. Identifying mismatches between the user's needs and tool features is important for refining the design and for exposing system constraints that must be captured in the requirements. This dual purpose of design deficiencies is illustrated well by the following example. A particular TMA feature had been designed to require several steps to access information. When exercised during shadow-mode operations, the feature was deemed unsuitable because immediate access to the information was needed. This not only identified a design deficiency but also exposed a system constraint that must be captured in the system requirements: immediate access to information for a particular traffic management decision.

Feature use in context is also instructive for defining and modifying system requirements. New technology can create new cognitive problems and information requirements for these problems must be defined.   Similarly, capabilities may emerge when the system is used in the operational environment that were not anticipated in the conceptual design; for example, the level of coordination and information exchange between the TRACON and Center with TMA. System constraints for such capabilities must be included in the system requirements.

It has been argued that the implementation of a system and its specification should not be kept separate, because, in practice, they are "inevitably intertwined." Models of system development that require their separation deviate from reality and restrict the development of effective systems (Swartout & Blazer, 1982). A similar argument can be applied to requirements definition. System implementation through field testing early during system development, can facilitate the evolution of system requirements for complex domains like ATC. As described in this paper, field assessments help highlight system constraints, enhance our understanding of the user's needs, and provide insight into the impact of new technology on existing operational practices. While systematic analyses, feasibility studies, and system modelling are necessary precursors for requirements definition, field development and assessment in the field can help augment the process.

CONCLUSIONS

The process for incorporating advanced technologies into complex aviation systems is as important as the final product itself. This paper described a process that is currently being applied to the development and assessment of an advanced ATC automation system, CTAS. The key element of the process is field exposure early in the system development cycle. The process deviates from current established practices of system development – where field testing is an implementation endpoint – and has been deemed necessary by the FAA for streamlining development and bringing system functions to a level of stability and usefulness. Methods and approaches for field assessment are borrowed from human factors engineering, cognitive engineering, and usability engineering and are tailored for the constraints of an operational ATC environment. To date, the focus has been on the qualitative assessment of the match between TMA capabilities and the context for their use. Capturing the users' experience with the automation tool and understanding tool use in the context of the operational environment is important, not only for developing a tool that is an effective problem-solving instrument but also for defining meaningful operational requirements. Such requirements form the basis for certifying the safety and efficiency of the system. CTAS is the first U.S. advanced ATC automation system of its scope and complexity to undergo this field development and assessment process. With the rapid advances in aviation technologies and our limited understanding of their impact on system performance, it is time we opened our eyes to new possibilities for developing, validating, and ultimately certifying complex aviation systems.

ACKNOWLEDGMENTS

We are grateful for the support of Barry Scott, Manager FAA Engineering Field Office at NASA-Ames Research Center and David Jones, Branch Chief ATC Field Systems Office at NASA-Ames Research Center. C. Halverson, E. Murphy, S. Nowlin, and P. Stager are acknowledged for their insightful comments on earlier versions of this paper.

REFERENCES

ATC Systems Field Office (1992). *Traffic Management Advisor (TMA) Reference Manual.* Moffett Field, CA: NASA-Ames Research Center.

Bentley, R., Hughes, J.A., Randall, D., Rodden, T., Sawyer, P., Shapiro, D., and Sommerville, I. (1992). Ethnographically-informed systems design for air traffic control. *CSCW Proceedings.* (pp. 123-129).

Christensen, J.M. (1958). Trends in human factors. *Human Factors, 1* (1), 2-7.

Davis, T.J., Erzberger, H., and Green, S.M. (1991). *Design and evaluation of air traffic control final approach spacing tool.* NASA Technical Memorandum 10287. Moffett Field, CA: NASA-Ames Research Center.

Department of Defense (1989). *Military Standard: Human Engineering Design Criteria for Military Systems, Equipment, and Facilities.* MIL-STD-1472D. Washington, D.C.: Department of Defense.

Erzberger, H. (1992). *CTAS: Computer Intelligence for Air Traffic Control in the Terminal Area*. NASA Technical Memorandum 103959. Moffett Field, CA: Ames Research Center.

Erzberger, H., and Nedell, W. (1989). *Design of Automated Systems for Management of Arrival Traffic*. NASA Technical Memorandum 102201. Moffett Field, CA: NASA-Ames Research Center.

Federal Aviation Administration. (1989). *FAA NAS Test and Evaluation Program*. ORDER No. 1810.4A. Washington, D.C.: Federal Aviation Administration.

Flexner, S.B. and Hauck, L.C. (1983). *The Random House Dictionary of the English Language*. Random House: New York.

Gould, J.D. (1988). How to design usable systems. In M.Helander, (Ed.). *Handbook of Human-Computer Interaction*. Elsevier Science Publishers BV (North Holland): New York. (pp. 757-789).

Harwood, K. (1993). Defining Human-Centered System Issues for Verifying and Validating Air Traffic Control Systems. In J. Wise, V.D. Hopkin, and P. Stager (Eds.), *Verification and validation of complex and integrated human machine systems*. Berlin: Springer-Verlag.

Hollnagel, E. and Woods, D.D. (1983). Cognitive systems engineering: new wine in new bottles. *International Journal of Man-Machine Systems, 18*, 583-600.

Hopkin, V.D. (1980). The measurement of the air traffic controller. *Human Factors, 22* (5), 547-560.

Hopkin, V.D. (1992). Human factors issues in air traffic control. *Human Factors Society Bulletin, 35* (6). Santa Monica, CA: Human Factors Society.

Hughes, J.A., Randall, D., and Shapiro, D. (1992). Faltering from ethnography to design. *CSCW Proceedings*. (pp. 115-122).

Hutchins, E. (1991). How a Cockpit Remembers its Speed. Technical Report, Distributed Cognition Laboratory, University of California at San Diego.

Johnson, and Baker (1974). Field testing: The delicate compromise. *Human Factors, 16* (3), 203-214.

Meister, D. (1985). *Behavioral Analysis and Measurement Methods*. New York, NY: John Wiley and Sons

Parsons, M. (1972). *Man Machine System Experiments*. Baltimore, MD: The Johns Hopkins Press.

Rasmussen, J. (1985). The role of hierarchical knowledge representation in decision making and system management. *IEEE Transaction on Systems, Man, and Cybernetics, 15*, 234-243.

Rasmussen, J. (1986). *Information processing and Human-Machine Interaction: An Approach to Cognitive Engineering*. Amsterdam: North-Holland.

Rasmussen, J. and Goodstein, L.P. (1988). Information technology and work. In M. Helander (Ed.), *Handbook of Human-Computer Interaction*. Elsevier Science Publishers BV (North Holland) : New York. (pp. 175-201)

Shneiderman, B. (1987). *Designing for the User Interface*. Reading, MA: Addison-Wesley.

Smith, S.L. and Mosier, J.N. (1986). *Guidelines for designing user interface software*. (Technical Report NTIS No. A177 198). Hanscom Air Force Base, MA: USAF Electronic Systems Division.

Suchman, L.A. (1987). *Plans, and Situated Actions: The Problem of Human Machine Communication.* Cambridge, MA: Cambridge University Press.

Suchman, L.A. and Trigg, R.H. (1991). Understanding practice: Video as a medium in reflection and design. In J. Greenbaum and M.Kyng (Eds.), *Design at Work.* Hillsdale, NJ: Lawrence Erlbaum Associates. (pp. 65-89).

Tobias, L., Volckers, U., and Erzberger, H. (1989). Controller evaluations of the descent advisor automation aid. In *Proceedings of the AIAA Guidance, Navigation, and Control Conference.* Washington, D.C.: AIAA. (pp. 1609-1618).

Van Cott, H.P. and Kincaid, R.G. (1972). *Human Engineering Guide to Equipment Design.* Washington, D.C.: U.S. Government Printing Office.

Whiteside, J., Bennett, J., and Holtzblatt, K. (1988). Usability engineering: Our experience and evolution. In M. Helander (Ed.), *Handbook of Human-Computer Interaction.* Elsevier Science Publishers BV (North Holland): New York. (pp. 791-817).

Weber, R.P. (1990). *Basic Content Analysis.* Newbury Park, CA: Sage Publications, Inc.

Woods, D.D., Roth, E. M., and Bennett, K.B. (1990). Explorations in joint human-machine cognitive systems. In S.P. Robertson, W. Zachary, and J.B. Black, (Eds.), *Cognition, computing, and cooperation.* Norwood, NJ: Ablex Publishing Company, (pp. 123-158).

20

Integrating Human Factors Knowledge Into Certification: The Point of View of the International Civil Aviation Organization

Daniel Maurino
Vincent Galotti
International Civil Aviation Organization, Canada

Human factors has matured into a core technology. This contention is best reflected in the attention the aviation industry has dedicated to this technology over the last decade. In fact, the aviation industry has experienced what has been dubbed the golden era of aviation human factors, both in research as well as in application, essentially through human factors training for operational personnel and the integration of human factors knowledge into accident prevention and investigation. Although skeptics and critics still exist in all segments of the international community who would state otherwise, the contribution of human factors knowledge to aviation system safety and effectiveness remains beyond question.

The most important step, however, has yet to be taken. No matter how potentially applicable the research, no matter how performance enhancing the training, no matter how sensible the investigation of accidents, these advances can only be deemed to be remedial measures. From the perspective of advanced technology systems safety, the major contribution of human factors will be fully realized only after the huge amount of existing and available human factors knowledge is incorporated into the certification process of such systems; that is, before the systems are in operation. The incorporation of human factors requirements into the certification process of new, advanced technology remains as the challenge of the future.

It has repeatedly been suggested that, unless awareness about a problem is first gained, for practical purposes, that problem does not exist. Therefore, this chapter advances a justification of why human factors requirements should be given the same consideration and weight as traditional hard-core requirements that exist during the certification process of equipment, procedures, and personnel that make up advanced aviation systems. Such justification is considered vital to secure the understanding and the necessary commitment of designers as well as regulators involved in certification processes. Second, an approach to the inclusion of human factors certification requirements is proposed, based on existing International Civil Aviation Organization (ICAO) regulatory requirements and guidance material as a basic starting point. An understanding of the role of ICAO in the processes of regulation is therefore necessary for the reader to identify how its mechanisms may be useful in attaining the required goals of human factors certification of advanced systems. Although general in nature, such a proposal is viewed as one feasible way to make progress toward the stated goal of human factors certification of advanced aviation systems.

The consideration of human factors requirements during the certification stage of advanced, new technology systems may be seen as resting on a three-legged stool. The first leg, the design of the technology that people in a system will utilize to achieve the system's goals, has traditionally attracted ergonomic considerations associated with equipment design, usually centered around "knobs and dials." More recently, this view has expanded to include other important human factors issues dealing with the cognitive, behavioral, and social processes of the human operators. Research and application in these areas must however be progressed, in particular, regarding the cognitive aspects of the human–machine system. The second leg of the stool, the procedures required to operate the technology, however, has not received equal attention. Procedures are not inherent to machines, but must be developed taking into consideration the operational realities of the context within which the technology will be deployed. The importance of proper human factors consideration in the design of procedures cannot be overstated. The third leg of the stool, certification of personnel who will operate the equipment, is very much underway. Although far from being complete, it can be safely asserted that human factors training for operational personnel is one area in which progress has overcome caveats and vacillations.

BACKGROUND

Over time, the contribution of human factors knowledge toward advancing the safety and effectiveness of sociotechnical systems, including aviation, has been hindered because of its piecemeal approach. Designers, engineers, trainers, and regulators have historically favored solutions that were biased by their professional backgrounds as well as by mind-sets. This often resulted in a state of affairs in which, although the individual components had been designed to the maximum level of available know-how and expertise, the aviation system as a whole received only partial benefits when considering the full potential of the implementation of new technology. Furthermore, an unlimited trust in technology-driven approaches toward system development often overlooked the fact that the human component of the system still remains as the old "Mark I" version, with essentially the same limitations as existed 5,000 years ago. In the simplest terms, technological design has often been guided by the possibilities afforded by state-of-the-art knowledge and engineering feasibility rather than by enduser validation and functionality. Indeed, the shortcomings of an over-reliance on technology to overcome system safety deficiencies have clearly been identified by research as well as the investigation of major sociotechnical systems' catastrophes. However, rather than revising the wisdom of this approach, the record suggests that the industry has renewed its commitment to technology with a blind euphoria over technical systems claimed to be absolutely safe. One cannot help but reflect on the *Titanic*: Such marvelous technology could not possibly sink. History, however, indicates otherwise.

When pursuing safety and effectiveness in aviation, we have tended to think in individual rather than in collective terms. At the operational level, this is reflected by the pervasiveness of remedial actions addressed toward individuals rather than to the system as a whole. A good example of this assertion is the omnipresence of the pilot error clause in accident investigation literature, which would seem to be a relic from World War II. Even with the introduction of human factors knowledge into existing protocols for accident investigation, the focus still often remains on those at the tip of the arrow. It is a common mistake to narrow down an investigation of human factors in accidents and incidents to human performance, to the behavioral aspects as they relate to the performance of operational personnel involved in the actual occurrence. Such investigations should rather broaden their perspective to include overall system performance, including the social aspects of the system performance and the cognitive elements of the human–technology system.

From the perspective of design, the picture does not vary substantially. Technology has been introduced piecemeal, producing excellent individual examples of equipment that is quite remarkable. However, because of the pervasive lack of appropriate macroanalysis at the time that individual technical designs are introduced into a system, the interface with other system components including, first and foremost, the human, is sometimes rather clumsily accomplished. This less than optimum interface has demanded, often times, significant adaptation efforts to accommodate a new piece of equipment into a system already in operation. Such adaptation often involves trade-offs and compromises that, at the end of the day, diminish the benefits of the new design in terms of its contribution to safety and effectiveness.

The introduction of high-level automation into flight decks and air traffic control exemplifies the shortcomings discussed in the previous paragraph. Some would consider it naive to accept that the design and introduction of automation routinely progresses beyond the level of microdesign and microanalysis. The consequences of an absence of macroconsiderations are well documented, and technology intended to reduce human error has in many cases merely displaced it. Likewise, technology intended to alleviate workload often increases it at the most inappropriate times and reduces it during other times so that it may actually foster boredom and complacency. The absence of macroergonomic considerations has led many to contend that the introduction of automation into many advanced aviation systems (e.g., new flight deck design, automated air traffic control systems, intelligent reasoning tools, etc.) reflects a regrettable, although quite preventable, failure of the human factors profession. The good news is that lessons have been learned, and these lessons will hopefully be translated into the design of future global communications, navigation, surveillance/air traffic management (CNS/ATM) systems. CNS/ATM is briefly described in a later section of this chapter because of its importance to the future of aviation and the criticality of addressing human factors certification concerns at an early stage in its development.

ICAO'S ROLE IN REGULATION AND IMPLICATIONS OF CERTIFICATION

Regulation remains the vehicle to ensure that the hard-learned lessons do not vanish into the labyrinths of sectorial interests or into the fragility of human memory. The literature available on how to proceed in securing safety and effectiveness in new technological systems accumulates rapidly. The justification behind new approaches is well documented and beyond challenge. Seldom does a week go by without one major human-factors-related event taking place in the industry's agenda. It could safely be stated that industry has pursued the goal of awareness and education about human factors to the maximum extent that can reasonably be expected. Although there will always be room for relevant fora to exchange ideas and foster feedback, it is time to progress beyond the stage of just talking and exchanging stories. One way to accomplish this progression is to ensure knowledge emanating from research and exchanged through workshops, seminars, and so on, be implemented at the worldwide level. To accomplish this goal, it is imperative to establish and introduce a requirement to include human factors knowledge in the certification process of equipment, procedures, and personnel through legislation that is binding to the larger community. Such a development would be a measure of assurance that, by virtue of the imperatives of regulation, a macroapproach to high-technology systems design will replace existing microapproaches.

Having been developed by the ICAO, an important aspect of this chapter is developing an understanding of the importance of ICAO Standards and Recommended Practices (SARPs) relevant to the certification processes. It is therefore important to discuss the role and processes of ICAO toward the attainment of these goals for the purpose of clarifying the virtues of international standardization and cooperation. A coordinated effort would certainly enhance a wider level of implementation and acceptance of the important work currently being developed concerning human factors certification of advanced systems.

For certification in aviation to have significant relevance, it must have worldwide application. Considering the global aspect of CNS/ATM systems and the goal of a seamless, global ATM system, based on satellite technology, the need to efficiently integrate all of the associated elements at the worldwide level is obvious. Whether discussion refers to air traffic control procedures, pilot licensing, aviation training, or maintenance procedures, it is essential that there be international agreements among nations for regulations to be effective, worthwhile and able to achieve all potentialities. Consider the following example: In an afternoon's flight, an airliner can cross the territories of several nations in which different languages are spoken and in which different legal codes are used. In all of these operations, safety must be paramount; there must be no possibility of unfamiliarity or misunderstanding. In other words, there must be international standardization and agreement among nations in all technical, economic, and legal fields. To accomplish these goals, the nations of the world established ICAO to serve as the medium through which this necessary international understanding and agreement can be reached. It acts as the mechanism whereby global coordination and harmonization are achieved.

The main accomplishments of ICAO associated with regulation, and thereby certification, have been the agreement of its member states on the necessary standardization for the operation of safe, efficient, and regular air services. This standardization has been achieved primarily through the adoption by the ICAO Council, as Annexes to the Chicago Convention, of specifications known as International Standards and Recommended Practices (SARPs). The 18 Annexes so far adopted cover the whole spectrum of aviation. A standard is any specification for physical characteristic, configuration, material, performance, personnel, or procedure, the uniform application of which is recognized as necessary for the safety or regularity of international air navigation and to which contracting member states will conform in accordance with the Convention. In the event of impossibility of compliance, notification to ICAO is compulsory. A *recommended practice* is similar to a standard. The application of a recommended practice is only recognized as desirable as opposed to necessary, in the interests of safety, regularity, or efficiency of international air navigation, and to which contracting states will endeavor to conform in accordance with the Convention (ICAO, 1990).

AN OVERVIEW OF THE ICAO ANNEXES

This overview is developed so that an understanding of the Annexes and their application toward the development of certification standards for Human Factors may be achieved. The Annexes to the Convention on International Civil Aviation have reduced many of the complexities of air transportation to everyday routine. They govern the standards of performance required of pilots, air traffic controllers, maintenance technicians, and flight dispatchers among others. The ICAO licensing requirements must be met by all contracting states and apply to all the operational personnel just mentioned. International specifications also exist for the design and performance of aircraft themselves and the equipment aboard. The rules of the air, by which pilots fly, were formulated by ICAO. These include both visual flight rules and instrument flight rules. The weather reports, so vital to the safety of international travel, are provided to pilots and airport staff by a worldwide network of meteorological stations. The aeronautical charts used for navigation throughout the world are also specified by ICAO, and all the symbols and terms on these charts are standardized for uniformity so that no confusion can ever arise. Even units of measurement used in aircraft communications are standardized so that no pilot can be confused by these. Aircraft telecommunications systems, radio frequencies, and procedures are also ICAO's responsibility. The way in which aircraft are operated is regulated by ICAO so that an international level of safety is maintained. Uniform rules of airworthiness of aircraft are ensured by internationally agreed certification processes. Without that certification, no aircraft can be accepted for flight. Even the requirements for aircraft registration and their identifying marks are based on international standards. To facilitate free and unimpeded passage of aircraft and their loads, ICAO seeks to speed up customs, immigration, public health, and other procedures for passengers, crews, baggage, cargo, and mail across international boundaries and to ensure that essential facilities and services are provided at international airports. ICAO has made significant progress in reducing aircraft noise by adopting stringent noise limitations for aircraft engines. This continuing effort has already resulted in a whole new generation of quieter aircraft. Safeguarding international civil aviation against unlawful seizure of aircraft, sabotage, and bomb threats has received special consideration during recent years. The standards, recommended practices, and guidance material on airport security have resulted in a marked decrease of such incidents. Finally, ICAO has recently introduced a comprehensive set of recommendations for the safe transport of dangerous goods to be applied uniformly all over the world. This latest Annex shows how remarkably well the Chicago Convention meets the continuing standardization needs, so intrinsically tied to certification of civil aviation (ICAO, 1982). A list of the currently existing Annexes is given in Appendix B.

(CNS/ATM) SYSTEMS

A discussion about CNS/ATM is necessary as it can be identified as a major and complex set of new systems that will have an influence on all new systems developed for civil aviation. An opportunity exists for an early human factors input that should include human factors certification requirements. The human factors profession should seek to ensure that all involved with the development and implementation of CNS/ATM systems are made aware of the possibilities regarding human factors certification along with the traditional Human Factors involvement. Human factors awareness should be sought after by all involved, especially those involved in international civil aviation and the development of international regulations.

The CNS/ATM systems concept was developed by the ICAO Future Air Navigation Services (FANS) Special Committee, which was established at the end of 1983 to study, identify, and assess new concepts and new technology in the field of air navigation (including satellite technology) and to make recommendations for the development of air navigation for international civil aviation over the period of the next 25 years. The Tenth Air Navigation Conference, held in Montreal in September 1991, endorsed the findings of the FANS Committee, which identified the major elements of the planned future system. These elements are communications, navigation and surveillance, which will be increasingly accomplished through the use of satellite technology. A resulting ATM system would have the capability of resolving the shortcomings of the present air navigation system and of alleviating some of the air traffic congestion problems experienced in many parts of the world. The implications for human factors certification of advanced systems are evident and offer a further direction of study and discussion for future gatherings (ICAO, 1991).

The Tenth Air Navigation Conference endorsed the view that the planning and implementation of improved ATM capabilities should include considerations of human factors impacts and requirements. They further stated that the many goals listed for the future ATM System should be qualified in relation to human factors. The Tenth Air Navigation Conference recommended that work by ICAO in the field of human factors include studies related to the use and transition to CNS/ATM systems and that ICAO encourage member states to undertake such studies. It also developed a list of considerations that are recommended for use when determining human factors aspects in relation to CNS/ATM systems. These are listed in Appendix A for discussion purposes. It is certain that human factors will be considered in all phases of implementation of the new systems. It is important to consider ways in which human factors certification can be introduced into this important development.

DISCUSSION OF HUMAN FACTORS AND THE CERTIFICATION PROCESS

As discussed elsewhere in this chapter, there are two possible alternatives to address human factors issues in contemporary systems: during the design stage of the system, or after system implementation in the operational context. If following the first alternative, the tools are (a) technology design mindful of human factors issues, and (b) standard operating procedures that reflect the needs and constraints of the operational context in which the technology is going to be deployed. If following the second alternative, the tools are (a) technology re-design (a very rare situation because costs are usually prohibitive), and (b) training to adapt human behavior to cope with less than optimum interfaces. The second alternative has been traditionally favored by aviation, and within this alternative training has been the tool most frequently resorted to, because it is apparently cheaper than redesign.

The decision of which alternative to pursue has significant financial implications, beyond the obvious safety-based considerations. Involving human factors expertise during technology design might incur additional expenses, but the costs are paid only once in the system's lifetime. Furthermore, in the medium to long term, initial costs will be recovered through the most efficient system performance (which will demand, *inter alia*, less training). Coping with flawed human–technology interfaces through training means increased expenses throughout the entire operational life of the system: Not only will it be necessary to increase training, but every time that operational feedback identifies an interface shortcoming that was ignored or unforeseen during design, additional training will be necessary. This means constantly paying for more training on a routine basis. Furthermore, training becomes the dumping ground for poor interface design.

In order to maximize safety and cost-effectiveness of CNS/ATM systems, the proactive management of human factors issues should be a normal component of the processes followed by designers, providers, and users of these systems. The time to address human factors issues is during technology design, before technology is deployed into operational contexts. This will allow anticipation of the negative consequences of human error rather than regrets about its consequences. Furthermore, flight deck operational experience shows that, once technology is deployed, it is too late to address human factors issues arising from flawed human–technology interfaces, and such issues become endemic, hindering safety and efficiency of the system.

ICAO supports a proactive approach to the management of human factors issues in CNS/ATM systems. The proposal for human factors SARPs for inclusion in the Annexes to the Chicago Convention discussed in this section, developed under the activities of its Flight Safety and Human Factors Program, must be viewed under the light of the previous discussion. These SARPs proactively address human factors issues relevant to the certification processes of equipment, procedures, and personnel.

Consider what has been achieved concerning personnel licensing. In 1988, a revision to Annex I included a requirement to the effect that each applicant for a license must demonstrate appropriate knowledge regarding: "human performance and limitations relevant to...(the license being issued)" (ICAO, 1988). Annex I also includes an augmented requirement for the demonstration of certain skills in a manner that dictates increased attention to particular aspects of human performance. Thus, for example, the holder of an airplane transport pilot license "shall demonstrate the ability to: exercise good judgement and airmanship, understand and apply crew coordination and incapacitation procedures; and, communicate effectively with other flight crew members" (p. 14).

This requirement effectively guarantees that, during the certification process (i.e., licensing and training) of operational personnel, human factors considerations must be duly accounted in the form of demonstration of human factors knowledge and skills. Other isolated efforts existed in the certification of procedures (Degani &Wiener, 1994). These efforts were valuable but were, nevertheless, random efforts toward pursuing an avenue of action that deserved serious consideration and the necessary framework for institutionalization that only regulation—international regulation—can assure. The initiative here should belong with the international community, and it is further contended that it should be pursued as such and within the context of existing structures.

From the perspective of ICAO, the inclusion of human factors requirements, following the licensing process of operational personnel as outlined in Annex I (example given earlier) is viewed as one possible way to proceed. The Annexes are well-established and internationally accepted documents covering all aspects of aviation. As such, they provide an ideal structure. ICAO contracting states must observe and abide by their provisions if they are to be accepted as members of the international community. It is difficult to identify a more solid foundation on which to build the much-needed framework necessary for legislating human factors. The ICAO Annexes are subjected to periodic revisions to ensure the relevance and applicability of their contents. If each of these Annexes were analyzed by dedicated groups of experts, then it would be possible to pinpoint if and where the inclusion of human factors requirements in such documents would be appropriate. Should it be deemed feasible, the drafted requirements would eventually be included in the Annexes as they progress through their routine revision cycles. At this level of analysis and drafting, the requirements would be broad in nature and scope. Annex I stands as a leading example to follow. Once the basic requirements are legislated, even in the broadest terms, the responsibility will by force switch to the international research, design, and training communities, which would then have to find the means and tools to implement such requirements into practice. The ICAO Annexes would establish the policies, and the international community would devise the appropriate procedures to achieve such policies. This combination has worked quite satisfactorily and has produced consistent and successful results throughout the implementation of Annex I (i.e., certification of personnel).

THE ICAO INITIATIVE

With this approach in mind, ICAO, during 1997, decided the development of further amendments to the Annexes and other documents with regard to the role of human factors in present and future operational environments. Underlying this proposal was an assessment of the adequacy of existing materials in Annexes and other documents to the role of human factors in CNS/ATM systems. This assessment is part of a long-term strategy by ICAO to address the human factors-related issues in CNS/ATM systems. This section presents the amendments to include human factors-related SARPs developed by ICAO.

The development of SARPs, which reflect the role of human factors in operational environments, and in particular in CNS/ATM systems, was a long-standing concern of ICAO. As discussed in previous sections of this chapter, such development was hindered in the past by the theoretically biased and stand-alone nature of human factors endeavors. Because of this theoretical orientation, the relationships between human error in operational environments and some basic aviation processes such as design, certification, licensing, training and operations were not always understood to the extent necessary to introduce human factors knowledge into the development of safety regulation. Over the last 5-year period, however, applied research has provided the foundation for operationally oriented human factors endeavors and the integration of human factors knowledge into the aviation processes mentioned earlier. This in turn has provided the necessary information to enable the development of safety regulations addressing the role of human factors in present and future operational environments.

The development of SARPs that address the role of human factors responds to the need to anticipate human error in operational environments by applying human factors knowledge during the design stage, before systems achieve operational status. Human factors knowledge can be proactively applied into systems design to make it resilient to the consequences of human error before consequences are beyond recovery. In technical terms, systems should be designed to be error resistant, error tolerant and error evident. One way to accomplish this is by including human factors requirements during the development and certification process of equipment, procedures, and personnel.

The proposal follows previous regulatory activities in regards to human factors by ICAO. In addition to Annex 1 (already discussed), Annex 13 (Aircraft Accident and Incident Investigation) requires that the format of the final report includes organizational and management information, (i.e, pertinent information concerning the organizations and their management involved in influencing the operation of the aircraft), since November 1994. Annex 6 (Operation of Aircraft) includes a standard regarding initial and recurrent training on human factors knowledge and skills for flight crews since November 1995.

THE NEW HUMAN FACTORS REQUIREMENTS

The new requirements introduce generic requirements on the observance of human performance and human factors principles into Annexes 1, 3, 4, 5, 6, 8, 10, 11, 14, 15 and 16 (see Appendix B). Definitions for human performance and human f actors principles are proposed and included for each Annex as required, as follows:

- Human Factors principles: Principles that apply to aeronautical design, certification, training, and operations and seek safe interface between the human and other system components by proper consideration to human performance.
- Human performance: Human capabilities and limitations that have an impact on the safety and efficiency of aeronautical operations.

Annexes 2, 7, 9, 12, and 18 are not included in the proposed amendment because the review undertaken did not indicate the feasibility of developing human-factors-related SARPs for inclusion in these Annexes .

The table included in Appendix C provides an overview of the ICAO Annexes that include new human-factors-related SARPs.

QUESTIONS FOR DISCUSSION

In addressing the questions put forward by the organizers of the workshop on human factors and certification held in 1993, some early and tentative answers can be advanced. They are simple and general and need refinement. It is felt, however, that often simple realities must be reasserted to further progress. Second, they are viewed as a bridge to foster sound discussion and an open exchange of information. Third, they are submitted as a preliminary attempt to provide a foundation to help orient discussion of a subject that appears, at first, abstract and controversial and on which professional and cultural biases and preferences may have influence. Short answers to the questions on which discussion during the workshop focused are proposed here.

- Who should have the authority to perform the human factors certification process when considering advanced aviation systems? Certification should remain a responsibility of the national authorities. The certification process could be facilitated if placed under the general umbrella of the ICAO human factors SARPs, as internationally accepted. Such requirements provide a basic framework and have been established with the support of the appropriate sectors of the international community. The approval process (even if only an endorsement) by ICAO would enhance the credibility of the requirements at the worldwide level.

- Where or how should certification be accomplished? Within the general guidelines discussed in the answers developed in this chapter. The review of the ICAO Annexes and the subsequent inclusion of human factors certification requirements by groups of experts from different states, called on by ICAO, allows national authorities to include the human factors certification requirements during the normal certification processes dealing with the "hard" components of an advanced system.
- What should be certified? As a minimum, those human-factors-related aspects included in the ICAO Annexes.
- Why should it be certified? The arguments advanced in the background discussion of this chapter and at the workshop as a whole are considered as the justification and answer to this question.

CONCLUSION

It is appropriate here to repeat the analogy described in the introduction to this chapter: The consideration of human factors requirements during the design stage of advanced, new technology systems may be seen as resting over a three-legged stool. The first leg, the technology that a system will utilize to achieve its goals, has traditionally attracted ergonomic considerations associated with equipment design, usually centered around knobs and dials. Lately, this view has expanded to include the so-called other important aspect of human factors study that deals with the behavioral processes of the human operators. Study in this area must be furthered. The second leg of the stool, the procedures to operate the technology, however, has been largely unaddressed. Procedures are not inherent to the technology, but must be developed. The importance of proper human factors consideration in the design of procedures cannot be overstated. Last, the third leg of the stool, the certification of personnel who will operate the equipment, is very much underway, but far from being complete. The real quest now, however, is to integrate these three legs into an indivisible one.

Finally, and most important, the workshop and its topic were extremely timely in that we are at the dawn of the most ambitious development ever undertaken in international civil aviation. This would allow us the rather unique opportunity to put theory into practice in the near future by ensuring that the concepts developed and furthered by the workshop and the follow-up are implemented in the design and certification of CNS/ATM systems described earlier in this chapter. Now is the time to incorporate human factors requirements during the certification processes of these systems. This might act as a test of the feasibility of these ideas. Such endeavors represent a challenge for the research, engineering, training, operational, and regulatory communities. However, there is certainly more to be gained by attempting to meet the challenge rather than refraining from progress by decrying the difficulties involved.

REFERENCES

Degani, A., & Wiener. E.L. (1994). *On the design of flight deck procedures.* (NASA Contractor Rep. No. 177642). NASA Ames Research Center, Moffet Field, CA.International Civil Aviation Organization. (1982). *The Convention on International Civil Aviation: The first thirty five years.* Montreal, , Canada: Author.

International Civil Aviation Organization. (1988). *International standards: Personnel licensing (Annex 1).* Montreal, Canada: Author.

International Civil Aviation Organization. (1990). *International standards: Rules of the air (Annex 2).* Montreal, Canada: Author.

International Civil Aviation Organization. (1991). *Report of the tenth air navigation conference.* Montreal, Canada: Author.

APPENDIX A

LIST OF CONSIDERATIONS WHICH ARE RECOMMENDED FOR USE WHEN DETERMINING HUMANS FACTORS ASPECTS IN RELATION TO THE CNS/ATM SYSTEM

1.) The level of safety targeted for the future system should be defined not only with reference to various system statistics, but also with reference to error-inducing mechanisms such as human capabilities and limitations as well as important individual cases.

2) Definition of system and resource capacity should include reference to the responsibilities, capabilities, and limitations of Air Traffic Service (ATS) personnel and air crews who must retain situation awareness and understanding to carry out all of their responsibilities.

3) Dynamic accommodation of three-and four-dimensional flight trajectories to provide user-preferred routings while an ultimate goal for users may initially be restricted by human capabilities and the need to organize the flow of air traffic in an orderly manner to provide separation. The transition period will need careful research and evaluation on human factors aspect.

4) Provision of large volumes of potentially relevant information to users and ATS personnel should be limited to what is absolutely necessary and mediated by methods that effectively package and manage such information to prevent information overload while providing information pertinent to particular operational needs.

5) Human–computer dialogues serving flight "air and ground requests" should be consistent in form and style with the ways in which air crews and controllers plan and negotiate.

6) A single airspace continuum should be free of operational discontinuities and inconsistencies between kinds of airspace and kinds of facilities that affect responsibilities and activities of air crews or ATS personnel at functional boundaries.

7) Organization of airspace in accordance with ATM procedures should also be readily learnable, recallable, and, to the maximum practical extent, intuitively understandable by air crews and ATS personnel.

8) Responsibilities of pilots, air traffic controllers and system designers should be clearly designed prior to the implementation of new automated systems and tools

(e.g., conflict resolution advisories, data link,   Automated Defendant Surveillance (ADS), etc.).

In addition to the analysis and assessment aimed at the specific concerns just outlined, evolution of the ATM should be accompanied by systematic pre- and post implementation evaluations of its more general human factors impacts. These assessments should encompass its effects on aircrew and ATS personnel workload and performance, as well as implications for their selection, training, career progression, and health.

APPENDIX B:
DESCRIPTION OF ICAO ANNEXES

1)   Personnel Licensing

2)   Rules of the Air

3)   Meteorological Service for International Air Navigation

4)   Aeronautical Charts

5)   Units of Measurement to be used in Air and Ground Operations

6)   Operation of Aircraft

        Part I - International Commercial Air Transport-Aeroplanes

        Part II – International General Aviation-Aeroplanes

        Part III- International Operations-Helicopters

7)   Aircraft Nationality and Registration Marks

8)   Airworthiness of Aircraft

9)   Facilitation

10)   Aeronautical Telecommunications

11)   Air Traffic Services

12)   Search and Rescue

13)   Aircraft Accident Investigation

14)   Aerodromes

15)   Aeronautical Information Services

16)   Environmental Protection

17)   Security

18)   The Safe Transport of Dangerous Goods by Air

APPENDIX C
ICAO Annexes Including Human Factors-Related Standards and
Recommended Practices (SARPs)

Annex	Part and Chapter
PERSONNEL LICENSING	CHAPTER 2 LICENSES AND RATINGS FOR PILOTS
	CHAPTER 3 LICENSES FOR FLIGHT CREW MEMBERS OTHER THAN LICENSES FOR PILOTS
	CHAPTER 4 LICENSES AND RATINGS FOR PERSONNEL OTHER THAN FLIGHT CREW MEMBERS
METEOROLOGICAL SERVICES FOR INTERNATIONAL AIR NAVIGATION	2.2 Supply and use of Meteorological information
	4.1 Aeronautical meteorological stations and observations
AERONAUTICAL CHARTS	2.1 Operational requirements for charts
UNITS OF MEASUREMENT TO BE USED IN AIR AND GROUND OPERATIONS	3.3 Application of specific units
OPERATION OF AIRCRAFT	PART I INTERNATIONAL COMMERCIAL AIR TRANSPORT AEROPLANES

CHAPTER 4 FLIGHT OPERATIONS
4.2.5.   Checklists
CHAPTER 6 AEROPLANE
INSTRUMENTS, EQUIPMENT, AND
FLIGHT DOCUMENTS
6.1.   General

VOLUME II   CHAPTER 5 AERONAUTICAL MOBILE SERVICE
Communications Procedures 5.1 General
Including Those With PANS
Status

VOLUME IV   CHAPTER 2   GENERAL
Surveillance Radar and   2.2 Human factors considerations
Collision Avoidance Systems

AIR TRAFFIC SERVICES   CHAPTER 2   GENERAL
                              2.2.   Service to aircraft in the event of an emergency
                              CHAPTER 4   FLIGHT INFORMATION SERVICE

AERODROMES   CHAPTER 9 EMERGENCY AND OTHER
VOLUME I AERODROME   SERVICES
DESIGN AND OPERATIONS

AERONAUTICAL   CHAPTER 3   GENERAL
INFORMATION SERVICES   3.4.8 Human factors considerations

ENVIRONMENTAL   PART V   CRITERIA FOR THE
PROTECTION   APPLICATION OF NOISE
                                                  ABATEMENT OPERATING

VOLUME I   PROCEDURES
AIRCRAFT AND NOISE

21

Improving Air Traffic Control by Proving New Tools or by Approving the Joint Human-Machine System?

Marcel Leroux
CENA, France

From the description of a field problem (i.e., designing decision aids for air traffic controllers), this chapter points out how a cognitive engineering approach provides the milestones for the evaluation of the future joint human–machine system.

The European air traffic control (ATC) system has entered a deep crisis. This system is unable to face a tremendous increase in demand. This is not only the consequence of its inertia; such inertia is normal for any complex system. Short-term measures to optimize the present tools appear to be insufficient; these tools have already reached the limits of their evolution capability. A large discussion on how to enhance ATC methods and tools is open. Very ambitious goals are assigned to the future systems; for example, the French CAUTRA V project plans to double the capacity of the system by the year 2005 and to significantly increase safety. Numerous ambitious projects exist, but none of them has already proven its efficiency or even its feasibility. ATC automation is short of effective solutions.

Obviously, major technology improvements (flight management system, Data Link, four dimensional-navigation, computational power) must be extensively used; but in the mean time, full automation cannot be a solution, at least for the next two or three decades. Human controllers must remain in the decision-making loop. As automation cannot replace human operators, it must assist them. This is a paradox of automation, as needed by the ATC system: As long as full automation feasibility and efficiency will not be proven (i.e., as long as we will need controllers to make decisions, even in an intermittent way) it is essential to preserve the controllers' skills. Whatever the tools that will be designed, human controllers must exercise their skills continuously.

Human operators are a factor of flexibility, of capability to deal with unexpected situations, of creativity, and of safety, thanks to their capability to compensate for the machine's failures or inadequacies. To preserve these capabilities, we may have to automate less than possible from a purely technological point of view.

However, in the meantime, the human operators are a factor of error. From this observation, and for years, system designers thought that the more human operators were put on the fringe, the more the risk of error would decrease. In fact, they add another kind of difficulty to the supervision of the initial system: the difficulty of understanding the behavior of the automatisms that partly monitor the system. Thus, automation makes the operators lose their skills, as they know less about the initial system. It creates additional sources of error and, as reported in numerous examples, the consequences of these errors are much more important than the previous ones. Rather than eliminating human operators with the consequences of depriving the joint system of major benefits and of increasing the risk of errors, it seems more sensible to design a system that is error tolerant. Such a system cannot be designed only from the technical advances: We must automate in a different way than suggested by the use of technology alone.

The consequences are very important for the design of the future system as well as for its validation. Intuitively, we must not only validate the machine components of the joint system, but we must also verify that the human–machine cooperation works well. As we need to take advantage of the human as a factor of safety and flexibility, we must prove that this requirement is fulfilled. Some aspects of validation can be less critical than with fully automated systems.

VERIFICATION AND VALIDATION FROM A COGNITIVE ENGINEERING POINT OF VIEW

Cognitive Engineering

Cognitive Engineering as a Method of Designing New Tools. Cognitive engineering has arisen from the expression of a need by numerous system designers: the need to understand what really makes the task difficult for the operators and how this difficulty impairs human performance to define the most effective aids (De Montmollin & De Keyser,1985; Hollnagel 1988; Rasmussen, 1986). Cognitive engineering is about the multidimensional open worlds that are the effective working context of the operators. Its aim is to understand and describe the present mental activity of operators, given their present tools, and how these mental mechanisms decay under time pressure, fatigue, and stress. Cognitive engineering also enables the anticipation of how new technologies will modify the activity of operators.

We must not only elicit the knowledge of operators, but, first of all, we must understand how this knowledge is activated and utilized in the actual problem-solving environment. The central question is not to identify the domain knowledge possessed by the practitioner but rather to point out under which conditions this knowledge is (or is no longer) accessible. This is the problem of defining a validity domain for human performance. Cognitive engineering must also identify and predict the sources of error and the mechanisms of error (Hollnagel, 1991).

When several agents can act on the system, under which conditions can they cooperate efficiently under time pressure? What are the mental resources that are involved? What is the cognitive cost of cooperation?

Then we have to point out how the present tools are inadequate and determine how the operators compensate for the deficiencies of their tools. Thus, we have to examine how tools provided for an operator are really used by the operators.

*Cognitive Engineering as a Method of Validating These Tools.*This whole analysis enables us to explain the cognitive model of the operator. Such a model is central to defining a global approach to the design of effective decision aids. In the meantime, this model provides a guideline for validating the joint human–machine system. Validation has no meaning per se. We have to validate according to some criteria which one has to specify. Cognitive engineering enables one to point out the weak points of the system as well as those of the human-machine interaction. Thus, it enables us to transform high–level validation requirements into relevant criteria to test the joint human–machine system. It determines which aspects of the machine or of the human–machine interaction must be verified closely so as to guarantee an effective performance of the whole system or to prevent error. Then we can determine or assess the gains along these dimensions.

We must verify that the new joint human–machine system preserves the sources of strong performance and really improves the weak points, from a safety point of view as well as from capacity. This suggests that we must assess the performance of the new system with reference to the previous one in real conditions, that is, whatever the variability of the real world is.

In the meantime, some crucial questions one should always ask are as follows: Does the new system preserve the sources of good performance of the operator? Does it preserve the capability of the operator to deal with unanticipated situations or machine deficiencies? These questions become more important as we consider cognitive tools.

Cognitive engineering provides a relevant frame to answer basic questions such as these: Who has to determine validation criteria? When do we have to experiment so as to verify these criteria? How and where to experiment? As cognitive engineering is an iterative process, at any cycle after designing new tools, the experiments enable one to determine how the global human–machine system evolves, how the bottlenecks in operator's activity evolve, disappear, decay, or are created, what kinds of problem are solved; and what kinds are created by the new system and the new human–machine cooperation philosophy; and what are the consequences on operators' training.

A FIELD STUDY: THE CASE OF AIR TRAFFIC CONTROLLERS

Method

The following is the description of en route air traffic organizer (ERATO), a project that is aimed at defining an electronic assistant for en-route air traffic controllers. This electronic assistant will include several decision aids.

At first, we have to elicit the cognitive model of the controllers. This model explains the mental mechanisms that are common to all controllers and enable them to process data and to make real time decisions. These mental processes are analyzed for the executive controller, the planning controller and then for both controllers as a whole, to assess the consequences of cooperation on mental load on global performance. The main goal remains to describe the mental mechanisms involved in the decision-making process and how these mechanisms evolve and decay under time pressure.

Decaying processes become bottlenecks in data processing and decision making. The bottlenecks assessment is a diagnosis phase: We have to point out the sources of poor and good performances of the air traffic controllers, given their actual working context. As long as the situation is not too demanding, controllers can compensate for these bottlenecks, but in very demanding situations these bottlenecks may severely impair the controllers' performance. The assessment of bottlenecks then enables one to specify the basic functions of effective decision aids. The specification of the interface also depends on the cognitive model. Prior to this specification is the definition of the working method in an electronic environment :

- What do controllers actually need to build an effective mental representation of traffic in a cooperative way and under time pressure?
- How should the system support decision-making processes?

This strong interaction between the specification of the tools and the definition of the working method is critical throughout the iterative process of defining the joint human–machine system.

This distinction enables the organizing of the problems associated with the specification of human–machine interaction into a hierarchy. The human–machine interaction must meet the following conditions, cited according to their criticality:

- Enable operators to exercise all the mental mechanisms that enable them to build the relevant mental representation of the system to be monitored.
- Enable operators to cooperate in an efficient way.
- Enable efficient inputs into the system.

These three points are necessary, but too often the third is the tree that hides the forest, although its single purpose is to ease the first two.

To define a logical representation of the model, we need to combine different laboratory logic to build a logical tool adapted to formalize controllers' knowledge. The design of the function defined earlier involves the use of an expert system that models large subsets of controllers' knowledge. The expert system provides the two electronic assistants with relevant data, so we have to face the problem of the integration of the expert system to real-time functions to design the two electronic assistants.

What Really Makes the Task Difficult

Data that depend on the time factor, have to be processed

- For their value.
- For their accuracy: When observing air traffic controllers, we found that they spent a lot of time and a lot of cognitive resources in eliminating ambiguity. A major reason controllers are often unable to make a clear assessment of the situation is based on their representation of predicted time intervals: Unless we sink into pure fatalism, we do not anticipate that an event will happen "at" a given time, but "about" a given time. The difference is fraught with consequences for the operator.
- For their availability: All the data necessary to make a clear assessment of the situation are not available at a given time; some of them may be definitely unattainable. The operator may have to make decisions in a state of partial ignorance.
- For their flow: The controllers must deal with any cases comprising the dual aspects of both time-dependent information processing and real-time decision making as: (a) dynamicity versus inertia of the system or of any subsystem, (b) lack of data versus data overflow, (c) prolonged low workload versus tasks overflow. We must also consider the question of how operators can adapt to sudden transitions from any of these aspects to the dual one, for example from a situation of lack of data to a data overflow.
- For their presentation: Data presentation is technology driven and uselessly bulky. Tasks and objectives are not well defined and may severely compete. Risk is so important that the controller has to guard

against errors from all the actors (including the machines) in the system.

A Rapid Overview of the Controller's Cognitive Model

Making Decisions in a State of Partial Ignorance. The controller anticipates, according to normal or routine behavior of the aircraft, called the *default world*, with reference to the default logic that models this kind of reasoning. This default behavior is illustrated by controllers when they use sentences such as, "Normally this aircraft going to Paris Orly will start descent about 30 NM before this fix." Controllers do not know the top of descent, but from their experience, they know that "this will normally happen about here," so they will ignore all potential conflicts that should happen if the given aircraft should start descending earlier to focus all their activity on the most probable conflicts. This is an efficient means of narrowing the range of contingencies to be examined and to increase efficiency: at first all the aircraft are processed as if their behavior always remain consonant with the "normal" behavior.

However, to process exceptions (i.e., to guarantee safety), controllers monitor sentry parameters. As long as these parameters remain in a normal range, all the previous diagnoses or decisions that are inferred from the default world remain valid. But if a sentry parameter drifts outside the expected range, then all the previous plausible inferences have to be revised; some additional conflicts can be created, due to this abnormal behavior. In normal situations, this way of reasoning is an efficient and safe way of making decisions in a state of partial ignorance. But we can observe that, in very demanding situations, the monitoring task may no longer be performed by the controllers. Thus, when outside its validity domain (in too demanding situations), this mechanism may become a major source of errors.

Making Decisions in a Fuzzy and Uncertain Environment. Very often, diagnosis is the result of an ambiguity-elimination mechanism. Even when remaining in the default world, the controller is often unable to assess a definite diagnosis. Controllers spend a large amount of time in ambiguity elimination processes. Allowing a doubt is a luxury for the controller; the mastery of doubt is an art. This is performed by associating one (or two) relevant parameter(s) to each undecided situation. To avoid a scattering of resources, these parameters will remain the only ones monitored.

Each conflicting situation, certain or potential, triggers several resolution frames. These frames are a part of the knowledge base common to all controllers. For example, let us consider the following situation: two aircraft converge over the same point, and the second climbs through the level of the first. This canonical situation can trigger three resolution frames :

- Clear the climbing aircraft directly to the requested level, under radar monitoring.
- Radar vectoring.
- Clear the climbing aircraft to a safety level until both have flown over the fix.

A part of the controller's activity is devoted to choosing the best frame. Each of these frames may be more or less demanding. In demanding situations the cognitive cost becomes a basic criterion to choose a frame. Of course while resolving a problem, the controller may have to shift from an inoperative frame to a more relevant one.

According to the assessment of the workload, the controller can instantiate a resolution frame in a more or less efficient way. The controller can also abandon a more elegant frame to shift to a more efficient one: This is the consequence of his or her own resource management policy, according to the problems occurring at a particular time.

All of these mechanisms are a part of the real-time data process. This process results in a problem driven organization of the raw data set.

Memorization Problems. The danger associated with forgetting a conflict situation, and the great number of data acquisition processes and the rapid changes in the status of these processes as time goes by make memory management a very demanding task for the controller. To solve a given conflict, the controller may have to perform actions within far-off time intervals. These time intervals may be very short; if a controller misses the right time span to act on traffic, the very nature and complexity of the problem may change rapidly. The only way to avoid this is to frequently monitor the position of the relevant aircraft. Obviously this mechanism is very costly. The controller must shift frequently from one problem to another. At each shift, he or she has to restore the resolution context. When conflicts are complex and under time pressure, this may become a critical task and lead to tunnel-vision problems.
The controller must keep in mind:

- Relevant traffic requirements, to be sure that his or her current activity complies with them.
- The decisions to act.
- The triggered frames.
- The active goals.
- The associated data acquisition tasks, as well as the targeted time spans to perform these tasks and their logical consequences.

Although data organization allows very efficient processing of information, it also furthers chunking. Early studies on controllers' memory show that experienced controllers are able to keep in mind a greater amount of data than beginners. Data relevant to conflict resolution are more easily memorized.

Cooperative Activity: Most of the previous tasks can be performed by two controllers, successively or in parallel. Mental mechanisms involved in cooperation are an essential part of the model. Efficient cooperation between the two controllers relies on three factors. They must have:

- The same skills, knowledge and training.
- The same representation of effective traffic requirements.
- Simultaneously available cognitive resources to exchange information.

When demand increases, the two latter conditions may decay so much that cooperation may no longer be effective. Numerous near misses have been reported due to cooperation failure in overly demanding situations. One controller did not even know that some tasks were urgent and important, whereas the other controller thought that these tasks were normally performed. This points out the limits of cooperation based on implicit task delegation.

Conclusion: Validity Domain of All the Mental Processes

All these mental mechanisms have a validity domain. We can easily observe how their efficiency decays under stress, time pressure, or fatigue.

For example, in demanding situations, the sentry parameters are less monitored. This may lead to errors when an abnormal behavior is not detected soon enough. We can also observe that the assessment of a given situation, including conflict detection and resolution assessment, needs a few tenths of a second, whereas it can lay for 7 to 8 minutes in very demanding situations. In this case, the controller is confronted with problems associated with numerous shifts from this conflict to concomitant ones, as described before, and the risk of error (forgetting a relevant aircraft, choosing the wrong resolution frame, etc.) is high. The validity domain of the mental processes directly depends on the number of aircraft that have to be processed by the controller. This is the reason we focused on a problem-driven presentation of the information.

The efficiency of the mental processes also depends on the capability of the operator to focus his or her attention on the relevant problem at the right time.

JUSTIFICATION OF THE TOOLS

This analysis, which corresponds to Phases 3 and 4 of the project, is the key point of the approach. It explains the reasons each tool has been designed and the improvements of the joint human–machine system that are expected. It also defines the criteria to experiment with this system. At this level, there is a deep symbiosis between design and validation. Of course, this can no longer be true during the experiments.

Guidelines

For the next decade, the nature of the data to be processed by the controllers will not significantly change, only the volume. Therefore, all the mechanisms that are inherent in the nature of data to be processed must be preserved in the new environment. However, the cues that trigger mental processing are associated with the physical environment, so they will disappear. It is necessary to make sure that the new environment will enable the operator to obtain a relevant set of efficient cues.

Cognitive tools can be specified :

- Either to improve the efficiency of cognitive resources, (or to economize cognitive resources): the information filtering functions, the extrapolation function, the memorization aids.
- Or to manage them in a more efficient way: the reminder.

Information Filtering

Justification The aim of problem-driven information filtering is to reduce the number of aircraft to be considered at one time. By splitting an overly demanding situation into several subsets of aircraft, we can expect that the controllers will have the capability to process these subsets of aircraft very efficiently. As we do not provide the controllers with the results of an automatic conflict detection and resolution, they will have to use all their mental mechanisms to assess the situation. This should preserve their skills and their capability to deal with any unanticipated situation more properly than they do now. The expected gain is that, as they will be working on appropriate subsets of aircraft, these mental mechanisms will be much more efficient than now. Thus, we will have to verify that this human–machine cooperation philosophy enhances:

- The way they anticipate in a state of partial ignorance.
- The associated sentry parameters' monitoring processes.
- The ambiguity elimination processes: The definite assessment of the situation should be made earlier than now.
- The choice of the relevant resolution frame should be made earlier than now and instantiated in a more "elegant" way
- The cooperation between controllers should be improved. The information filtering is supposed to enhance the definition of the mental representation. Both controllers' mental representations of the situation should remain consistent over time, as they will be able to update it very easily.

*Design Problems.*The role allotted to the expert system is to provide the electronic assistants with adequate data to show how to organize the raw data in a problem-driven way. Information filtering techniques are under dispute (De Keyser, 1987). The point are how to make sure that the operator will not need data that is hidden by the system. Data retention and access should not be a source of errors.

To answer these precise questions, the expert system must not only encode an exhaustive model of the controller; we must also carefully define its role in the system.

*Description of the Expert System.*The first version of the expert system included about 3,000 Prolog first order rules. It processes the same set of data as the controllers have to process now (i.e., the information from the strips) and, when available, the radar information. It includes two main modules. The first computes the default representation of each aircraft. From this representation, the second module associates to each aircraft its relevant environment, called the *interfering aircraft subset* (IAS).

This environment is composed of:

- The subset of all conflicting aircraft. These conflicts may be certain or potential. This subset is not determined by means of a pure mathematical computation, but rather according to the current expertise of controllers.
- The subset of all the aircraft that may interfere with a "normal" radar resolution of the conflict; that is, all aircraft that may constrain conflict resolution. A normal resolution is a solution that is consistent with the current know-how of controllers.

The relevant environment of an aircraft is typically a problem-driven filtering of information. The IAS represents the relevant working context associated with an aircraft. Such an environment embodies traffic requirements and all information that may be useful to fulfill these requirements. The number of rules is explained by the need to represent current knowledge of the controllers to make sure that information filtering really meets the controller's needs.

The definition of the relevant environment is "according to the traffic requirements, provided the EC works normally, he may need all, or a part of, the displayed data, but he will in no way need any other data."

*Discussion..*The discussion on the exhaustiveness and the relevance of data filtered by the expert system is central.

- The first answer includes in taking into account the default behavior of the aircraft in a more "prudent" way than the controller. This will result in the display of some aircraft that may not be relevant for the controller. In the most demanding situations (more than 25 aircraft in the sector), the most numerous IAS never include more than 12 aircraft. If one or two additional aircraft are displayed this is not really a problem. In all cases, the number of aircraft displayed as a result of information filtering remains lower than the maximum efficient processing capability (about 15 aircraft), where as the initial number was above this figure.

- The system detects all potentially abnormal behavior of an aircraft, in order to advise the controller as soon as possible and to update information filtering accordingly. In future versions, this mechanism should be performed using FMS/Data-Link capabilities.

- These two first answers do not really solve the problem. The knowledge elicited in the expert system defines a set of normal behaviors of the controllers. However, whatever the number of rules can be, it is impossible to represent the whole knowledge of all the controllers. Should we be able to do this, we should have to deal with controllers' errors or creativity. The solution defined in ERATO consists of considering the expert system as a default representation of the controllers. To guard against the consequences of human error or creativity (i.e., unexpected behavior), a monitoring process exists. This process is inspired by the natural sentry parameters' monitoring process of the controllers. This monitoring process will detect any discrepancy between the actual position of all aircraft and any of the possible positions that could result from a normal behavior of the controller. When necessary, this process will trigger an alarm to advise the controller that the previous information filtering is no longer relevant and has been updated. We have to make sure that this mechanism is efficient in demanding situations, and that the controller is not overwhelmed by the warnings. In other terms, the expert system must be accurate enough.

This monitoring process associated with the expert system allows the electronic assistant to smoothly adapt to operator error and creativity. Such information filtering is error tolerant.

These results are used by several functions of the electronic assistant: simulation of problem's resolution, extrapolation, memorization aid, data transfer from one device to the other, and cognitive resources management aid. The problem-driven information filtering allows the controller to concentrate on well-formulated problems to operate in a more efficient and creative way. This function substitutes a set of easily manageable problems for the initial complex situation. The basic information filtering will be used by all the functions of the electronic assistant.

The Extrapolation Function

Justification. This function will enable both controllers to improve problem formulation. As long as the aircraft are not in radar contact, the controller has to process data from the strips. In demanding situations, it is easy to observe that the strips are not read in an exhaustive way, which can cause severe errors. This function will display the situation at any future time, taking into account the default behavior of all the aircraft. It substitutes a graphical display for an alphanumeric representation of data. This will improve the choice of the right resolution frames. It will also enable the controller to assess more easily where uncertainty lies, and consequently to point out more efficiently the parameters that are relevant to eliminate ambiguity.

Design. is commonly admitted that operators spend a significant part of their activity in compensating for tool deficiencies. An ill-adapted interface can significantly devalue the results of information filtering. We have suggested that very often the referential used by the controller is not a temporal one but a spatial one: the question "When will you act on this aircraft" is answered "There." Thus, the interface will enable the controller to drag an aircraft along its trajectory with the mouse; all the other aircraft will move accordingly. This interface responds in the way the controller anticipates. In some situations, the controller refers to a temporal referential. The interface will display simultaneously the time corresponding to the simulated position of the aircraft. However, should this interface have only one of these referentials, the controller should would have to mentally convert distances into time intervals. In demanding situations this could represent a significant additional workload.

Simulation Functions. These functions will allow the controller to experiment with different resolution frames and to answer questions such as "What would happen if I climb this aircraft to...," or "Is it better to set course directly to..." The expert system will deliver the simulated information filtering. These answers will be updated until the controller has made a decision. For the time being, the controllers have no tool that assists them in performing this task.

Memorization Aids

The controller will have the ability to indicate the trajectory section where he or she intends to vector an aircraft. When the aircraft flies over this point (or abeam), a warning will be triggered. Then, after having consulted the filtered information, the controller will initiate his or her decision. This should solve both problems of keeping in memory the "right time to act" and context-resolution real-time updating.

Improving the Management of Cognitive Resources: The Reminder

*Justification.*We observed how it may become difficult for the controller to focus complete attention on the right problem at the right time. The reminder consists of a specific window of the electronic assistant where each problem will be associated with a label. A problem is defined as a conflicting situation involving two or more aircraft. The labels are positioned according to urgency. The display of the relative urgency of problems should enable the controller to avoid wasting cognitive resources on nonurgent and unimportant tasks as the short-term situation decays. In normal operations, this should allow the controller to objectively manage all cognitive resources.

The aim of the reminder is to show what the traffic requirements are and their urgency to the two controllers. There are several ways to split a given situation into relevant problems. This variability can be observed for several different controllers as well as for a given controller, according to his or her cognitive resources management philosophy. The more the situation is felt to be demanding, the more the controller will split it into "little" problems and solve these problems in a very tactical way: with short-term solutions. If the controller feels that the situation is mastered he or she will consider these elementary problems as a part of a whole and solve them in a more strategic way. Statistically, about a quarter of the problems may be more or less broken down, whereas the other three quarters are always described in the same way by controllers.

Design Problems The reminder will propose labels by default. Most of these labels will correspond to the effective needs of the controller; some will not. Thus, the controller will have the capability of editing these labels.

When a situation can be described as a single problem or as several problems, the reminder will propose the simplest situation for three reasons:

- It corresponds to the need of the controller when the situation is the most demanding. If the controller wants to concatenate some subproblems, the situation is probably not too demanding, and there is time to do this properly. This is a means of avoiding clumsy automation: The interface assists the operator in the most demanding situations.
- It is easier to concatenate than to split labels.
- This enables more information to be pointed out on each subproblem.

VALIDATION TECHNIQUES

Classically we define dependability as that property of a computing system that allows reliance to be justifiably placed on the service it delivers (Laprie, 1987).

We can point out four classical methods regarding dependability procurement or dependability validation: fault avoidance, fault tolerance, fault removal, and fault forecasting. These definitions can be applied to a complex heterogeneous human–machine system such as the ATC system and to any of its machine components. In the first case, the users are the airlines (or their passengers), where as in the second case the user is defined as the controller or any subsystem.

The specification of decision aids relies on a philosophy of future human–machine cooperation, whether this philosophy is clearly defined or not. The central question is to make sure that this cooperation fulfills the initial requirements regarding capacity and safety. To answer this question, we have to choose the right parameters to be evaluated and then determine the minimum set of experiments to get a significant amount of data (Woods & Sarter, 1993). Some basic questions must be answered about the robustness of the joint human–machine system: Is it error tolerant (Foster, 1993)? Does its organization allow a quick and easy correction of errors (Reason, 1993)?

The design of decision aids implies an analysis, either implicit or explicit, of operator deficiencies and of the most effective means to compensate for these deficiencies. The ultimate step of the verification and validation process should be the verification of these initial assumptions (Hopkin, 1993).

The validation of the electronic assistant is threefold (Leroux, 1993).

- The cognitive model has been verified through the analysis of the behavior of air traffic controllers during simulations of demanding traffic (Leroux, 1991). The observation of the controllers and their comments enabled us to point out the mental processes and the bottlenecks as described in the model.
- The philosophy of human–machine cooperation and the specification of the electronic assistants have already been presented to more than 180 fully qualified controllers.

The expert system is used to provide the various functions of the electronic assistant with adequate data. To validate this knowledge-based module, we have to check that the outputs are acceptable by controllers in real conditions; that is, that they always include at least the minimum set of information. Although the validation of the knowledge-based module has not been carried out, it has been successfully confronted to the most usual problems. However, we will have to validate very carefully the monitoring process, as it guarantees the relevance of the outputs. We must prove that this mechanism detects all discrepancies between the observed state of the world and the expected one. The trust of the operator on the machine relies on the efficiency of this mechanism. Thus, the validation of the knowledge-based module will be twofold: from a pure dependability point of view and from a human factors point of view. This will be performed by two different teams with different techniques to experiment. Validation of the Problem-Driven Filtering Algorithms.

This validation was carried out in 1993, using five very demanding traffic samples from five different upper sectors. These experiments involved 44 FPL controllers from the five French control centers.

The protocol consisted of interviewing controllers at identified moments of the simulations to ask them the following questions on target aircraft (about 25 per traffic sample) :

- For each aircraft, describe what its behaviour will be inside the sector.
- Describe all the conflicts (certain or possible).
- For each of these conflicts, describe the resolution frames that can be used.
- For each of these resolution frames, describe the aircraft that are, or may be, a constraint.

The results were then merged and compared to the results of the algorithms.

The first result is that the algorithms are generic to all French upper sectors, required only very minor adjustments. The second result is that the rate of filtering is correct. Despite the fact that the number of aircraft in the sector was between 30 and 40 (about twice the present peak of demand), on 122 target aircraft, there were only 4 IAS higher than 15 aircraft (15 aircraft is considered by controllers as their maximum efficient processing capability), whereas 92 IAS were less than 8 aircraft.

The first level of validation of the Electronic Assistant will consist of the following:

- Experimenting the interface of each function.
- Verifying that it really improves the target bottlenecks.
- Making sure that it does not decay some sources of good performance. The criteria for experimenting with these functions are directly issued from the cognitive model, as previously described.

Then we will have to assess the validity domain of each function: Are they really efficient for situations where the controller needs an effective aid? After that, we will have to validate the electronic assistant as a whole to analyze if the function as a part of a whole loses properties or acquires unexpected ones. Finally, we must assess the performance of the joint human–machine system and answer questions such as these:

- How is this cooperation philosophy accepted by controllers?
- How will it modify the activity of the controllers?
- Does it enable them to work in a more efficient and creative way?

- Does it provoke a loss of vigilance or of skill?
- Does it improve the global performance from capacity and safety points of view?
- What are the consequences on training?
- Does it enable a progressive and "soft" integration of technological advances in avionics?

The First Evaluation Campaign October 1994-April 1995

The Role of the Cognitive Model in Defining the Experimental Protocols The cognitive model provides a guideline for evaluating the joint human–machine system. It enables the transformation of high-level validation requirements into relevant criteria to test the joint human–machine system. It determines which aspects of the machine or of the human–machine interaction must be verified closely to guarantee an effective performance of the whole system or to prevent error. Then one can determine or assess the gains in these aspects.

We must verify that the new joint human–machine system preserves the sources of good performance and really improves the weak points from both a safety and a capacity point of view. This suggests that we must assess the performance of the new system with reference to the previous one in real conditions; that is whatever the variability of the real world is, in demanding and very demanding situations. The experiments should enable one to determine how the joint human–machine system evolves; how the bottlenecks in the operator's activity evolve, disappear, decay, or are created; what kind of problems are solved and created by the new system, and what the consequences are for operator training.

The experimental protocol was based on the comparison of the performance of controllers in demanding or very demanding situations, in the present environment and in the future one.

The experiments involved nine teams of two controllers. Each team of two controllers was confronted with four traffic simulations, two of them using a conventional environment (paper strips and a radar display), the two others using the electronic assistant. The average duration of a simulation was 45 minutes. A video record of the controllers' activity and a computer record of all the actions on the system were made. Then controllers had to fill in a questionnaire on the functions: For each function they had to answer questions on the relevance of the function (regarding safety, capacity, cooperative activity, etc.), its usability, the function as a (potential) source of error, and their opinions on planned improvements of each function. Lastly, they had to comment on their own activity in a semidirected way, using both the video replay and the computer replay of the simulation. The interview was orientated on the workload as it was felt by both controllers, and focused on the consequences for conflict detection, conflict resolution, and cooperation, to get data on resource management and to compare with equivalent situations in a conventional environment. The average duration of the interviews was about 5 hours for each traffic simulation.

Results

As mentioned previously, the experiments have feedbacks at all levels of the project. Whenever a function is not satisfactory, it must be assessed whether it comes from the working method, the interface design, the algorithms, the specification itself, or from bottleneck assessment. Whatever the reason is, we have to refine the cognitive model. Different opinions of controllers in the questionnaire were set against the present state of the cognitive model to improve it and to understand the genuine reasons behind those diverging opinions.

A statistical analysis of actions on the interface, with reference to the number of aircraft, problems, and so on, is cross-checked with the comments to analyze how each function is really used over time when the demand varies. Combined with the analysis of the video record and of the interviews, this statistical analysis enables us to understand how new tools will modify controller activity, to verify if the decision aids enable them to work in a more efficient and creative way without provoking a loss of vigilance or of skill, and to make sure that they improve the global performance of the joint man-machine system, from capacity and safety points of view.

The Reminder. The reminder is the main factor in increasing productivity, but its efficiency relies on efficient cooperation between controllers. Three agents can act on the reminder: the algorithms and both controllers. The experiments showed that the role of each agent needed to be clarified. The refining of the reminder has been an exemplary illustration of the methodology used by the team.

The starting point was the confidence that the reminder was central both in building the mental representation of traffic and as a support to cooperation.

From these initial assumptions we worked with controllers to determine what the most efficient working method with the reminder should be. From that working method, we inferred the specification of the second version of the interface of the reminder. A mock-up was designed and connected to the whole interface; enabling us to test the working method and the interface under realistic conditions and to refine both.

The Cues. The electronic environment deprives the controller of all relevant cues that were familiar to him or her in the previous environment. This is not only a matter of nostalgia. These cues trigger all the mental processes that enable the controller to build his or her mental representation of traffic. Building this representation is not a deterministic process; it is guided by the detection and the processing of these cues. So it is critical to verify that:

* The interface provides relevant cues.
* The controller can easily detect and process them.

This task is central to the interface specification. It is made more complex in an electronic environment because of the saturation of the visual channel. Thus we have to define a global policy of data presentation on the interface, and in fine, a working method relying on relevant visual circuits on the interface. This definition process will make use of more advanced studies on the attention and the focus of attention.

The Second Evaluation Campaign October 1997-May 1998

This campaign has involved eight teams of three controllers, each team for 3 weeks. The aim was to verify that :

* The updated version of ERATO enables controllers to build their mental representation of traffic in a more efficient way.
* The immediate consequence is a major increase in capacity.
* The transition from the present system to the new one was possible from an operational point of view.

The controllers were full performance level, and none of them knew ERATO before these trials. They were trained for two weeks. The third week was dedicated to the trials.
The trials were focused on:

* How the system supports cognitive activity in demanding situations.
* How the system supports cooperation in demanding situations.
* The usability of the human–machine interface.
* The fatigue: Are people able to control more than the present peak of traffic for 8 hours a day?

or the cognitive activity and the cooperation trials, each team of controllers was confronted with two traffic simulations. The average duration of a simulation was 90 minutes. A video record of the controllers' activity and a computer record of all the actions on the system was made. Controllers had to comment on their own activity in a semidirected way, using both the video replay and the computer replay of the simulation. The average duration of the interviews was about 4 hours for each traffic simulation.

The usability trials consisted of interviews on the relevance of the functions and of the data that are shown by the system, the ease of the processing of these data, and of specific tests.

The fatigue trials consisted of a normal working day, in which the traffic was increased, to be at any time more demanding than the most important peak of traffic during the year 1997.

The data coming from the video records, the audio records and the computer records are encoded as predicates that enable precise analysis of the interaction among individual activity, cooperative activity, and the use of the interface. It also enables us to cross-check the opinion of controllers on the functions with the way they really use it.

Even if the data processing is in progress, it is now clear that with appropriate training, all the controllers succeeded in getting acquainted with the system in less than 10 days. For some of them 5 days was enough. Second, the acceptance of the system is total, even if we must improve its usability. The informal results of this trial campaign are so positive that we can now plan a final test on real traffic in early 2000, which will precede the operational deployment by the year 2001.

CONCLUSION

The cognitive engineering approach applied to ATC proves to be successful. The backbone of this approach is the cognitive model of air traffic controllers. The activity of the design team is driven by this model, to infer the specification of the tools, to search for relevant criteria to evaluate the new joint human–machine system, or to improve the cognitive model itself.

Identifying at the outset what is critical in the operators' activity makes it possible for the designer to focus on relevant questions during any validation phase. The validation phases must happen all along the design process, and they must begin before the design process during the preliminary research phases. The validation must cover all topics, from the usability to the cognitive activity or the cooperative activity. It must include the transition problems. From this point of view, the validation of training is an essential part of the validation of the joint man–machine system, as training must not be seen as a way of compensating for design inadequacies, but as the continuation of the principle-driven design of the system.

REFERENCES

De Keyser, V. (1987). How can computer-based visual displays aid operators? *International Journal of Man–Machine Studies*, 27, 471–478.

De Montmollin, M., & De Keyser,V. (1985). Expert logic vs. operator logic. In G. Johannsen, C. Mancini, & L. Martensson (Eds.). *Analysis, design, and evaluation of man-machine systems*. Ispra, Italy: IFAC.

Foster, H. D. (1993). Resilience Theory and System Valuation. In J. A. Wise, V. D.Hopkin, .& P. Stager (Eds.), Verification and Validation of Complex Systems: Human factors issues. (pp. 35–60) . Berlin: Springer-Verlag.

Hollnagel, E. (1988). Information and reasoning in intelligent decision support systems. In E. Hollnagel, G. Mancini, & D. D. Woods (Eds.), Cognitive engineering in complex dynamic worlds. London: Academic Press.

Hollnagel, E. (1991). The phenotype of erroneous actions: Implications for HCI design in Weir, & Alty (Eds.), Human–Computer interaction and complex systems. London: Academic Press.

Hopkin, V. D. (1993). Verification and validation: Concepts issues and applications In J. A. Wise, V. D.Hopkin, .& P. Stager (Eds.), Verification and Validation of Complex Systems: Human factors issues. (pp. 9–33) . Berlin: Springer-Verlag.

Laprie, J. C. (1987). Dependable computing and fault tolerance at LAAS: A summary. In A. Avizienis, H. Kopetz , J.C. Laprie (Eds.), New York: Springer-Verlag.

Leroux, M. (1992). The Role of Verification and Validation in the Design Process of Knowledge Based Components of Air Traffic Control Systems. In J. A. Wise, V. D.Hopkin, .& P. Stager (Eds.), Verification and Validation of Complex Systems: Human factors issues. (pp. 357–373) . Berlin: Springer-Verlag.

Rasmussen, J. (1986). Information processing and human–machine interaction: An approach to cognitive engineering. New York: North–Holland.

Reason, J. (1993). The Identification of Latent Organizational Failures in Complex Systems. In J. A. Wise, V. D.Hopkin, .& P. Stager (Eds.), Verification and Validation of Complex Systems: Human factors issues. (pp. 223–237) . Berlin: Springer-Verlag.

Woods, D.D. & Sarter, N.B. (1993). Field Experiments for Assessing the Impact of New Technology on Human Performance. In J. A. Wise, V. D.Hopkin, .& P. Stager (Eds.), Verification and Validation of Complex Systems: Human factors issues. (pp. 133–158) . Berlin: Springer-Verlag.

22

Certification for Civil Flight Decks and the Human-Computer Interface

A. J. McClumpha
Defense Evaluation and Research Agency

M. Rudisill
NASA Johnson Space Center

This chapter addresses the issue of the human factors aspects of civil flight deck certification, with a focus on the pilot's interface to the automation. In particular, three questions are asked that relate to this certification process: (a) Are the methods, data, and guidelines available from human factors adequate to address the problems of certifying as safe and error tolerant the complex automated systems of modern civil transport aircraft, (b) do aircraft manufacturers apply effectively human factors information during the aircraft flight deck design process, and (c) do regulatory authorities effectively apply human factors information during the aircraft certification process?

PROBLEMS WITH AUTOMATION

Progressive automation is a feature of the modern civil cockpit and the trend toward more information, greater complexity, and more automated aircraft has the potential to isolate significantly the pilot from the aircraft and decrease understanding and awareness. Billings (1991) reported that the dominant cause of aircraft accidents is human error and that the most important purpose automation can serve is to make the aviation system more error resistant and more error tolerant. It is generally accepted that future cockpit developments in the civil flight deck environment will have substantial automation to provide cost-effective air transport. However, it is widely reported that the dominant factor in many civil aircraft accidents in the recent past has been the automation, and that the automation has had a debilitating contributory role (Sheridan 1990). For example, 65% to 70% of civil jet transport accidents are attributed to human error and of that the majority were controlled flight into terrain ("Human Factors Research," 1989). Accidents that are caused by controlled flight into terrain are of particular concern to human factors because a primary contributor appears to be the pilot interface to the aircraft automation ("Confusion Over Flight Mode," 1993; "French Government Seeks A320 Changes," 1993).

Pilots criticize flight deck design. Poor human factors are cited (although often not explicitly) as a contributory factor in a wide range of aircraft accidents. Often these problems are caused through the pilot's interaction with the automation. The Air Accident Investigation Bureau (AAIB) report of the Boeing 737-400 accident at Kegworth, U.K. (Department of Transport, 1989) noted that some time was lost as the copilot attempted unsuccessfully to program the flight management system (FMS) to produce the correct flight instrument display for landing at East Midlands Airport. From the cockpit voice recorder it is inferred that the first officer had selected the route page and was entering the correct information (for the destination airfield) but as an en-route point. He did not notice the inadvertent selection or understand the limitations of the available selections with respect to this information. An absence of appropriate feedback within the FMS allowed the error to remain. The first officer had failed to select the arrival airfield page; this page is similar to the en-route data page in terms of data layout and data entry. In addition, there was no clear, unambiguous indication (e.g., a title) of the selected page within the FMS. The Airbus A320 FMS is also often criticized for not providing easy and efficient information access and selection. Pilots report having to select up to five different pages to obtain information necessary to complete a single action sequence. Both examples are illustrations of violations of well-recognized and explicit guidelines applicable to the design of human–computer interfaces (HCIs).

The results of recent surveys addressing attitudes toward glass cockpits have shown that pilots report an erosion of flying skills, increased workload at critical times of flight, and a sense of being out of the loop as a result of the automation (McClumpha, James, Green, & Belyavin, 1991; Weiner, 1989). An analysis of the comments made by the respondents of the Royal Airforce Institute and Aviation Medicine (RAF IAM's) survey of pilots' attitudes toward automated aircraft highlighted two main types of comments about the HCI to civil flight deck automation (Rudisill, 1993). One set of comments relates to specific systems and specific problems with the HCI of that aircraft. Although this information is particular to an aircraft type, it indicates human factors problems that could be used to help develop the human factors assessments required within a certification program. The other set of comments, however, relate to the broader nature of the automation and the difficulty pilots experience in interacting with the automation. For example, pilots report difficulty in knowing what the automation is doing, what the boundaries or limits of performance are, and how to intervene effectively if problems arise. Pilots also report that many of these types of problems only emerge after some considerable line flying or with specific experience in unusual situations. It is exposure to these types of situations that forces pilots to appraise their ability to understand and feel competent with the automation. These issues present considerable difficulty for a manufacturer in terms of how to design aircraft systems and interfaces for the pilot. Automation also presents difficulties for a regulatory authority in terms of how to certify as safe and error tolerant these aspects of the design.

Billings (1991) stated that only a subset of conditions and failures can ever be evaluated in a system as complex as modern civil flight decks. It is not, therefore, surprising that Wiener and Curry (1980) concluded that the rapid pace of automation is outstripping the pilots' ability to comprehend all the implications for aircraft performance. They reported that an uneasy balance now exists between, on the one hand, accepting aircraft in which all the implications and potential for debilitating behavior are unknown and, on the other hand, training to support the crew tasks when operating these aircraft.

CIVIL AIRCRAFT DESIGN AND REGULATORY AUTHORITIES

Regulatory authorities are responsible for providing certificates of airworthiness for all aircraft operating within their airspace. To that end and in that role they act as the arbiter of acceptability for the aircraft. They assure, among other things, that the aircraft performs to the requirements for engine power and thrust and that the airframe structures have the appropriate mechanical integrity. They assess, to stringent requirements, the ability of the aircraft to be operated to a safe landing under circumstances unlikely to be met in normal flying. Therefore, the essence of regulatory authority certification is whether the aircraft meets the minimum acceptable standards that are identified as requirements for certification. This is achieved by the authority defining a number of requirements and criteria and applying them during a systematic evaluation of the aircraft. Regulatory authority certification assesses whether the aircraft achieves the minimum acceptable standards that are identified as requirements for certification.

Aircraft manufacturers take great care in the design process leading to certification. The Boeing 757 had 3,942 hours of simulator missions flown before certification in 1982. The Boeing 767 program involved 5,348 hours of simulator flying and the 747–400 involved 5,981 hours. A total of 369 pilots from 29 airlines flew the 747–400 simulator and nearly 200 nonflying personnel were involved in the design process ("Human Factors Research," 1989). A regulatory authority may require more than 1,000 hours of simulator flying and more than 100 actual landings before certificates of airworthiness are issued.

There is no doubt that certification of civil aircraft by regulatory authorities is approached thoroughly, diligently, and carefully and that aircraft manufacturers are responsive to certification requirements and that they design for safe, error-tolerant systems.

Aircraft design (and flight deck design, in particular) is evolutionary. This evolutionary approach carries with it two strong implications from a human factors perspective. One is that a flight deck will not be substantially different to previous versions. This will help ensure that pilot training for that aircraft will be kept to a minimum (i.e., cost-effective). The second and more worrisome implication is that flight deck design appears to have evolved without systematic adherence to any rigorous human factors scientific base and that this situation has been perpetuated because of the evolutionary nature of aircraft design. Norman (1991) stated his position with regard to this issue quite strongly. He argued that cockpit procedures, flight instruments, and regulations are "guided more by instincts, anecdotes, and reactions to individual incidents rather than by systematic, scientific analysis of the issues". There does appear to be a failure to comply with a basic tenet of human factors. This tenet dictates that the user interface (which in this context includes procedures, instruments, and regulations) should be derived from an understanding of the tasks, procedures, and information requirements of the pilots. This understanding and information should then be mapped to the interface design in a way that exploits the pilot's sensory, perceptual, physical, and cognitive processes. However, certain flight situations (e.g., engine fire) do require specific actions by nominated crew members in clearly defined ways and there is a legal responsibility on the crew to follow set operating procedures. Unfortunately, even in this situation, nowhere is it explicit that these procedures, instruments, and regulations have been derived from an understanding of the pilot's task, procedures, and information requirements and mapped to the design of the interface in a way that supports the cognitive activities of the pilot.

Existing certification requirements do, however, establish the need for specific information to be available to the crew. Unfortunately, neither the form nor the content of this information, nor the pilot's ability to access it easily and efficiently to meet task needs, appear to be established as requirements for certification. A recent research program at the NASA Langley Research Center has been investigating the information pilots use, for how long, and when. It is curious that fundamental research of this nature is being undertaken as recently as 1991. It could be argued that there has been a fundamental shift in the pilot interface on flight decks from one that is essentially electromechanical to one that is essentially software based and automated, and consequently this research needs to be undertaken. Nevertheless, the pilot's task has not changed appreciably as a result of this change in technology. Therefore, the information requirements of pilots have remained essentially the same and should have served as the basis for design, independent of the implementation technology.

There also does not appear to be a requirement levied on the aircraft designer by the regulatory authority for an aircraft design to conform to detailed and explicit requirements for good quality human engineering. The specific number of hours flown or tested in a development and certification program is not necessarily an indication of good design. What is examined, found, and remedied during the design process is of greater interest. However, information on the form and content of specific human factors examination does not appear to be readily available. More specifically, the pilot interface to aircraft systems does not appear to have been subjected to a rigorous and critical examination based on existing human factors and human–computer interaction methods, standards, or guidelines. There does not appear to be a requirement to evaluate systematically the aircraft for adherence to good human factors.

The certification authority, as the arbiter of civil aircraft acceptability, has the responsibility for ensuring that the pilot interface provides the necessary facilities for safe flight. Besides ensuring aircraft conformance to the traditional requirements for aspects such as appropriate propulsion and structural integrity, they should establish conformance for the pilot interface to requirements for good quality human engineering based on established human factors design principles. Mechanisms and processes to evaluate human factors systematically do not appear to be presently in place within these authorities, and it is suggested that they should be put into place in such a way as to help establish flight deck design based on sound human factors principles.

The certification process has the flexibility to allow a manufacturer to request a derogation to the certification requirements. Derogations or relaxations to requirements can be given at the discretion of the regulatory authority, but only when it is considered safe to do so. The question asked by a regulatory authority is whether the aircraft is likely to be as safe as a previous version. Unfortunately, relaxations to human factors requirements are strong candidates for derogations because of the often subjective and generally implicit nature of the human factors evaluation. There is also a concern with the possible cumulative effect of many, but separate derogations on pilots' performance. The possible effect of any derogation on the pilots' ability to operate the aircraft safely should be considered as part of the derogation analysis. However, as a result of the generally subjective nature of the human factors assessment, particularly with regard to the HCI, it is perhaps not surprising that human error is identified as the major contributory factor in aircraft accidents.

We have described the exacerbating effects of progressive automation and the apparent absence of systematic human factors practice in the design of civil aircraft flight decks. The apparent absence of systematic evaluation of human factors by regulatory authorities for aircraft certification has also been commented on. Two questions remain: Is information available within the human factors community to support the processes of flight deck design and aircraft certification? What is required to achieve the goal of incorporating human factors into the design and certification processes?

ROLE OF HUMAN FACTORS

Human factors contains concepts, principles, and methods that are sufficiently objective and advanced to meet the need for certifying aircraft to specific standards, requirements, and criteria. This information is derived from the scientific base of research from experimental psychology, human factors, and physiology. It describes human sensory, perceptual, and motor performance, and, to a lesser degree, cognitive performance. This empirically based information is potentially available for the analysis and certification of sensory and perceptual characteristics, manual operation, and the physical effects (e.g., noise, vibration, anthropometry, biomechanics) operating within the pilot interface. For example, anthropometric design standards can precisely define acceptable reach requirements for appropriate percentiles of the population. It would be highly unusual for an aircraft flight deck to be designed that prevented the pilot from correctly reaching required controls and displays. Regulatory authorities would not certify such a design. A detailed knowledge of vision and character readability allows the equally precise definition of alphanumeric font size, based on the seated eye position of the pilot and copilot. Reach envelopes, workspace layout, display readability, and the acceptable limits of force functions can all be predicted, established, evaluated, and certified with the specific consequence of derogations predicted. Many references are available for the design of these system characteristics and these could be adapted for use in certification programs (e.g., Boff & Lincoln, 1988; Department of Defense, 1989).

The situation is rather different for design and certification of the aircraft HCI on today's automated civil aircraft. Automation per se and software-based interfaces in general have changed the locus of control within the cockpit from the physical domain to the cognitive domain. Therefore, much of the HCI today on the flight deck is related to human cognitive performance. It is this particular domain of performance that presents the true challenge for certification programs that deal effectively with human factors.

To date, regulatory authorities have generally based examination on theory and models and then tested to criterion performance. Hence, mathematical modeling is used to predict the likelihood of system failure. There are two problems with this approach when applied to human factors certification, especially certification of the HCI. First, mathematical modeling will be used to predict the failure of individual elements of the system to a 10–8 error rate. Assessments of human error as a result of the system design are the domain of the certification test pilots who will assess how critical a potential error is likely to be. An assessment may suggest that a system has a low error rate and is therefore acceptable. What does not appear to be assessed is the contribution of the error potential of an individual system component with all other aircraft systems, or the effects of combinations of errors created through poor interface design. It is unclear if mathematical modeling is an appropriate technique for modeling and hence predicting failures of the pilot–vehicle interface. Indeed, the approach generally taken during HCI design involves designing the HCI by following established good design practices, rather than modeling the HCI system after design and then assessing error potential.

It is our contention that regulatory authorities do not yet possess the understanding to deal effectively with the demands that cognitive systems require for effective review and assessment. A typical fallback position of a regulatory authority is to adopt the blame and train philosophy. Human error is often treated as a training issue rather than as a reflection of underlying design deficiencies (cf. recommendation from AAIB report on Kegworth accident). However, the difficulty here is that the current methods and practices used by regulatory authorities do not appear to address the HCI aspects of the system design. Furthermore, they do not appear to support this in a way that can lead to improved understanding of the potential problems and establish standards and requirements for design that can be used by manufacturers. The cognitive aspects of system performance require quite different methods and techniques from those applied to the classic domains of human engineering. It is the cognitive aspects of the effect of the system that require attention and impose different methods for certification. However, with the status and form of the science of HCI a substantial change in philosophy of flight deck HCI, design certification by regulatory authorities is required. These issues are dealt with in the following sections.

Contributions of Cognitive Psychology

In a paper addressing the role of cognitive psychology in the design of useful and usable systems, Landauer (1991) addressed the current limitations of theory. He suggested that a theory of human–computer interaction good enough to allow prediction of system characteristics for design is likely to be impossible because the behavior of systems and the cognition that supports user interaction with the systems are complex. He argued that laws, such as Hick's law, Fitts' law, and laws of visual and auditory perception have indeed had an impact, albeit minor. The point to emphasize from this is that a theory that allows predictions and design decisions from domains other than classic human factors is unlikely to be available for sometime, or indeed ever.

The data, methods, tools, guidelines, and standards that would serve as the basis for analysis of more cognitive aspects of pilot operations and design of the HCI to support them do not yet exist in a form readily-usable for certification. In a sense, this is understandable given the complex nature of the underlying psychological domain. As has already been mentioned, sensory and perceptual processing and manual control are complex behaviors, but they are also more amenable to direct investigation. Systematic investigation of cognitive processing and development of coherent theories has been more recent. A base of objective data and sufficiently complex unifying theories of cognition and an appreciation for how this information should be reflected in an HCI are now being formalized.

However, what is presently available for use in HCI design and certification is an explicit representation of what constitutes good HCI design practice with supporting methods, and validated guidelines and standards based on substantial supporting research. This work occurs in several forms: (a) a software design model that explicitly identifies HCI design tasks and requires iterative prototyping and review; (b) HCI guidelines and standards that embody principles of good HCI design derived from empirical research; and (c) instruments and methods, such as HCI checklists and usability testing tools, that may be used for HCI evaluation and certification. It is this approach that can provide the foundation for effectively integrating HCI knowledge into the civil flight deck design and certification processes.

HCI and the Design Model

A feature that is often incorporated within the military weapons system procurement cycle is an explicit model of the design that includes consideration of the HCI within the design cycle. These models require that the HCI design is iterative, with prototyping and formal assessments integrated throughout the design cycle. Requirements for acceptance test procedures are also identified early. The basic, general model of this procurement is concept studies, feasibility, project definition, development, build, and in-service support. Human factors can, in principle, be integrated into this cycle at each stage with differing specificity and requirements at each stage. For example, NATO Standardization Agreement 3994 AI (1999) is an example of a procedural guide for the integration of human engineering in system design and evaluation.

At present, the civil transport aircraft model of procurement and production is different with different roles and responsibilities for system designers, airlines, and regulatory authorities. There appears to be no explicit requirement for HCI prototyping and review by the user population. Certification typically occurs late in the design and development process and certification approval will be sought within the final phase of the development program. Furthermore, certification is formally called up late in the development schedule. If a problem is identified at this stage, it is usually too late either to adequately address the problem or implement a redesign. However, it is noted that certifying authorities can participate in reviews during the aircraft design process. In this way, design features that would prevent certification of the aircraft could potentially be noted and corrected.

However, the international nature of aircraft manufacturing, purchasing, and regulation exacerbates the problem. Typically, a regulatory authority from a different nation than the airframe manufacturer will not formally interact with the manufacturer until the point at which certification is required. Experience from work with military systems procurement strongly suggests that this approach is likely to be problematic and problems identified at this stage of certification are usually too costly to remedy. The model of development within civil flight deck design needs to be updated to include explicit consideration of the HCI. Aircraft regulatory authorities need to be more actively involved in the design and development process.

HCI Guidelines and Standards

A trend within the human factors industry has been the development of standards for HCIs. In recognition of the importance of the user's interface to the system, Apple Computers requires all applications for their products to adhere to their HCI guidelines and standards. More recently, other HCI guidelines and standards have been published and are beginning to come into wide use in commercial applications (e.g., OSF Motif, Open Look, Presentation Manager).

A design practice that has emerged in projects involving software-based systems is the provision of both project-specific and general HCI guidelines and standards. The purpose of these guidelines and standards is to provide a human-factors-based, well-designed, common HCI across all user interfaces to the system. These guidelines and standards can be levied on all system components, even if the components come from several suppliers. However, this well-recognized and explicit HCI design practice does not appear to be integrated either into the airframe manufacturers' flight deck design approach or into the regulatory authority's acceptance testing and certification procedures. There is evidence to show that many designers are not aware of the many relevant HCI design guidelines and standards. There is also evidence that many designers view guidelines and standards as a hindrance that constrains design rather than as an aid that provides structure and boundaries to good design practice.

Within airframe manufacturers, the problems are further compounded because a flight deck will consist of a mix of avionic equipment from a disparate range of suppliers. Each separate avionic subsystem supplier is unlikely to have either an in-house guidance for the HCI design or its own in-house proprietary standard for the look and feel of the system. Major airframe manufacturers integrate many subsystem components into a coherent whole, yet little or no effort is made to create a common look and feel across equipment. Often a flight deck will have a proliferation of HCIs, a unique one for each piece of avionic hardware (e.g., there are cases of two different keyboard layouts on the same flight deck). Many general HCI design guides and standards are available that could be used for flight deck HCI design and certification (e.g., Boff & Lincoln, 1988; Department of Defense 1985,1989; Section 5.15; NASA, 1991; Smith & Mosier, 1984).

HCI Evaluation Methods: Checklists and Usability Testing

Many HCI guidelines and standards can be represented as high-level HCI evaluation criteria embodied within checklists. HCI checklists help to formalize design evaluation by domain experts (i.e., human factors HCI design practitioners). For example, Ravenden and Johnson (1990) identified nine checklist criteria: visual clarity, consistency, informative feedback, explicitness, appropriate functionality, flexibility, control, error prevention and correction, and user guidance and support. More than 100 questions are asked in the checklist and the answers help to provide a standardized and systematic method of enabling those evaluating the interface to identify problem areas. Shneiderman (1986) identified 20 major items in a short form of a generic user-evaluation questionnaire for interactive systems. The Shneiderman questionnaire identifies a range of questions with bipolar semantically anchored items that require users to evaluate aspects of system performance. These examples are representations of generic forms of audits that can be used in a variety of ways.

Usability testing has become one of the most well-established methods of product evaluation used by human factors engineers. A recent series of studies funded under ESPRIT Project 5429 (1992: or MUSIC; or Metrics for Usability Standards in Computing) examined and identified methods and tools for measuring the usability of products that require human–computer interaction. It is not the intention of this chapter to review HCI procedures of this form but simply to point out that considerable work has been underway for sometime in this area, that the information is available, and that it is sufficiently refined to be directly applicable to the civil aviation context.

ORGANIZATIONAL IMPLICATIONS

The point has been made that regulatory authorities will need to adjust to the different requirements that the cognitive elements of the system impose for effective design assessment and hence certification. At an organizational level, though, the form and function of human factors also need to be addressed. Rouse (1991) argued for the concept of human-centered design and made two important points. He suggested that not only are the concepts, principles, and methods in human factors not sufficiently advanced, but more important, most currently available principles have little impact on design. This, he argued, is the result of the massive influence of the organization as a whole and the structure within which human factors is allowed to function. This has obvious parallels to Billings' (1991) human-centered approach to automation. Whereas Billings refers to users primarily as pilots, Rouse took a system orientated view and referred to users as any stakeholders within the ultimate system (e.g., regulatory authorities, airlines, maintainers, sales staff, pilots, etc.). Rouse argued that human factors will have greater impact when these wider implications are considered. To this end we believe that regulatory authorities require a substantive change in philosophy about human factors. Four organizational changes are required.

The first organizational change is the recognition and integration of the developments within human factors in general, and HCI design in particular, into the aircraft design and certification processes. This is a key message of this chapter. There is good evidence that pilots, engineers, and system designers do not typically understand the benefits that can be accrued from a formal human factors evaluation. Regulatory authorities need to recognize the unique contribution of human factors and formally establish the mechanism for its incorporation into the certification process. It should also be recognized that the science base of HCI can be of a qualitatively different form to other more entrenched areas of certification. Regulatory authorities can set requirements by specifying good quality human factors and HCI design, even if these are set at a high level. Designers will have to work to meet those requirements or provide substantial and appropriate data for a relaxation to the requirement. If this approach is adopted, the examples of poor HCI design in civil cockpits may, in time, begin to decline. A better representation of the aviation and, in particular, nonaviation related but immensely applicable, standards and guidelines needs to be achieved.

The second organizational change concerns the model used for flight deck HCI design and the stage at which certification occurs. The detailed form and content of human factors support to certification should be defined and established. Regulatory authorities should commence a review of the technicalities and implications of early sight of an aircraft's design. In addition, consideration should be given to allowing regulatory authorities from nations other than that of the aircraft manufacturers to review proposed aircraft designs during the design process.

The third organizational change involves the end users. The line pilots who operate these systems daily and who must interact with their aircraft through the interface provided by the manufacturer need to be involved in some formal way in the HCI review process. Neither the test pilots of an aircraft manufacturer nor airline managers, even those who are pilots, can adequately represent the typical line pilots' operational requirements.

The fourth and perhaps most ambitious requirement has to be the incorporation of expert human factors practitioners as sign-off authorities within the regulatory authority and the manufacturer's design team to ensure that human factors is given appropriate weight.

CONCLUSIONS

Being pragmatic, it is assumed that the way aircraft are designed is not going to change unless regulatory authorities establish requirements and criteria for design verification that force changes within the aircraft manufacturers' organization and within the regulatory authority. To help achieve these changes, integration of knowledge from domains of human factors other than those specifically related to aviation needs to occur. This knowledge is in the form of an approach to design that explicitly incorporates human factors and HCI considerations and uses empirically based design guidelines and standards. In addition, the use of formal assessment methods and tools for usability testing and certification assessment need to be used. This chapter has mentioned the potentially valuable contributions from HCI studies and usability assessments and the form and content of that intervention have been described.

In addition, a number of organizational changes are required to allow these data and methods to be applied to aircraft design and certification. Norman (1989) argued for a complete revision to the process of aircraft design. Such a revision is probably untenable. The effective management of human factors in the civil aircraft certification process is the more tenable option and probably implies "a little and often". We would recommend that a better appreciation of the nature and form of HCI assessment be established within the regulatory authorities. In particular, an appreciation is needed of the type of assessment that is demanded by cognitive systems. There is also a strong case for early examination of systems and subsystems in the design process by the regulatory authority. Effective requirements setting of the HCI aspects by these authorities is clearly needed. Finally, formal recognition of the unique contribution of human factors in the form of sign-off authority should be established. These are the approaches needed to ensure the appropriate integration of human factors in the certification process of civil flight deck design.

Our position in this chapter is that a number of changes are required to the aircraft design process and the aircraft certification process that will enhance the design of the flight deck HCI. These changes include the use of a software design model that explicitly considers the HCI, the use of empirically based guidelines and standards during HCI design, and the use of instruments and methods, such as checklists and usability testing tools, that may be used for HCI evaluation and certification. It is this approach that can provide the foundation for effectively integrating HCI knowledge into the civil flight deck design and certification process.

REFERENCES

Billings, C. (1991). *Human centered aircraft automation: A concept and guidelines*. (NASA TM 103885).

Boff, K. R., & Lincon, J. E. (Eds). (1988). *Engineering data compendium,* (Vols. I–III). New York: Wiley.

Confusion over flight mode may have role in A320 crash. (1993, February). *Aviation Week and Space Technology.*

Department of Defense. (1985). Human engineering guidelines for management information paper systems (DOD-HDBK-761). Washington, DC: Author.

Department of Defense. (1989). Human engineering design criteria for military systems, equipment, and facilities (Mil-Std 1472D). Washington, D.C: Department of Defense.

Department of Transport. (1989). Report on the accident to Boeing 737-400 G-OMBE near Kegworth, Leicestershire on 8 Jan 1989 Rep. No. 4/90. London: her Majesty's Stationary Office.

ESPRIT Project 5429. (1992). *Metrics for usability standards in computing.* Teddington, UK: National Physical Laboratory.

French government seeks A320 changes following Air Inter crash report. (1993, March 2)*Aviation Week and Space Technology.*

Human factors research aids and glass cockpit design effort. (1989, August 7). *Aviation Week and Space Technology.*

Landauer, T. (1991). Let's get real: A position paper on the role of cognitive psychology in the design of humanly useful and usable systems. In J Carroll. (Ed.) *Designing interaction* (pp. 60-73). New York: Cambridge University Press.

McClumpha, A., James, M., Green R., & Belyavin, A. (1991). Pilots' attitudes to cockpit automation. Proceedings of Human Factors Society 35th Annual Meeting.(pp. 107-112). San Francisco.

NASA. (1991). Human computer interface guide (SSP 30540). Reston, VA: Space Station Freedom Program Office.

NATO Standardization Agreement 3994 AI, Edition 1. (1999). *Application of human engineering to advanced aircrew systems design.* Brussels, Belgium: NATO.

Norman, D. (1989, July). *The "problem" of automation: Inappropriate feedback and interaction, not "overautomation."* (Rep. No. 8904). , University of California , San Diego: Institute of Cognitive Science.

Norman, D. (1991). Cognitive science in the cockpit. In *CSERIAC,* (Vol II, No. 2.)

Ravenden, S., & Johnson, G. (1990). *Evaluating usability of human–computer interfaces: A practical method.* New York: Wiley.

Rouse, W. (1991). Design for success: *A human-centered approach to designing successful products and systems.* New York: Wiley.

Rudisill, M. (1993). *Pilot comments concerning the pilot interface to automation: results from the RAF IAM survey*(Tech. Rep.) Unpublished manuscript.

Sheridan, T. (1991). Automation, authority, and angst–revisited. Proceedings of Human Factors Society (pp. 2-6) Meeting. San Francisco.

Shneiderman, B. (1986). Designing the user interface: Strategies for effective human-computer interaction. Reading, MA: Addison Wesley.

Smith, S., & Mosier, J. N. (1984). Design guidelines for user–system interface software. Mitre Corporation.

Wiener, E. (1989). *Human factors of advanced technology (glass cockpits) transport aircraft.* (NASA CR 177528). Mofflet Field, CA: Ames Research Center.

Weiner, E., Curry, R. (1980). *Flight deck automation promises and problems.* (NASA TM 81206). Mofflet Field, CA: Ames Research Center.

23

Some Inadequacies of Current Human Factors Certification Process of Advanced Aircraft Technologies

Jean Pariès
Managing Director, Dédale, France

Automation-related accidents or serious incidents are not limited to advanced technology aircraft. There is a full history of such accidents with conventional generation aircraft. However, this type of occurrence is far from sparing the newest "glass cockpit" generation, and it even seems to be a growing contributor to its accident rate. Nevertheless, all these aircraft have been properly certified according to the relevant airworthiness regulations. Therefore, there is a concern that with the technological advancement of air transport aircraft cockpits, the current airworthiness regulations addressing cockpit design and human factors may have reached some level of inadequacy. This chapter reviews some aspects of the current airworthiness regulations and certification processes related to human factors and focuses on questioning their ability to guarantee that the intended safety objectives are met, particularly for advanced cockpit design.

CURRENT CERTIFICATION PRINCIPLES

Certification References

According to Article 31 of the Convention on International Civil Aviation , any aircraft involved in international operation shall hold a certificate of airworthiness delivered or validated by its state of registry. Furthermore, Annex 8 to that convention sets the minimum international degree of standardization. However, airworthiness certification has been achieved through the implementation of national codes of airworthiness containing the full scope of requirements considered necessary by the states to reach the target safety level. Today, merely two airworthiness codes form the potential reference for any transport category aircraft certification: the U.S. Federal Aviation Regulations (FAR) 25 from the Federal Aviation Administration (FAA), and the European Joint Aviation Requirement (JAR) 25 from the European Joint Aviation Authorities (JAA). Due to the pressure of highly internationalized business and the implications of airworthiness regulations on the competition between manufacturers, these two codes only display minor differences, and most of the amendments to these codes are closely harmonized. They can therefore be referred to as a single reference for this discussion.

It is fair to note that most of this chapter was written in 1994. Since that period, the certification of human factors in aircraft design has started to be the focus of several international initiatives. Particularly, a FAA Human Factors Task Force report was issued in June 1996 (Abbott, Slotte, & Stimson., 1996) and JAA draft regulations addressing human factors in aircraft design were produced by the JAA Human Factors Steering Group Design and Certification Subgroup in February 1997. Both include recommendations to the airworthiness authorities. However, the current airworthiness regulations, which are referred to in the following discussion, are still in application. This chapter has intentionally not been updated with all the preceding initiatives, and should therefore be taken as reflecting the state of thinking in 1994.

Basic Purpose

The airworthiness requirements concerning the cockpit , as for any other subsystem of the aircraft, are not aiming at any " best possible" design. They intend to specify the minimum objectives to be matched by an applicant design. This is a very basic principle of any certification. As far as human factors in cockpit design and equipment are concerned, the minimum objectives currently set by the airworthiness code are more or less limited to the following:

- To guarantee that the minimum crew (i.e, after one crew member incapacitation) is still able to do the job without excessive workload or fatigue (ref. FAR/JAR 25–1523).
- To provide the crew members with acceptable comfort and protection against outside conditions, so that they can do their job without excessive effort, concentration or fatigue (ref. FAR/JAR 25–771).
- To provide the crew with sufficient visibility to the outside (ref. FAR/JAR 25-773).
- To minimize the risk of error in the use of controls, particularly through a standardization of the shape and movements of the primary flight controls (ref: FAR/JAR 25–777; 781);

- To minimize ambiguities in the information displayed by the instruments (ref FAR/JAR 25–1303, 1321,1322).
- To provide the crew with relevant alerting information about unsafe functioning states of any equipment or system, and to allow appropriate crew action (ref. FAR/JAR 25–1309).

However, these generic requirements can be completed by a set of special conditions, adapted to the specificity of each particular aircraft. These special conditions may well include extensive and detailed requirements for systems like CRT (cathod ray tube) display flight instruments.

Demonstration of Compliance

The methodology used to check the compliance of a proposed design with a relevant airworthiness requirement heavily depends on the explicit versus implicit nature of the requirement.

Explicit requirements are directly expressed in terms of design features. For example, JAR 25–781 quotes: "Cockpit control knobs must conform to the general shape ... in the following figures". In this case, the compliance of a proposed design is rather easy to check, and direct examination of descriptive material (drawings, scale models, mock-ups) can be used.

However, most of the human-factors-related issues are covered by implicit requirements, expressed in terms of general outcomes to be achieved. For example, JAR 25–777(a) quotes: "Each cockpit control must be located to provide convenient operation and to prevent confusion and inadvertent operation".

In the latter case, the methodology used to evaluate the ability of a proposed design to reach the objective is obviously the critical part of the certification process.

A first possible source of difficulty is the acceptable means of compliance with the objective (as interpreted). Consequently some regulatory requirements are complemented with advisory material, including interpretation guidelines or indications on acceptable means of compliance. Acceptable means of compliance more often than not are proven solutions, or sets of solutions, that have been shown by service history to be satisfactory. Even with the same reference airworthiness code, large variations generally occur in the selection of items subject to specific attention in the certification process of different aircraft. Great departures from previous designs will normally be given a closer look. This is true for technical designs, and for human factors aspects as well. However, this is rather easy to implement for technical systems: Plenty of technical experts are available in the industry, and they will reach an agreement (although not always easily) about what is new technology and what is not. It is much more difficult to assess what a new design or layout will change in the crew–cockpit interaction. A first reason may be that the experts are not available. A second reason may be that even when available, they do not have the models to assess that. The next section expands on this issue through test pilot judgment.

A second possible source of difficulty is the interpretation of the regulatory objective itself. For example, what is the exact meaning of "to minimize ambiguities in the information displayed by the instruments"? Let us take a first example. Glass cockpit displays provide more and more written information in English. How can we make sure that information is not ambiguous, when a significant number of pilots around the world are poor English speakers? A study conducted within Airbus training showed that the failure rate of pilots transitioning to the A310 or A320 was correlated to level of proficiency in English.

Now let us take a more challenging example. In many glass cockpit accidents, it appears that the crew lost their situation awareness (Paires, 1996). The airworthiness goal should therefore be to guarantee that the interface design can provide the crew with a safe situation awareness. The challenge is that situation awareness is the result of the cognitive control of a dynamic situation. It is about being able to filter those aspects of the situation that lead to correct actions. A critical condition for such a control is the proper management of the cognitive compromise (Amalberti, 1996) between several contradictory dimensions: in-depth understanding of details versus large-scale monitoring of the big picture, strategic anticipation of the future versus tactical reaction to real-time information flow, maximum (economic) performance versus optimal safety, short-term cognitive resource investment versus long-term fatigue management, and so on. Assessing the adequacy of a cockpit design from this perspective obviously reaches beyond checking that all the data are displayed in a conspicuous color. Many questions remain unresolved at this point. How do we consider the propagation of the consequences of undetected errors? How do we address inappropriate crew situation awareness through design? How do we acknowledge the frequency and the severity of errors? Should a design be rejected because of one single severe error, committed by one crew member one day or because the majority of pilots would make the same minor error?

The ultimate interpretation of the airworthiness specifications (as well as the specifications themselves) refers to more philosophical questions, such as what role should be allocated to humans regarding computers in future flight decks. At this level, there are no hard scientific models and no hard data, just convictions or philosophies.(Sarter & Amalberti, un press)

One can identify two opposite safety philosophies, which have very different implications for design, shown in Table 23.1

The Test Pilot Judgment Methodology

As a matter of fact, the main tool, when not the only one, currently in use to evaluate the human factors acceptability of a new design in a certification process is test pilot judgment. This judgment is based on regulatory and company test pilots' comparative experience in the subject cockpit and in existing designs. Experience in the subject cockpit follows from actual or simulated flight exposure (several thousands of hours for major test programs). Previous experience with existing designs will serve as a reference. In other words, test pilot judgment is based on extrapolations to a new design of expertise gained on the previous ones.

Furthermore, the certification process cannot wait until the first aircraft prototype is built to get started up. No manufacturer would take the risk of becoming involved in such highly expensive development programs without reasonable guarantee that the projected designs are certifiable. Consequently new designs are submitted to regulatory authorities to get some "certifiability" agreement well before a prototype aircraft is built. In this situation, pilot judgment cannot be exercised in a real cockpit, real flight context, but has to be exercised in a mock-up or any other form of simulated environment.

Finally, at the end of major certification programs, certification authorities are now calling for operational route-proving programs. These programs are the occasion for evaluating the aircraft in airline-type environments, including "natural" and artificially induced failures, with mixed crews being composed of airline pilots and manufacturer test pilots. They can currently include more than 100 flights.

DISCUSSION

The history of automation-related accidents or serious incidents includes conventional generation aircraft such as L–1011 Tristar , DC10, and B747. This type of occurrence even seems to be a growing contributor to the newest glass cockpit generation accident rate. This may be an incentive to question the current human factors certification process of advanced technology aircraft. The ability of a certification process to provide safety protection depends on several factors, including, but not limited to:

• The ability of the requirements to express relevant safety objectives.
• The adequacy of the methodology used to assess the compliance of the new design to the requirements.

As far as the human factors certification process of advanced technology aircraft is concerned, it seems that criticisms may be expressed concerning both aspects.

TABLE 23.1
Two Opposite Safety Philosophies

Philosophy 1	Philosophy 2
Aviation operations can be entirely specified through standardized procedures, programs, schedules, rules, nominal tasks, certification, selection, norms,ans so on	Aviation operations cannot be entirely specified through standardized procedures, programs, and the like. One reason is it includes humans.
Safety results from the nominal operation of the system .	Safety results from the dynamic control of the system.
Safety improvement will result from more specification (more extensive, comprehensive, and detailed procedures) and more discipline from the operators.	Safety improvement will result from a better respect of the "ecology" of the system and a better acknowledgment of its self-protection mechanism.
Deviations from nominal operation are both a cause of lower performance and the main threat for safety.	Deviations from nominal operation are both a necessity for adaptation to random dimensions of real life and a potential threat for safety.
Human operators are ultimately the only unpredictable and unspecifiable components of the system. They are the main source of deviation.	Human operators are up until now the only intelligent, flexible, and real-time-adaptable component of the system. They are a deposit of safety.
Automation, whenever feasible and reliable, will decrease deviation rate and therefore improve both performance and safety.	Automation will increase reliability, improve performance, make the operation more rigid, and create new problems in human–machine coupling.
Errors are nonintentional but regrettable deviations from standard actions. Errors are unfortunately inevitable.	Errors are non intentional but regrettable deviations from standard actions. Errors are unfortunately inevitable.
Errors are just as negative for safety as any other deviation. Any effort should be made to reduce the number of errors.	Uncorrected errors may be a threat for safety. However, self-error awareness is a critical governor of operator's behavior and food for risk management processes (regulator of confidence level).

Some Potential Biases in the Current Human Factors Certification Objectives

Airworthiness codes and their objectives are not floating by themselves in the vacuum. They are embedded in a global aviation safety system, including components such as personnel certification (selection, training, proficiency checking) or operational procedures certification. The design and functional characteristics of this safety system reflect specific theories about risk and safety in the aviation transportation system. These theories are far from being mere rational constructions, consistent with all the available scientific evidence. They also are historical and social outcomes, conveying the current fears and faiths of the aviation community.

The cutting up of the different safety codes and their role distribution is a first indicator of the background safety philosophy. This is particularly perceptible with the human error question. On the one hand, it is widely claimed that pilot error contributes to about 70% of air transport accidents. On the other hand, merely four paragraphs (in sections 777, 781, 1303, and 1309) of the airworthiness code, which includes about 330 of paragraphs, explicitly or implicitly address pilot error. This 1% or so score suggests an implicit assumption that pilot error is not really an airworthiness concern and consequently is a design concern, but is much more related to pilot qualification, procedures and operational regulations. As a matter of fact, this is the prevailing theory in the airworthiness world, and all the publications of the human factors researchers have had little influence on it until now, except perhaps for this ironic one: A shift has occurred from culpability-based theories (pilot should not make errors) to fatalism (pilots will always make errors, whatever the design, so we should substitute automation for pilot action as much as possible).

A second illustration is offered by the workload assessment focus in the certification of the glass cockpit generation during the first 10 years. The glass cockpit generation has progressively brought a drastic change to the previous pilot environment, including two-crew design, computer-generated displays, sophisticated automated flight controls, flight management computers. These changes presented a lot of challenges to pilots, such as auto-pilot active mode awareness, total energy awareness, crew communication, automation over-reliance, computer interface problems (Sarter & Woods, 1991, 1992, 1995). But in practice, one single question surpassed anything else: the workload question. A Presidential Commission was even set up in the United States in 1980 to endorse the concept. Since then, much effort, research and flight test time have been spent for evaluation, rating and judgment of workload levels during certification programs for the purpose of minimum crew complement assessment. But there is no real history of overload related accidents on glass cockpit aircraft. It is not the intention here to minimize the importance of minimum crew evaluation, but only to suggest that its perceived importance is also subjective, highly influenced by the politically highly sensitive nature of the acceptance of a two-crew member cockpit.

The Adequacy of the Compliance Checking Methodology

As already stated, the nearly only tool currently in use in a certification process to evaluate the acceptability of a new design in terms of human factors is test pilot judgment. This judgment is based on extrapolations to the proposed new design of expertise gained on the previous ones. Furthermore, as new designs are submitted to regulatory authorities to get some "certifiability" agreement well before a prototype aircraft is built, it often has to be exercised in a simulated environment.

This test pilot assessment methodology is by nature affected by some biases. First, it is based on the assumption that the experience of test pilots on previous aircraft is transferable to the new ones. This may not be true for great evolutions in human–machine interface design. Second test pilots are not a representative sample of the airline pilot population. They have a very specific knowledge of the aircraft, which leads to different mental models of the aircraft. They do not share the daily routine operations and the associated constraints; therefore the cognitive processes involved at the crew–aircraft interface are very doubtfully the same for test pilots and for airline pilots. Furthermore, typical figures for the number of individuals involved is about 10, or a few 10s, and the total exposure time is about 2,000. This is to be compared to the frequency of the safety critical events or combinations of events, which is more likely to be of the order of 1 per 100,000 hours. To make things worse, recent studies (Amalberti & Wioland, 1996) indicate that cognitive behavior evolves a lot during the first year of experience on a glass cockpit aircraft. The average experience needed for the training process to reach a maturity stage and stabilize the cognitive behavior is about 800 hours. This is by far a figure that no test pilot will reach during a typical test period with a new type of aircraft. Test pilot judgment will therefore be exercised within a cognitive frame that is significantly different from the average airline pilot situation. As a consequence, the types of errors associated will be different, and the conclusions reached through some of the certification processes may not be valid for line operations .

Furthermore, test pilots and designers should ideally refer to the behavior of an average pilot . But what is average pilot behavior? Is there such a thing? And how to assess something like a nonambiguous cross-cultural symbology? Helmreich, Merritt, and Sherman (1996), for example, showed that attitudes toward automation show major differences when compared among Europe, America, and Asia. Some pilots turn on automation when they are in trouble, although others turn it off. The rapid global appreciation of the aviation industry reveals the extent of these cultural differences. In the 1970s, 80% of the world market and all major manufacturers were located in the United States. The next millennium could see 60% of the market and major manufacturers in Asia.

Several methods can be envisaged to counteract these biases. Specific tests could be conducted at different milestones of the design process, early enough to allow modifications. Documenting the objectives, the assumptions, and the results of these tests would help to support and keep track of the design decisions. The participation of airline pilots in operational tests has often been presented as a basic condition for efficiency, which is probably true. However, experience indicates that poor design can result from airline pilot judgment as well as from test pilot judgment. How to better organize the participation of airline pilots is still difficult to tell. What pilot population should be involved? How should the individuals be selected? What should be expected from them: passive performance measurement or active opinion? When should they be involved in the design process? At the end of major certification programs?

Certification authorities are now calling for operational route proving experimental flights. These flights give the opportunity to evaluate the aircraft in airline-type environments, including natural and artificially induced failures, with crews being composed of airline and test pilots. Such programs have included more than 100 flights on some occasions. But most of the time they can only lead to modifications in the procedures or the training, because the design is already frozen and any design modification at that stage has a huge cost.

CONCLUSION

For years, aircraft have been designed and certified with indisputable success, although very little human factors was integrated into the certification process. The current aircraft airworthiness certification requirements addressing human factors issues are expressed in rather general terms, and they are subject to interpretation uncertainties. Ironically, however, technocratic creeds and naive models of human behavior, still criticized as such by series of current human factors papers, including this one, have led to the highest safety level of the transport industry.

Until now, the evaluation process for the certification of human-factors-related aspects of cockpit design has relied almost entirely on test pilot judgment. This method has proven satisfactory for the past years, although it is marred by several biases. However, the changes induced by the new design of the pilot–aircraft interface have recently increased the effects of these biases to a significant degree. Consequently, it seems that there is a need for the development of documented human factors assessment protocols to complement test pilot assessment methodology. This will first imply the clarification of a need for a redefinition of the objectives of the human factors certification program.

REFERENCES

Abbott, C., Slotte S., & Stimson D. (Eds.). (1996). *FAA human factors team report on the interfaces between flightcrews and modern flight deck systems.* Washington ,DC: Federal Aviation Administration..

Amalberti, R. (1996). *La conduite de systèmes à risques.* Paris: PUF.

Amalberti, R. (1993) Safety in Flight Operations in (Eds.). Wilbert B., Quale T. , *New Technology, Safety and System Reliability*, L. Erlbaum, Hillsdale NJ.

Federal Aviation Administration, (1998) Federal Aviation Regulations FAR–25
    Large Airplanes. Washington, DC: Author.
Helmreich, R., Merritt, A., Sherman P. (1996). *The flight management attitudes
    questionnaire: An international survey of pilot attitudes regarding cockpit
    management and automation,* University of Texas, Austin.
Joint Aviation Authority (1998). Joint Aviation requirements JAR–25 Large
    Airplanes. Hoofdorph, Netherlands: Author.
Pariès, J. (1994). Etiology of an accident: human factors aspects of the Mont Sainte-
    Odile crash. *ICAO Journal 49*(6). (pp. 37–41).
Sarter, N. , & Amalberti , R. (Eds.). (in press) *Cognitive engineering in aviation.*
    Mahwah, NJ: Lawrence Erlbaum Associates.
Sarter, N., & Woods, D. (1991). Situation awareness: A critical but ill-defined
    phenomenon. *I nternational Journal of Aviation Psychology, 1*(1);   45–57.
Sarter ,N., & Woods, D. (1992). Pilot Interaction With Cockpit Automation:
    Operational Experiences With the Flight Management System. *I nternational
    Journal of Aviation Psychology, 2*(4), 303–321.
Sarter ,N., & Woods, D. (1995). *Strong, silent and "out-of-the-loop": Properties of
    advanced automation and their impact on human automation interaction,*
    (CSEL Tech. Rep. No. 95–TR–01), Columbus University, Columbus, OH.
Wioland, L., & Amalberti, R. (1996, November). When errors serve safety : Towards
    a model of ecological safety.Paper Presented at Cognitive Systems Engineering
    in Process Contro*l*, Kyoto, Japan.

VII

ISSUES IN FUTURE DESIGN AND CERTIFICATION OF COMPLEX SYSTEMS

24

Advanced Automation Glass Cockpit Certification

René Amalberti
IMASSA, France

F. Wibaux
DGAC-SFACT, France

When the first version of this chapter was written in 1993, almost no specific human factors requirements or methodological standardization existed within the civil aviation regulations (Federal Aviation Administration [FAA] or Joint Aviation Authority [JAA]). Since that time, considerable efforts have been developed in the field. The improvements are threefold: First, a thorough human factors audit of the flight deck was issued in June 1996 by the FAA (with participation of the JAA). Second, mixed working groups involving authorities and industry have been created on both sides of the Atlantic and have framed the content of new guidance material and regulations. Third, researchers have made considerable progress in the comprehension of human error management and risk management that impacts the future regulations. The earlier version of this chapter has been revisited as a matter of these new advances in human factors in design and certification. A specific section has been added as a final part of the chapter to capture all the changes. The other three sections of the earlier version follow, all revisited and updated. The first of these sections debates human error and its relation with system design and accident risk. The second describes difficulties connected to the basically gradual and evolving nature of pilot expertise on a given type of aircraft, which contrasts with the immediate and definitive style of certifying systems. The third section focuses on concrete outcomes of human factors for certification purposes.

The goal of aircraft certification is simple: Guarantee that an aircraft fits the legal flight safety requirements when flown by qualified standard pilots, who are as representative as possible of end users. The crews taking part in the certification campaign are used to bring systems into play and are at times coappraisers, but they are never examined in their own right in the certification campaign. Thus, human errors observed during the certification campaign were classified in two categories. First, there are errors related to system design; that is, input errors, such as stick and throttle errors and inappropriate settings. These errors were taken into consideration in the certification process. Second, there are errors resulting from pilot attitudes, air traffic control (ATC) dialogues and clearances, or individual weaknesses related to general aviation know-how. This second type of errors was not considered to be relevant for aircraft certification purposes.

WHAT MODEL FOR HUMAN ERROR AND WHAT LINKS BETWEEN SYSTEM DESIGN AND ACCIDENT RISK?

Why Change the Current Procedures of Certification?

The present level of flight safety is very good. The risk of accident is about 1 per 1 million departures in industrial nations. However, this value has been virtually stable for 15 years. It was attained before the growth of automated aircraft and before human factors became a target objective of the FAA and the international aeronautical community.

In this context, why should the certification process bother with human factors? The answer is twofold. First, advanced automated aircraft have fewer dramatic failures, but the rate of accidents is not decreasing. Accident causes are increasingly due to the cognitive failure of crews. This introduces new problems of interference between system reliability and human reliability. In other words, the technical improvement of system safety based on automation could result in a negative outcome for human reliability. Moreover, because these technical changes are barely related to the previous experience acquired on standard aircraft, certifiers, whether they are expert pilots or engineers, must ask human factors specialists for scientifically based assistance to evaluate systems better. Second, with a constant rate of accidents, any increase in traffic volume will result in an increase in the total number of accidents. Moreover, the negative impact of each (rare) accident on customers (passengers and also companies) is multiplied by modern information networks. Naturally this is undesirable, and the improvement of this situation is the stated objective of the aeronautical community. The only solution to remedy this situation is to continue to increase flight safety. However, it is clear (given the good current level) that gains will not be easy in previously examined domains and also that new domains will have to be taken into consideration. Human factors is one of the most important of these new domains, aside from the future of ATC communications.

Let us examine how these objectives impact the relation among human error, system design, and accident risk, and hence impact certification procedures. Of course, the first problem to tackle is defining human error without ambiguity.

The Wariness About the Definition of Human Error

The ergonomics literature has proposed numerous definitions and classifications of human error from process control. The dominant definition considers human error as a deviation from the norms, whether these norms are written or assumed from practice. This definition is both the easiest to use and the most debatable. Most recent cognitive ergonomics field studies (Amalberti, 1992; deKeyser, 1986; Rasmussen, 1986) show that novices continuously interpret norms to make the job feasible with their limited resources and knowledge and that experts interpret norms still further with routines, shortcuts, and violations. To sum up, norms are never respected, although they serve operators as references both to give and to limit degrees of freedom in adapting to the ongoing situation. Human operators are experts in piloting this derivative of norms to fit the situation requirements with minimum workload.

In contrast, another way to define error would be to consider error as a deviation from operator intention. This approach is probably less biased than considering error as a deviation to norms. However, operator intentions are hard to evidence, especially after a lengthy delay between action and analysis, and in the end human error still continues to be considered as a deviation from norms during the process of certification.

Experience acquired in participating as a neutral observer in the minimum crew campaign of advanced automated glass cockpits shows that ambiguities in human error definition lead the certification team, poorly trained in human factors, to make numerous mistakes in classifying and interpreting the crew errors. These mistakes are threefold: first, errors immediately corrected by the crew tend to be ignored. However, many of these errors point to incorrect system design, especially when they are repeated almost systematically. Second, deviations from norms are too often considered as errors although they are not. In many cases, they represent pilots' attempts to conserve resources or to manage the system and the task more conservatively. Third, some deviations where there are no specific procedural norms are ignored, although they potentially endanger the flight. This is typically the case of poor synergy and poor crew coordination that can result from the system design as well as procedure or input errors.

Modeling Relations Between Human Error and System Design

Ergonomics has always argued that systems should be designed primarily for end users (human-centered design). This is a very basic and central value of ergonomics. Nevertheless, the concept of the human-centered system, and in a certain sense of a "good system," has significantly shifted over the last 10 years with changes in technologies and ergonomics theories. Let us examine these changes.

A good ergonomics design has long been considered to be a design that prevents errors and facilitates good performance with as low a workload as possible. In the 1940s, the main interest was in unambiguous commands designed to minimize errors; for example, confusion between gear and speed brakes (Fitts & Jones, 1947). This type of ergonomics, which is dominant in United States, is termed *classic human factors*. The basic philosophy draws on the central idea that human error is avoidable if the design respects human limitations and capacities and leads to the concept of fault-preventing system design. It has been extensively and successfully applied to cockpit design and is currently used.

However, several factors have contributed to a recent decline of this approach: new technologies, new needs, and new ergonomics theories.

New Technologies and Classic Human Factors

Cockpit automation has enhanced flight performance in many domains such as precision approach, flight accuracy, engine performance, and pilots' situation awareness (with the introduction of map display). Automation has also mechanically reduced a great number of human errors resulting from improper power or stick settings and system handling simply because these tasks are no longer pilot dependent.

However, the drawbacks of automation for human behavior are as numerous as the advantages just described. Cockpit automation and its consequences for cockpit layout have considerably reduced the benefits expected from a simple human-factors-based system design. The flight management system (FMS) with its undifferentiated and multiplexed keyboard, is a blatant example of this (Pelegrin & Amalberti, 1993; Sarter & Woods, 1992). This design is the source of many input errors in programming systems.

Criticisms have also been directed to information displays. The ability to display, aside from classical dials, much more information in various new forms (such as drawings and text) has led designers to a series of poor ergonomics solutions regarding the capacities and limitations of human perception. Perception time tends to be increased with the use of textual information, perceptual feedback is reduced in peripheral vision (due to loss of motion of sticks and throttles and also due to cockpit architecture that requires the other crew member to move less), and auditory and kinesthetic sensations are also reduced (due to computer program smoothing system reaction to improve passenger comfort).

But of course the expected benefits of automation for ergonomics are elsewhere. Situation awareness has been improved with the use of map displays, primary flight display, and ECAMs or EICAS.

To sum up, the new cockpit layout (glass cockpit) is assumed to enhance the pilot's situational awareness, but the solutions chosen to reach this goal are clearly to the detriment of classic sensorimotor human factors.

Evolution of Theories: Cognitive Ergonomics, Another Way to Consider Ergonomics

The change in goals for cockpit design prompted new developments in ergonomics theories in the 1980s. This is the domain of cognitive ergonomics, which draws heavily on the European ergonomics tradition. The value of such ergonomy is pilots' cognitive modeling focusing on their strategies and know-how. Numerous field studies have shown the advantages of this type of operator's cognitive model (Amalberti, 1992; Bainbridge, 1989; de Keyser, 1986; Hollnagel, 1993; Reason, 1990). The following section summarizes the main characteristics of one pilot's cognitive model.

Professional pilots generally have satisfactory procedural knowledge of their work domain and remarkable reasoning capacities, but they are resource limited and cannot use all the knowledge and the reasoning capacities they would like to in time-related situations. The true task of pilots is to develop strategies to get the job done with respect to this resource limitation bottle neck.

Solutions Call for Planning, Anticipation, and Risk Taking

Because their resources are limited, pilots need to strike a balance between several conflicting risks: an objective risk resulting from flight context (risk of accident) and a cognitive risk resulting from personal resource management (risk of overload and deterioration of mental performance).

To keep resource management feasible, the solution consists of decreasing outside risks, simplifying situations, only dealing with a few hypotheses, and schematizing reality. To keep outside risk within acceptable limits, the solution is to take as many preflight actions as possible to simplify the flight. Any breakdown in this fragile and active equilibrium can result in unexpected situations, in which pilot performance may be decreased.

Evidence shows that human errors result from internal characteristics of cognitive models and are not suppressible (Reason, 1990; Senders & Moray, 1991). Because resource limitations force the pilot to make a series of compromises between what the situation should ideally require and what he or she is capable of doing, errors are the logical consequence. Moreover, expertise results from experiencing errors (Anderson, 1985), and errors are generally profitable when the pilot receives immediate feedback from his or her errors.

New technologies confirm this general picture. As Wiener and Curry (1980) and Bainbridge (1989) pointed out, automation does not reduce the number of global errors, but merely changes error types. There are more routine errors and representation errors.

The consequences for ergonomics and certification purposes are twofold. First, the concept of fault-tolerant system design replaces the one of fault-preventing system design. Fault-tolerant system design does not aim at limiting local errors, but merely at improving the pilot's awareness, giving as clear feedback as possible of error, possibly correcting the immediate consequence of error when the flight is endangered; that is, logical testing on FMS inputs or safety envelope of flight laws (alpha floor). Second, it is clear that in this theoretical framework, it is no longer satisfactory to measure human performance as a simple error rate. More complex approaches are required to efficiently serve the certification process.

Should Error Analysis Be Restricted to Human–Machine Interaction?

Standard certification procedures do not deal with crew errors that are not directly related to system design. However, new technologies could lead to a change in this position. The interdependency of any component of the aeronautical system, aircraft, crews, ATC, or maintenance, makes the analysis of causality between design and consequence of design much more complex than on a simple system. Any change in system philosophy influences the way operators carry out the task, even for actions not directly related to the system interface. This is typically the case for crew coordination in glass cockpits.

Glass cockpit layout is generally assumed to make crew coordination more difficult. The reasons are threefold. First, as already described, communications require more and more central vision and active vocal dialogue to read the written information and to remedy the relative deprivation of sensory inputs. Second, new cockpit architectures, with independent access to information and commands, facilitate desynchronization. Pilots can display modes, change modes, or change parameters on their channel, which the other crew member is totally unaware of. Third, problems of language emerge because more and more information is written. The standard language of aviation is English, although most pilots in the world have a different native language and do not speak perfect English (Pelegrin & Amalberti, 1993).

What emerges from these various difficulties is that many situations of poor coordination in a glass cockpit can be related to system design although they are not directly related to a specific action on the interface. This level of causality challenges the philosophy of the system and calls for complex corrections. It is easy to understand that designers are very reluctant to consider that these errors are related to system design and prefer to pass on the problem to trainers.

Relation Between Human Error and Accident Risk

Safety is a central concern of aircraft certification. System-failure classically serves to measure system reliability, and human error serves as an equivalent measure of human reliability. Measures could be quantitative or qualitative (type of error), but it is explicitly admitted that a good design and a safe system would provoke fewer errors than an unsatisfactory design, and would therefore result in fewer accidents.

This is only partially true. We have seen that human error is not totally avoidable. Moreover, it is important to remember that the accident-rate is 1 accident per 1 million departures, and that there are an average of more than 2 human errors per flight that are not detected and corrected immediately by crews (Amalberti, 1998; this value comes from numerous flights made in 1992 on glass cockpits in the observer position). Thus, even though human responsibility appears to have risen in glass cockpit accidents, the relation between human error and accident is far from being trivial.

The key point is that the relation between system design, human error and the risk of accident is not linear from great risks to no risk. Obvious bad system design and unadapted regulations or training will cause numerous human errors and will increase the risk of accident. However, even if the design, training, and regulation are perfected, numerous human errors and a nondecreasing risk of accident will remain. Without strong relations between the two arguments, remaining accidents are poorly linked to human errors. They are better linked to a matter of circumstances, a dramatic combination of unexpected events in which human error can occur but is not seen as a decisive factor. This picture specifically applies to rare accidents arising in the context of high reliability.

For certification concerns, this nonlinearity between human error and risk of accident can become a problem. The central concern is to have enough references on human error theories to clearly separate what should be a normal rate and type of human error from an abnormal rate and type of human error due to poor system design, training, or regulation. This is typically a domain in which human factors could improve the current process of certification.

The Evolving Nature of Pilot Expertise and the Immediate and Definitive Style of Certifying Systems

Although computer technology clearly enables software modifications during and after the end of certification, aircraft philosophy and most subsystem designs are considered to be stable and definitive at the beginning of the certification process. This is not the case for pilots' expertise. Most official pilots in charge of flying the machine during the certification process have less than 200 hours of experience on the machine (see Fig. 24. 1). This is far from having stabilized expertise in the glass cockpit. Results from field experiments (Pelegrin & Amalberti, 1993; Sarter & Woods, 1992) all indicate that pilot expertise for flying an aircraft with a glass cockpit shifts significantly up to 800 flight hours and perhaps more.

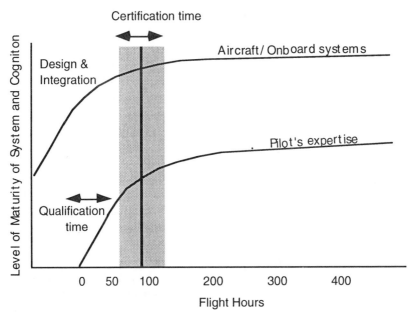

FIG. 24.1. Pilot's expertise on the type related to design and certification process.

These values are almost double the values observed for those required for expertise in a standard cockpit.

The problem is that behaviors change with experience. Errors also change in nature. The lengthy period required to stabilize expertise in glass cockpits creates difficulties for pilots becoming used to the system. Pilots cannot easily grasp the enormous possibilities of the system. With pilots of up to 400 flight hours, the main risks are overconfidence or excessive doubt concerning their own capacities, which, respectively, result in engaging the system in unknown domains or hesitating to make the right choice of action.

Once pilots gain confidence with the system, routine errors and violations are multiplied. The risk of accident still exists but changes in nature. It is clear that some relations between human error and system design will only emerge in experienced pilots.

This picture raises a key question about certification: Do we test all aspects of system design and system safety with novices? If not, do we have to consider various levels of pilot expertise? A positive answer would result in envisaging a double-track certification, one initial test comparable to what is done currently and one operational complement administered after a few months of experience on the system.

Another solution would be to use official pilots who are already experts on similar machines (machines of the same family) to gain time and experience. In this case, novice official pilots in glass cockpits will also be required to represent this class of pilots and their specific problems.

Note that, in any case, the choice of official pilots who represent the future range of company pilots and the composition of crews is a key for efficient certification; that is, take into account the pilots' level of experience, form a representative panel of pilots, avoid crews made of two captains, and so on. This area could improve greatly in the future.

PRACTICAL OUTCOMES: HOW TO IMPROVE CERTIFICATION WITH HUMAN FACTORS CONSIDERATIONS

When to Bother With Human Factors in the Certification Process?

A good answer would be anytime the human factors perspective gives a plus to the standard approach that is already being used. The objective of integrating human factors should not be to make a revolution in certification, but merely to support and improve the current way of doing certification.

The expected benefit is threefold. First, it must be in terms of the identification of undetected system weaknesses. Second, it is in terms of giving a rationale for pilots' problems and relations to system design in the risk of accident. Last but not least, an aircraft can no longer be viewed as an isolated system. Certification typically concerns the integration of aircraft into operational conditions. Therefore, outcomes of certification must concern system design as well as pilot training and regulations so that subcomponents of the macrosocial system might be included in the evaluation. For this specific concern, psychosociology can effectively support certifiers to make the relevant decisions.

What to Certify?

The complexity and the novelty of systems lead one to consider that the integration of human factors in certification should overpass the simple evaluation of system performance and the reliability of an end product. It should extend to design procedures, based on general principles that have proven to be efficient, and it should begin with prototype evaluation at a period when changes are effectively possible.

What to Measure?

The starting point should be to analyze pilots' activities and detect human errors. However, the analysis of these human errors must vary according to the goal: assessing the risk of accident or testing a pilot's ease with the system design. This is the reason no unique classification of human error can figure out all questions raised by certification. Multiple classifications are required. Further analysis would concern the assessment and possible measurement of mental reasoning, mental workload, communications, and crew coordination, to sum up all cognitive activities that serve pilots and set up a relevant representation of the ongoing situation.

Who Makes the Evaluation?

We saw in previous sections of this chapter that the selection of the panel of official pilots and engineers participating in the certification campaign is a key factor in obtaining relevant results. One could suggest that this panel should be as large as possible to grasp a great variety of opinions, and also to avoid making certifiers codesigners due to a (too) long relationship the certifier develops with designers. Whatever the panel, it seems useful to require a minimum human-factors background for people in charge of certifying.

What Limitations for Human Factors?

Many human factors aspects of cockpit automation are beyond the traditional certification process. We have seen that some of them could be easily better taken into consideration. Yet, numerous others that relate to psychosociology, work organization, careers, trades, or companies will also be beyond human factors investigation during the certification campaign. However, they should be crucial to system acceptability and risk acceptability, but this is another story.

Assuming that there is the integration of a human factors specialist in the certification process, another clear limitation should be the legal responsibility of this specialist in case of accident. Human factors cannot answer all the cockpit problems, either because of the lack of knowledge or because of the lack of time to apply relevant methodology. Therefore it will probably be required to specify in writing what domains are relevant for human factors actions and what mix of responsibilities will draw on human factors specialists and on other certifiers.

What Should Change?

In most cases, the modifications in system design required by the certification are small. The reason is obviously the financial cost. Therefore, when problems are observed, they tend to be solved by putting effort into pilot training and regulations.

Software technologies have changed this picture a little by introducing a greater level of flexibility, but it is clear that the underlying system philosophy remains unchanged.

Even though this is an acceptable outcome and even though experienced pilots rate the system as very good (this is the case of the modern glass cockpit), human factors specialists worry about this increasing reliance on training solutions. What will occur for flight safety if we continue to produce opaque systems that require more than 1,000 flight hours before pilots are experienced? Is this realistic?

Similarly, the presence of various generations of aircraft poses unsolved problems at this time: What will occur with pilots flying successively old and new aircraft? What will occur with multiqualified pilots flying almost identical aircraft with just a few differences, in particular as regards cross-crew qualifications? In both cases, the questions overlap systems certification, introducing new types of questions to investigate and new constraints in forming the panel of pilots called for testing the system.

Systematic flight analysis of current modern aircraft is a fantastic tool to anticipate most troubles pilots will have with the next technology. This is a central direction for improvements for the entire aeronautical community, and it can be useful (for certification purposes) to ask the designer to take into account lessons from the previous design.

But again, any envisaged modification in a new machine will have to be investigated with a human factors perspective not really for itself, but for the possible negative consequences it will introduce when flying similar systems with biqualification.

Finally, one should remember that certifiers do not have to overpass the mandatory mission they are paid for (assessing that the system fits safety and minimum requirements). Once these minimum requirements are established and respected, designers will be free to create and certifiers will not have to officially judge a design in terms of being good or bad. In the context of a free market there is competition between manufacturers and the success or failure of a product remains a decision of customers.

1993–1998: Five Years of Intense Activity in Human Factors for Design and Certification

The period between 1993 and 1998 must be considered as a historical transition for the setup of future regulations for design and certification of large airplanes. During this period three changes occurred. First, the diagnosis of human factors problems has been refined and redrawn in a more comprehensive technical form for the industry. Second, the dialogue among authorities, research, and industry has been installed on a collaborative basis, overcoming the initial jams and reluctances of each party. Last, research has continued to progress in a more applied fashion that permits the offering concrete solutions and methods for the purpose of design and certification.

The Interfaces Between Flight Crews and Modern Flight Deck Systems, Report of the FAA HF Team

In November 1994, the FAA tasked a group of 16 experts from FAA, JAA and Academy to audit the implementation of human factors in modern flight decks. The group audited manufacturers, companies and researchers all around the world and issued a report in June 1996 (Abbott et al., 1996). The main results of this report are threefold. First, the human factors problems of automation are not the problems of a specific black box, (e.g., the FMS or the autopilot), nor are the problems of a given aircraft or a specific manufacturer. Automation is a vast movement with a change in culture and habits that results, and most problems lie in the global harmony of the human and technical components, regulations, and organization of such a vast system. This is the reason most safety problems in aviation today are systemics rather than locotechnical. This is the reason also why an accident can be reputed to be caused by a local technical or human deficiency, but the solution to prevent such a future accident is never only at that local level; it must address system organization.

CLOSING NOTES

The international aeronautical community aims at introducing human factors more efficiently in the certification loop because of the desire to reduce the risk of accident and because of new technologies that have negatively impacted on human performance in a few domains. Analysis shows that this improvement cannot be made without a reconsideration of the concept of human error before (flight analysis), during, and after the certification phase (feedback and accident analysis).

One last and chronic source of misunderstanding in bridging knowledge between human factors specialists and engineers is that engineers superimpose human error and system failure. This makes no sense. Humans are intelligent and flexible. They can be perfectly adapted, whatever the complexity of the situation, and can ensure a very high level of safety. They learn from errors, cover billions of domains, and can adapt to unknown domains. However, errors are always possible and they always occur. These errors are poorly predictable and tend to occur at times, in areas and with people that nobody would have predicted. On the other hand, machines are rigid, unintelligent, and repetitive, and failures are predictable and curable. Because of their stability, machine reliability appears to be easily modeled and also more reliable than human reliability. The result is that engineers give a systematic priority to machines to the detriment of pilots because they feel this is the only way to improve and control safety.

All human factors findings show that they are wrong. Human and machine reliabilities are simply different and must work in synergy to reach a better level of reliability. Unfortunately, the solutions chosen at this time to increase system reliability interfere with human reliability and lower this human reliability.

Thus, it would not be realistic to discuss a very detailed point in the interface, although fundamental points are being ignored. A French maxim perfectly summarizes this point: "It is not good that the tree hides the forest."

The ideas expressed in this chapter are aimed at launching debates. They are not firm directions decided on SFACT (Service de la Fornation Aeronautique et du controle technique), but only preliminary thoughts. Future decisions of French official services will take into consideration, in addition to some ideas expressed in this chapter and other technical ideas, all legal, international, and sociotechnical aspects of the problems that have not been mentioned in this chapter.

REFERENCES

Abbott, K., Slotte, S., Stimson ,D., Amalberti, R., Bollin G., Fabre, F., Hecht, S., Imrich T., Lalley, R., Lyddane, G., Newman, T., & Thiel, G. (1996, June) *The interfaces between flighcrews and modern flight deck systems*, (Report of the FAA HF Team), Washington DC: Federal Aviation Administration.

Amalberti, R. (1992). Safety in risky process-control: An operator centred point of view. *Reliability Engineering & System Safety, 38*, 99–108.

Amalberti, R., Paries, J., Valot, C., & Wilbaux, F. (1998). Human Factors in Aviation: An introductory course. In K. M. Goeters (Ed.) *Aviation Psychology: A science or a profession* (pp. 17–43). Adelshot, UK: Ashgate Avebury Limited.

Anderson, J. (1985). Development of expertise, In *Cognitive psychology and its implications*(pp. 235–259). New York: Freeman

Bainbridge, L. (1989). Development of skill, reduction of workload. In L. Bainbridge & Quintinilla (Eds.), *Developing skills with new technology*, London: Taylor & Francis.

de Keyser, V. (1986). Technical assistance to the operator in case of accident: Some lines of thought. In E. Hollnagll, G. Mancini & D. Woods (Eds), NATO ASI Series Vol F21 (pp. 229–254). Berlin: Springer-Verlag.

Fitts, P., & Jones, R. (1947). *Analysis of factors contributing to 460 "pilot error" experiences in operating aircraft controls*, (Rep. No. TSE AA–694–12). Wright-Patterson AFB, Dayton OH.

Hollnagel, E. (1993) *Reliability of cognition: Foundations of human reliability analysis*. Amsterdam: Elsevier

Pelegrin, C., & Amalberti ,R. (1993, April). *Pilot's strategies of crew coordination in advanced glass-cockpits; a matter of expertise and culture*. Paper presented at the Second Flight Safety International Congress, Washington, DC.

Rasmussen, J. (1986). *Information processing and human-machine interaction*. Amsterdam, North Holland.

Reason, J. (1990). *Human error*. New York: Cambridge University Press.

Sarter, N., & Woods, D. (1992). Pilot interaction with cockpit automation: Operational experiences with the flight management system. International Journal of Aviation Psychology., 2(4), 303–321

Senders, J., & Moray, N. (1991). *Human error.* Hillsdale, NJ: Lawrence Erlbaum Associates.

Wiener, E., & Curry, P. (1980). Flight desk automation: Promises and problems. *Ergonomics, 23,* 988–1011.

25

Beware of Agents when Flying Aircraft: Basic Principles Behind a Generic Methodology for the Evaluation and Certification of Advanced Aviation Systems

Denis Javaux
University of Liège, Department of Work Psychology, Belgium

Michel Masson
Dédale, France

Véronique De Keyser
University of Liège, Department of Work Psychology, Belgium

There is currently a growing interest in the aeronautical community to elicit the effects of the increasing levels of automation on pilots' performance and overall safety.

The first effect of automation is the change in the nature of the pilot's role on the flight deck. Pilots have become supervisors who monitor aircraft systems in usual situations and intervene only when unanticipated events occur. Instead of "hand flying" the airplane, pilots contribute to the control of aircraft by acting as a mediator, instructions being given to the automation.

By eliminating the need for manually controlling normal situations, such a role division has reduced the opportunities for the pilot to acquire the experience and skills necessary to safely cope with abnormal events (Bainbridge, 1987).

Difficulties in assessing the state and behavior of automation arise mainly from four factors :

- The complexity of current systems (e.g. Billings, 1991; De Keyser & Javaux, 2000; Javaux, 1997; Javaux & De Keyser, 1997) and consequent mode-related problems (Degani, 1996; Degani, Shofto, Kirlik, 1999; Javaux, 1998; Sarter & Woods, 1995; Vakil, Hansman, Midkiff, & Vanek, 1995).•
- The intrinsic autonomy of automation that is able to fire mode transitions without explicit commands from the pilots (e.g., Sarter & Woods, 1992) .
- The bad quality of feed-back from the control systems displays and interfaces to the pilots (e.g., Javaux, 1998; Norman, 1990 ; Sarter & Woods, 1994).
- The fact that the automation currently has no explicit representation of the pilots' current intentions and strategy (Onken, 1992a , 1992b; Sherry, Kelley, McGobie, Feary, ALkin, Polsan, Hynes, Palmer, 1997).

The conjunction of those factors induces a large set of crew–automation interaction problems that pose questions for the current research: difficulties in anticipating computer-generated mode changes; difficulties in assessing the implications of changes to previously given instructions; difficulties in reacting to unanticipated events and to command changes; difficulties in finding, integrating, and interpreting relevant data for situation assessment; and difficulties in building extended and refined mental models of how automation is working and how instructions have to be input (Javaux, 1998; Sarter & Woods, 1992, 1994).

For pilots, the consequences of those difficulties are an increase in cognitive workload and the development of unofficial strategies to override or "hijack" the automation, in an attempt to satisfy official goals (Amalberti, 1992).

As a result, certification is facing a range of new and complex problems that challenge the aeronautical community to predict and account for all kinds of pilots–automation interaction patterns arising from the introduction of new and sophisticated technologies in cockpits. For an attempt at predicting the behaviors, errors, and mental models of pilots faced with automated systems, see Javaux (1999).

> The rapid pace of automation is outstripping one's ability to comprehend all the implications for crew performance. It is unrealistic to call for a halt to cockpit automation until the manifestations are completely understood. We do, however, call for those designing, analyzing, and installing automatic systems in the cockpit to do so carefully; to recognize the behavioral effects of automation; to avail themselves of present and future guidelines, and to be watchful for symptoms that might appear in training and operational settings. (Wiener & Curry, 1980 cited in Billings, 1991, p. 67).

In particular, this chapter tries to characterize the added complexity and problems (Javaux, 1997, 1998) created by the introduction of autonomous agents (intended as automated resources) in new generations of aircraft.

As an example of the potential for catastrophic consequences of these problems, we would like to refer to the China Airlines B747–SP accident 300 miles northwest of San Francisco, on February 19, 1985, using the accident report proposed by Billings (1991) :

The airplane, flying at 41,000 ft. enroute to Los Angeles from Taipei, suffered an inflight upset after an uneventful flight. The airplane was on autopilot when the n. 4 engine lost power. During attempts to relight the engine, the airplane rolled to the right, nosed over and began an uncontrollable descent. The Captain was unable to restore the airplane to stable flight until it had descended to 9500 ft.

The autopilot was operating in the performance management system (PMS) mode for pitch guidance and altitude hold. Roll commands were provided by the INS, which uses only the ailerons and spoilers for lateral control; rudder and rudder trim are not used. In light turbulence, airspeed increased. As the airplane slowed, the PMS moved the throttles forward but without effect. The flight engineer moved the n. 4 throttle forward but without effect. The INS caused the autopilot to hold the left wing down since it could not correct with rudder. The airplane decelerated due to the lack of power. After attempting to correct the situation with autopilot, the Captain disengaged the autopilot at which time the airplane rolled to the right, yawed, then entered a steep descent in cloud, during which it exceeded maximum operating speed. It was extensively damaged during the descent and recovery (1991, p. 98).

As noted by the author, the National Transportation Safety Board (NTSB) concluded that :

The probable cause was the captain's preoccupation with an inflight malfunction and his failure to monitor properly the airplane's flight instruments which resulted in his losing control of the airplane. Contributing to the accident was the captain's over reliance on the autopilot after a loss on n. 4 engine. The Board noted that *the autopilot effectively masked the approaching onset of loss of control of the airplane.*(p. 98)

Without stating too much about the concepts that are developed in the following sections, yet in contrast to the first elements of analysis retained by the NTSB, this chapter claims that this accident's main contributing factors are flaws in the design of the information and control systems combined with the presence of agents that operate independently of any pilot's control action but without adequate feedback.

More precisely, as revealed by this incident, the breakdown of the pilot–automation system onboard this aircraft—which is typical of a design current known as technology-centered automation—is mainly due to a lack of controllability of the automatic systems involved, coupled with a lack of visibility and predictability of those systems' status, effects, and interactions over the considered flight phase, and an engine failure.

Assuming certification has among its major goals to guarantee the passengers' and pilots' safety and the airplane integrity under normal and abnormal operational conditions, we suggest it would be particularly fruitful to come up with a conceptual reference system providing the certification authorities both with a theoretical framework and a list of principles usable for assessing the quality of the equipment and designs under examination.

This is precisely the scope of this chapter.

THE MULTIPLE RESOURCES OF AUTOMATION

We consider automation to be a tool or resource – a device, system or method by which the human can accomplish some task that might be otherwise difficult or impossible, or which the human can direct to carry out more or less independently a task that would otherwise require increased human attention or effort. (Billings, 1991, p. 7)

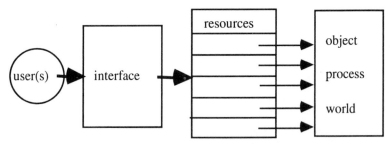

FIG. 25.1. A simplified diagram of automated control of automation

Four components define the classical automated control situation (Fig. 25.1) (e.g. Sheridan, 1988):

- A set of users, operators, or pilots with some goals or tasks to achieve.
- An object, a process, or a world, characterized by a set of state variables, on which the users want to act.
- A set of automated resources that possess the capability to change the state of the world on the behalf of the user.
- An interface that provides the user with the means to activate and control these resources.

It is clear from everyday life experiences (Norman, 1988) that resources can display very different behavioral characteristics, and that this influences the way we use them as well as the type and amount of knowledge we need to do this efficiently.

The following three essential categories of resources can be identified according to their different behavioral characteristics.

Functions constitute the simpler type of resources and affect the state of the world in a straightforward way. Their effect only depends on the state of the world prior to their activation. Moreover, this effect can be described by a simple state transition: the state of the world before and after the activation of the function. Functions are thus usually extremely predictable resources (e.g., manual seatbelt and nonsmoking sign activation, manual throttle control, etc.).Functions affect objects, processes, or worlds by means of open control loops.

Functional patterns constitute the second type of resource. The behavior of functional patterns is also only dependent on the state of the world prior to their activation. Nevertheless, contrary to functions, their effects are not described as a simple state transition, but as sequences of successive states. Predictability of these patterns is still high, but it requires more information than with simple functions (e.g., landing gear extraction and retraction, flaps retraction). The cognitive complexity of the prediction is higher (Javaux, 1997; Javaux & De Keyser, 1997). Functional patterns like functions affect objects, processesor worlds by means of open control loops.

Agents finally can also be described by sequences of successive states. Nevertheless, with agents sequences are not only influenced by initial conditions but also by conditions varying during execution of the sequences themselves (e.g., agents range from low-level automatisms, such as attitude stabilizers, to high-level pilot aiding devices, such as, the performance management system or the flight management system). Agents affect objects. processes or worlds by means of closed control loops.

In more automated systems, the level of animacy of *machine agents* has dramatically increased. Once activated, systems are capable of carrying out long sequences of tasks *autonomously*. For example, advanced Flight Management Systems can be programmed to automatically control the aircraft from takeoff through landing. (Sarter & Woods, 1995, p. 6)

As suggested by the previous examples, automated functions, functional patterns, and agents are present in most technological contexts. Processes and agents are especially useful in task-intensive situations and have come to progressively replace functions in modern airplanes. Several reasons account for that evolution.

The first is that, as any human operators, pilots are limited in their perceptual abilities (e.g, they cannot fine control the airplane's attitude without assistance or manually react to events in milliseconds) and in their capacities to process information. Compare the classical concepts of bounded rationality (Simon, 1957) and short term (Miller, 1956) or working memory (Baddeley, 1986; Baddeley & Hitch, 1974 ; Reason, 1987) limitations in cognitive psychology.

> The capacity of the human mind for formulating and solving complex problems is very small compared with size of the problems whose solutions is required for objectively rational behaviour in the real world—or even for reasonable approximation of such objective rationality. (Simon, 1957, cited by Reason, 1987, p. 76)

Some external resources can be introduced to cope with these limitations, but it should be clear that purely functional resources cannot suffice in highly dynamic situations such as piloting an airplane. Because of humans' limited bandwidth input-output channels and because of their limited and rather slow processing capabilities, it is not possible to ensure correct coordination and activation of multiple functional resources. Computerized agents, on the other hand, because they can be considered as functions with autonomy, display that ability to coordinate, at least locally, several specialized functions (Maes, 1989). Agents integrate the logic behind functional integration and activation (acting on the process through functions—see the notion of competence—or recursively through simpler agents).

Producers (airplane designers) and consumers (commercial airlines) have extended the scope of the tasks expected from the global system of crew, airplane, air traffic control (ATC). The necessity to enhance safety and performance while flying in densely crowded airspace is among the main motivations for the introduction of agents in airplanes. As a result, the complexity of the flying task has grown to such levels that it has become necessary to extend the perceptive, motor, and processing capabilities of the crew. The task itself has been broken down into simpler primitive subtasks that have been allocated to specialized agents.

Thus, there has been a continuous trend in aeronautics to introduce more and more automation into cockpits. However, some problems related to this approach have been described by several human factors experts, like Sarter and Woods (1994):

New automation is developed because of some payback (precision, more data, reduced staffing, etc.) for some beneficiary (the individual practitioner, the organization, the industry, society). But often overlooked is the fact that new automated devices also create new demands for the individual and groups of practitioners responsible for operating and managing these systems. The new demands can include new or changed tasks (setup, operating sequences, etc.), and new cognitive demands are created as well. There are new knowledge requirements (e.g., how the automation functions), new communication tasks (e.g., instructing the automation in a particular case), new data management tasks (e.g., finding the relevant page within the CDU page architecture), new attentional demands (tracking the state of automation), and new forms of error or failure (e.g., mode error). (p. 17)

This kind of role sharing and interaction pattern has the long term effect of removing the pilot from the loop and decreasing system awareness especially as feedback on automation status and behavior may be poor and difficult to obtain.

Although our goals are not to ignore these very important problems, we would like to draw the attention to the problems specifically related to the interfacing of functions and agents in modern aircraft. We especially believe that some of the problems encountered in modern glass cockpits appear because the agent character of some resources has not been sufficiently recognized, and that in some cases agents have been interfaced as if they were mere functions.

As shown later, the main question behind usable interface design is, how to provide the user with the necessary knowledge and the means to interact with the resources to act and react in the world according to goals or tasks. We first show how the approach adopted by the classical human–computer interaction (HCI) community regarding the interfacing of functions on a static setting has succeeded in its attempts to answer this question; how it has provided designers with principles to support evaluation, certification, and design methodologies; and, in the end, end-users with highly usable forms of interfaces. We then show how such a strategy could be applied to interface agents in dynamic worlds.

In the end, we will have provided the reader with two sets of principles, respectively for functions and agents interfacing, that could influence the way evaluation and certification of interfaces incorporating these two types of resources are performed.

INTERFACING FUNCTIONS ON STATIC WORLDS: CLASSICAL HCI

The now-classical domain of HCI has proven its ability to solve interfacing problems with powerful computerized tools. Such successes must be related to three factors:

1. Cognitive theories of human-computer interaction have been produced.
2. Some general principles that interfaces have to verify have been defined, either as a subproduct of the cognitive theories of the interaction, or of empirical data (controlled experiments, analysis of errors, etc.).
3. Some generic forms of interfaces conforming to these principles have been designed and have received a wide acceptance.

Cognitive Theories of Interaction

Cognitive theories of interaction between users and computers have existed for some years now. Strongly influenced by the early attempts of artificial intelligence to produce models of problem-solving and planning (such as general problem solver, Newell & Simon, 1972), nearly all rely on the same approach and assume that the user achieves goals by solving subgoals in a divide-and-conquer fashion (Dix, Finlay, Abowd and Beale, 1993): goals, operators, methods, and selection rules (GOMS; Card, Moran & Newell, 1983), Cognitive Complexity Theory (Kieras & Polson, 1985), Task-Action Grammars (Payne &Green, 1986).

The GOMS model, which has served as the basis for major research in cognitive modeling applied to HCI (Dix et al., 1993), considers for example that an interaction situation can be described in terms of GOMS

- *Goals* are the user goals; what has to be achieved.
- *Operators* are the basic operations the user can perform to affect the system state (represented as state transitions).
- *Methods* describe how alternative sub-goals decomposition can help the user to reach the same goal.
- *Selection rules* attempt to predict which methods the user will use to reach goals depending on the user itself and the state of the system.

In such models, the computer is clearly considered as a static setting; that is, one whose state only changes as an effect of the actions of the user considered as the application of operators.

To illustrate how the distinction between a static object or world and its related operators or functions encounters personal experience, we analyze how the file management problem is treated on most personal computers.

Interface designers confronted with the file management problem have to define ways to represent files as they appear on some physical support (e.g. a hard disk) and provide users with the means to manipulate them. Files are usually organized on this support according to a hierarchical structure (a tree). This structure is static; it remains as it is unless the user attempts a modification. Files and directories can be moved, copied, or deleted. Individual files can be transformed thanks to applications (word processors, spreadsheets, etc.) that change their internal structure. All these operations are under the control of the user.

The desktop metaphor (Booth, 1989) elegantly solves this problem:

1. *Static objects*:The desktop metaphor is a possible alternative to the problem of representing static objects. Files and directories are represented by icons or names in lists. Files that are in use are represented by open windows

2. *Functions or operators*: Most of the functions are accessible directly on the desktop (direct manipulation interface; Hutchins, Hollan, & Norman, 1986). File displacement and deletion are operated through moves of the mouse or function activations through menus. Activation of an application on a specific file is possible through double-clicking commands or menus.

General Principles

Thanks to the coherent framework provided by the analogy with problem-solving or planning on static objects or worlds, it is possible to produce a structured and theoretically sound set of principles about properties of usable interfaces. These principles rely on four underlying ideas.

1. To act efficiently on a static object, the user must have access to some knowledge about the object itself and the functions that can be applied.

   - The user must be able to assess the current state of the object.
   - He or she must know which operators or functions can be applied to this state.
   - What transition will occur if an operator or function is applied.

Without this information, the goals cannot be reached (one would say, in terms of problem-solving or planning theory, that the problem cannot be solved).

2. Part of this knowledge is related to the static objects and the other part concerns the functions themselves.
3. The knowledge required to interact with static objects can be distributed within the interface–user system. Well-designed interfaces provide the user with a lot of knowledge about the current state of the object (visibility), the functions that can be applied (availability), and the related transitions (predictability). To paraphrase Norman (1988) in such interfaces, information is in the world. In badly designed interfaces, the

current state of the object is not visible and a lot has to be remembered (in short-term memory). It is hard to tell which functions can be applied or what their effects will be. In such interfaces, information is in the head.

4. Principles (necessary principles) that warrant the presence and availability of the necessary knowledge can be stated. Secondary principles, considered less important, can be proposed to indicate how to make the interface more usable or how to support the user in his/her tasks.

Principles for Static Objects

Visibility—Can I See What I Need to See? The goal of this principle is to ensure that the user might have a full and accurate representation of the current state of the object under control.

Interpretability—Do I See What I'm Supposed to See? It is not sufficient for the user to have access to a representation of the object. A representation conveys some meaning about some real situation that is abstracted into symbols that have to be interpreted by the user. This principle ensures that the user correctly interprets the representation. Some simpler but nevertheless essential principles usually support interpretability: consistency or coherence of the symbols and of their interpretation, and familiarity and generalizability of the symbols.

Flexibility—May I Change the Way I See? The possibility of tuning the representations, in particular to modulate the informational flow according to the bandwidth of the human cognitive processing limitations and the particular needs of the current situation, is an especially desirable property of usable interfaces.

Reliability—Is This Thing the Real Picture? This one deals with a critical feature of any interface. It must be reliable, and the user must be confident with the information it provides or the resource it helps to use. When applied to object representation, the reliability principle wonders whether the representation presented to the user constitutes an accurate and reliable representation of the object and how this can be assessed by the user.

Learnability—Can I Avoid Headaches? This principle is important because the way users accept new products or interfaces is influenced by their learnability. In the case of a static object representation, how easily can the user learn the rules that help to correctly interpret the representation? Once again, simpler principles such as consistency, familiarity, and generalizibility strongly contribute to facilitating learnability.

Principles for Functions

Availability—What Can I Do? To apply functions on the static object as if they were operators, the user must be in a position to decide which functions can be applied on the object. General availability refers to the complete list of functions provided by the interface. Local availability concerns the limited list of functions applicable to specific states of the object. Knowledge concerning both types of availability should be accessible to the user.

Accessibility—How Can I Do It? Once the user has gained some knowledge about which functions can be applied on the object and has chosen one or a sequence of them to apply, he or she has to specify it for the interface. Knowledge about how to access functions and how to activate them on the correct objects should be available to the user. Consistency, familiarity, and generalizability are once again among the simpler principles that help the user to access functions.

Predictability—What Will Happen? Predictability is without any doubts the essential principle to conform to. In problem-solving and planning models, the ability to predict how the state of the world will change when an operator interfaces with a machine is crucial for resolution or task satisfaction. The user must possess the necessary knowledge to be able to generate plans or sequences of actions on the interface that in the end will meet its goals. Modes, if any, have to be made visible to the user, because they influence, by definition, the way functions behave and thus constitute a threat to predictability (Degani, 1996).

Feedback—How Is It Working and What Are The Effects? Feedback is essential because it permits the user to assess that the intended state has been reached and hence that the activated function has been applied correctly. Feedback is thus associated with error detection, but also with the ability to learn (see learnability principle) the necessary knowledge to predict functional effects (see predictability principle). Two forms of feedback are usually encountered. The first type concerns the visibility of the function status (progression bars) and helps to confirm that the access to the function has been successful (see accessibility principle). The second type of feedback ensures that the effects of the activated functions are visible. *Stricto sensu*, this second form of feedback is more concerned with visibility of the representations, and is thus not a pure functional principle.

Controllability—How the Hell Do I Stop This Thing From Erasing My Hard Disk? As dramatically stated by the previous sentence, controllability is a particularly desirable feature. Nevertheless, in general, interfaces provide a very limited set of interactions between a running function and the user (otherwise, it would be an agent). Control is usually limited to interruption (either temporary or definitive) of the function execution.

Flexibility—Can I Do it the Other Way? Users are not machines, and they are faced with very variable tasks. Moreover, users all differ. They have different backgrounds, different cognitive styles, and usually different goals. For such reasons, while not resorting to the major, necessary principles, flexibility is generally appreciated by users.

Automatibility - Can I Automate This Sequence of Operations? There are two facets to automatibility with obvious advantages. Machine-supported automatibility refers to the possibility for the user to define *macros*, automated sequences of functional activations, with or without parameters. Cognitive automatibility concerns the ability of the user to automate the motor and cognitive processes that support its access to functions. This form of automatibility is strongly conditioned by good visibility of the objects and easy and consistent access to functions.

Task Conformance—Does It Really Cover All My Needs? This principle concerns the scope of the available functions, regarding the nature of the task they are to perform. It can be considered from a general point of view (the global availability) or more locally according to specific situations (the local availability compared to the local task); that is, is the function available when needed?

Error Management—What if I Err? Users are fallible (Reason, 1990). Good interfaces have to take this into account and exhibit error resistance (prevent users from making errors; e.g., Masson & De Keyser, 1993) and error tolerance (help users to correct effects of errors through reversibility, escapability, recoverability of function applications).

Reliability—Is This Stuff Really Working as It Is Supposed To? Although extremely reliable systems, modern computers are nevertheless mere material human artifacts. Consequently, they suffer from design errors as well as from the usually hidden effects of the second law of thermodynamics. At the software level, bugs are present in any complex application. At the hardware level, problems and troubles sometimes occur due to heat, dust, fatigue, or even failure of a component. Interfaces should furnish the user with means to ensure that the functional resources effectively affect the state of the object as reflected by its representation and the different feedback mechanisms.

Learnability—Can I Avoid Headaches? Learnability of functions is essential. As already stated for object-related principles, it is generally a necessary condition for the acceptance of an interface. Several aspects can be learned and thus lead the user to eliminate exploratory or information-seeking behaviors. Every piece of the necessary knowledge related to the primary principles (availability, accessibility, and predictability) can be learned. Some rules that help the user to deduce such essential pieces of information from the representation of the object can also be abstracted and then greatly contribute to simplify the activation of the functional resources; hence the pervasive character of the consistency principle.

GENERIC FORMS OF INTERFACES

Classical HCI has also succeeded in its attempts to apply these principles to interface design. Graphical user interfaces (GUIs), and especially windows, icons, menus and pointers interfaces (WIMP; Dix et al., 1993), which constitutes the standard interface for interactive computer systems (Macintosh, Windows-based IBMs and compatibles, desktop workstations) have proven their usability to millions of end users.

Such kinds of interfaces indeed provide users with an excellent visibility over the current state of the objects or world manipulated (e.g., the desktop of the Macintosh) as well as consistent and familiar rules to interpret the representation (the desktop metaphor). Users habitually have the opportunity to tune these representations (e.g., different ways to display files in a directory) and this contributes to the interface flexibility. Moreover, such interfaces are highly learnable, especially because of their coherent and metaphoric nature.

GUIs and WIMPs equally perform at their best regarding functions. Availability is usually very well documented by the interface. This is at least true of the most used functions. Less common functions are not well known to users, especially in the case of very powerful tools such as word processors that provide users with hundreds of functions. Accessibility is extremely good, thanks to the mouse and its clicking commands and to menus, (which also contribute to availability). Predictability is good (at least for simple operations on the desktop) because of the coherence of the access rules and the already quoted metaphoric nature of the interface. Feedback is immediate, but restricted to objects visible on the desktop. Controllability is limited, but it is enhanced for functions that have destructive effects on the desktop. Flexibility is usually good, thanks to the several different ways to perform operations (directly on the desktop or through menus). Macros are provided as default features or they can be added thanks to dedicated applications. Task conformance is the principle where these graphical interfaces are at their worst: The possible scope of what can be done is somehow limited, especially if compared to very powerful command languages (e.g., Unix) dedicated to file management. Errors are handled differently by the manufacturers of common GUIs. Operations that imply a displacement of files between two places (a move operation) can generally be undone, but file deletion is sometimes an operation that cannot be reversed without specifically dedicated tools. Learnability, finally, is usually extremely good (perhaps it is in the end the main reason for the success of these interfaces within a computer-illiterate population) especially because of the so-praised consistency of the interface (even between applications) and the metaphor of the desktop.

INTERFACING AGENTS ON DYNAMIC WORLDS: HCI GOES TO THE REAL WORLD

We have carefully analyzed the approach followed by classical HCI to solve the problems related to the interfacing of functions in static worlds. Now we would like to see how such a strategy can be applied to interface agents in dynamic worlds. According to other authors (Kay, 1990; Laurel, 1990), the interfacing of agents is the challenge of tomorrow's HCI. We already have shown how agents constitute invaluable resources for users, operators, or pilots in their respective tasks. That is why manufacturers and designers have introduced them at several different levels of automation used in process control.

Cognitive Theories of Interaction

There has been for a few years an emergent interest about ideas related to the integration of a distributed work or processing force into a coherent and goal-oriented whole. Computer science, for example, has already produced numerous formal studies about parallel processing and synchronization problems. Distributed artificial intelligence aims at designing systems or societies of agents (multiagent systems) that collectively exhibit the ability to solve complex problems with more robustness than classical approaches (Maes, 1989). On the linguistic side, Winograd and Flores (Flores, Graves, Hartfield, & Winograd, 1988) developed linguistic-based theoretical perspectives for analyzing group actions. Coordination theory (Malone & Crowston, 1990) as a general and abstract theory tries to establish the connections between several different disciplines that are concerned with similar coordination phenomena. On the applied side, computer support to cooperative work is aiming at providing organizations or groups of users with better ways and tools to work together (Dix et al., 1993). As demonstrated by the next excerpts, concerns about agents and modeling human–agent interaction have even been expressed in aeronautics by human factors authors.

> Pilots were surprised when the aircraft did not respond as expected; they did not realize or understand why their instructions to the automation had not resulted in the desired change. In some sense, this is a good

example to show how pilots try to communicate with the system in a way analogous to communication with another human agent. They assume that entering the desired target value is sufficient for the system (as it would be for a human crew member) to understand that it is supposed to achieve this new target and how it is supposed to do so in detail.

(This) direction is to consider supervisory control of automated resources as a kind of cooperative or distributed multi-agent architecture (Sarter & Woods, 1995, p. 16).

Despite these efforts and remarks, there is nothing like a single and coherent body of theory about coordination between agents (Malone & Crowston, 1990), and it is hard to think of any integrated cognitive theory of interaction between humans considered as agents, or between humans and automated agents. Nevertheless, there is more and more awareness of the similarities between the problems encountered by researchers involved in these approaches to cooperative systems as is witnessed by the increasing number of workshops or conferences on the topic. On the cognitive side, expectations about future progress will rely on domains such as social or developmental cognitive psychology as well as psycholinguistics to produce a coherent and integrated theory of human interaction with agents.

General Principles

Designers faced with the problem of interfacing agents are still left without the sort of powerful framework they used to rely on when designing functional interfaces. Nevertheless, some important principles that interfaces with agents should verify can already be stated, thanks to extensions of the basic principles for functional interfaces, to reflections about the necessary knowledge required for usable interaction, and to recommendations formulated by analysts when incidents with such interfaces were reported. We will try to rely on these excellent studies of problems and incidents encountered with automation in modern glass cockpits as major sources for defining general principles.

On the epistemic side, it is at least clear from a formal point of view that more knowledge (distributed between the user and the interface) is needed to control a dynamic object world than a static one. Anticipatory behaviors, of which some researcher have shown the heuristic value (Van Daele, 1992), are only possible if the user, operator, or pilot has some knowledge or ability to predict how the controlled system will naturally evolve if no action is taken. Interfaces to dynamic worlds or objects should therefore provide the user with such knowledge or resource (see the predictability principle for dynamic worlds).

More knowledge is also needed to interact with agents than with functions. Agents can be of numerous different types and differ in terms of complexity (ranging from reactive to cognitive agents; Erceau & Ferber, 1991). Whatever the importance of such factors, the main difficulty with agents certainly comes from their flexibility (complex agents can exhibit different behaviors in similar situations) and from their autonomy (agents incorporate their own logic behind functional activation and act autonomously on the world). As a consequence, agents must be considered as generally less predictable resources than functions. The cognitive operation of predicting and controlling agent behaviors is more complex. The dynamicity and autonomy of the agents increases cognitive complexity (Javaux, 1997; Javaux & De Keyser, 1997).

Respective Scopes or Competencies and Cooperative Modes. A supplementary and rather essential distinction must be introduced before devoting some attention to the principles. It concerns the distribution of competence between the user, the operator or pilot, and the agent. With a functional resource, control behaviors while it is running are limited (controllability is low). To use a multiagent terminology, one would say that only two cooperation modes are possible: Either the job is done by the function or it is done by the user. The situation is quite different with agents. Because such resources display possibilities for extended controllability (controllability is high), they provide the capability for more complex cooperation modes.

Three classes of cooperation modes have to be considered:

1. The job is done by the user. The agent is not active or works on another task. This corresponds to the concept of direct manual control. According to Billings (1991) *direct manual control* is characterized by the pilot's direct authority over systems, manual control using raw data, unaided decision making, and manual communications.

    However, as pointed out by Billings (1991), no modern aircraft can be operated entirely on that mode: "Indeed, an aircraft operated even by direct manual control may incorporate many kinds of control automation, such as yaw dampers, a pitch trim compensator, automated configuration warning devices, etc." (p. 26).

2. The job is done by the agent. The user is not active or works on another task. This is precisely the meaning of the autonomous operation concept. As summarized by Billings (1991), autonomous operation is characterized by the fact that the pilot has no role to play in operation, that the pilot has normally no reason to intervene, and that the monitoring is limited to fault detection (ibidem, p. 26).

    Until the introduction of the A320 and MD11, very few complex systems were operated in a full autonomous fashion. In those new aircraft, however, major systems operate this way. For example, in the MD11, failure detection and subsystem reconfiguration are performed autonomously. Longitudinal stability and control wheel steering are also autonomous operations (see Fig. 25.2)

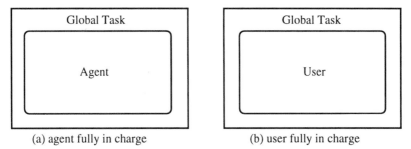

(a) agent fully in charge	(b) user fully in charge

FIG. 25.2. Agent and user information.

2. The job is done by both the user and the agent. Each of them has its own part of the task. Two situations or cooperation patterns have to be distinguished (see Fig. 25.3)

The first situation is that the two tasks are exclusive. This occurs for example when the agent and the user work on two different subsystems. An example of such a sharing pattern is given by Billings (1991):
The pilot may elect to have the autopilot perform only the most basic functions: pitch, roll and yaw control... he or she may direct the automation to maintain or alter heading, altitude or speed, or may direct the autopilot to capture and follow navigation paths, either horizontal or vertical... . In all cases however, the aircraft is carrying out a set of tactical directions supplied by the pilot. It will not deviate from these directions unless it is capable of executing them." (p. 28).
The second situation is that the two tasks share a common part. This could occur when the agent and the user do work on the same subsystems. In such cases, conflicts are likely to arise and resolution techniques have to be provided.
For example, in the 320, the flight control system incorporates an envelope limitation system that operates at all times and interacts with pilot's commands, to guarantee that safety barriers are not overcome. For example, bank angle, pitch and angle of attack cannot be exceeded by the pilot unless the flight control computer is turned off.

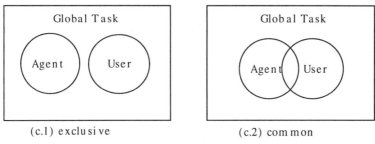

FIG. 25.3. Agent and user scopes.

Moreover, cooperation patterns with agents cannot solely be considered in a static perspective (they are fixed and cannot be changed over a task session or a flight). Dynamic (cooperation) pattern changes are also observed in modern cockpits (patterns change over the course of a task session, either through agent or user instruction).

> An important characteristic of *automatic flight-path control* is the high degree of dynamism. Transitions between modes of control occur in response to pilot input and changes in flight status. Automatic mode changes can occur when a target value is reached (e.g., when leveling off at a target altitude), or they can occur based on protections limits (i.e., to prevent or correct pilot input that puts the aircraft in an unsafe condition). (Sarter & Woods, 1992, p. 306)

For such reasons, as stated by Billings (1991), feedback (see the feedback principle) should be given to the user or pilot whenever an important mode change occurs:

> Automation should never permit a situation in which "no one is in charge"; pilots must always "aviate" even if they have delegated control to the autopilot. It is for this reason that autopilot disconnects are usually *announced* by both visual and aural alerting signals. (p. 85)

To confirm the importance of issues related to cooperation modes, Sarter and Woods (1992) also described how the pilot's inability to dynamically change modes can lead to some drastic measures:

> During the final descent, the pilots were unable to deselect the APPR mode after localizer and glideslope capture when ATC suddenly requested that the aircraft maintain the current altitude and initiate a 90° left turn for spacing. They tried to select the ALT HOLD and HDG SEL modes on the MCP to disengage the APPR mode and comply with the

clearance, but neither mode would engage and replace the APPR mode. They finally turned off all autoglide systems. (p. 311)

This leads us to some critical remarks about the way evaluation or certification of agent-based interface relying on principles should be performed.

- The analysis should begin with a very careful study of the possible cooperation modes between the user and the agent.
- It should detail who is in control of the cooperation mode changes and the relative scopes of the user and the agent within a given   cooperation mode (task migrability).
- For each possible cooperation mode, consider how the duality user–agent is positioned according to the principles. However, due to the very different ways the task is conducted in different cooperation modes, principles have to be applied with some nuances in mind and be related to the current specificities of the current mode.

In a short-response-time agent (e.g., a regulator), whose capabilities are far beyond those of the pilot, the cooperation mode is such that the task is exclusively under agent control. The main principles in this situation are reliability and the capability for the pilot to assess that the agent is working in its competence domain. Principles such as predictability, which used to be essential for functional resources, are hereby not necessary (e.g., the gyroscopic stabilizer of Maxim in 1891, the stability augmentation system of Wright in 1907, and their successors in modern autopilots).

Principles for Dynamic Worlds or Objects

Visibility—Can I See What I Need To See? Visibility of the objects or world acquires hereby a special status due to their dynamicity. With static objects where the world does not change spontaneously, the user or pilot can rely on short-term memory to maintain awareness and orientation. With dynamic objects or worlds, updating is necessary and this is only possible through predictions or observations of the future states. In a particularly complex dynamic context with heavy task constraints, the concept must even be extended to meet the notion of situation awareness (Sarter & Woods, 1991). In such situations, it is not enough to provide the user with the means to perceive the state of the world, but also to ensure that it will be attended to.

Situation awareness has recently gained considerable attention as a performance-related psychological concept. This is especially true in the aviation domain where it is considered an essential prerequisite for the safe operation of the complex dynamic system "aircraft." There are concerns, however, that inappropriately designed automatic systems introduced to advanced flight desks may reduce situation awareness and thereby put aviation safety at risk. (Sarter & Woods, 1991, p. 45)

The problem of the amount of information that must be visible is also addressed by Billings (1991): "How much information is enough? How much is too much? Though pilots always want more information, they are not always able to assimilate it" (p. 46).

As pointed out by Billings (1991), such a question should be answered (as suggested earlier) according to a clear consideration of the cooperative mode between the pilot and the agent and their respective involvement in the control task. "Pilots need much less information when subsystems are working properly than when they are malfunctioning" (p. 46).

Interpretability—Do I see What I'm Supposed To See? No significant difference with functional counterpart.

Flexibility—May I Change the Way I See? Flexibility applied to the representation of the dynamic object or worlds means that the user or pilot is capable of adapting this representation to its current or future needs. Such a possibility is present in several subsystems displays (e.g., changing the range on the navigation display on the OVD range) of glass cockpits.

Predictability—What Will Happen if I Stop Acting? The new principle is based as already stated on the heuristic value of predictory or anticipatory behaviors in most dynamic control situations (Van Daele, 1992). Good interfaces for dynamic worlds should provide the users with means to anticipate future states. Several examples of such an approach already exist in aeronautic contexts. On ATC radar control screens, airplanes can be represented with small tails indicating their speed and direction. This helps operators to anticipate their future trajectory. On traffic alert and collision avoidance system (TCAS) screens, vertical speeds of surrounding airplanes are represented by small arrows. In general any graphical representation of a trend indicator can contribute to the predictability of the dynamic objects.

Reliability—Is This Thing the Real Picture? The problem of reliability related to dynamic worlds and objects was perfectly stated by Billings (1991):

"It must be kept in mind that sensors, processing equipment or display generators can fail, and that when incorrect information is presented, or correct information is not presented, there is the potential for confusion in the minds of pilots" (p.40). An interface on a dynamic object should help the user to ensure that it functions correctly, both in its ability to display correct and accurate information about the real state of the monitored systems and in its ability to inform the user about future states (support to predictability principle). Redundancy of display equipment or availability of displays related to interdependent subsystems can help the user or pilot to ensure that the informational interfaces are functioning correctly thanks to comparison or interdisplay coherence checking.

Learnability—Can I Avoid Headaches? Here, again, there is no significant difference with functional counterpart.

Critical State Assessment - Is This the Right Place to Stand Still? This new principle concerns the peculiar problems associated with dynamic worlds. In such worlds, states are rarely equivalent. Some of them require special care or attention, either because the monitored system ventures within space regions where irreversible damages could be observed, or because its intrinsic dynamic could lead the user or pilot to lose control. Interfaces that provide critical state assessment support users and help them to enhance their performance.

Principles for Agents

Availability—What Can I Do? Availability as such refers to the capability the user has to decide whether a resource exists and is available. Users or pilots should be informed of the different agents they might use as resources (global availability) as well as when these can effectively be used (local availability), and in which cooperative modes.

Accessibility—How Can I Do It? Users or pilots willing to use an agent as a resource should have access to some knowledge about how to activate and configure it in the cooperative mode of their choice (if they are in control of this variable).

Predictability—What Will Happen? Predictability is, without a doubt, one of the principles that must be considered with extended caution when trying to interface agents. As previously stated, agents are autonomous systems. They consequently present less predictable behavior than functions (by definition, their behavior is influenced by external variables). Numerous examples of incidents related to a lack of predictability of some agents in glass cockpits have already been reported. Sarter and Woods (1994) realized a study through a questionnaire asking pilots to describe instances in which FMS behavior was surprising and to report modes and features of FMS operations that were not or poorly understood. One hundred thirty-five B–737–300 line pilots from an airline company participated in that survey (Sarter & Woods, 1994):

Pilots indicated that the algorithms underlying the calculation of VNAV are not transparent to them. They cannot visualize the intended path; therefore, they are sometimes unable to *anticipate* or *understand* VNAV activities initiated to maintain target parameters... several pilots reported that they have been *surprised* by VNAV when it failed to start the descent on reaching the top-of-descent (TOD) point. (p. 310)

The problem the user or pilot is faced with is one of agent modeling. Designers must ensure that they provide the user with a correct model of the agent. Two principal classes of models govern the theories about agents: mechanistic models and intentional models. In mechanistic models, the user relies on a kind of finite-state machine approximation of the agent whose behavior can be predicted thanks to the awareness of relations between some internal parameters or variables of the agent, the input it is actually processing and the resulting behavior. In intentional models of agents, the user predicts future behaviors of the agent on the basis of its goals or intentions. Consequently, and whatever the type of model held by the user (depending on the type and complexity of the agent), it seems essential that any important autonomous change that might modify the way the agent will behave in the near future (a change of mode in mechanistic models or a change of goals or intentions in intentional models) is reported to the user. Javaux (1998) presented a theory that attempts to explain why and predict how the pilots will actually build incomplete mental models of mode behavior on the B-737-303, the aircraft studied by Sarter and Woods. The models are incomplete or erroneous because mode behavior is inconsistent (Javaux, 1997), and because the pilots lack the experience of the most irregular and rare patterns of behavior (Javaux, 1998).

Scope or Competence Awareness—What Can This Thing Do and When? This new and essential principle concerns the competence of the agent: what it (can) do and in which circumstances. With purely functional resources, competence awareness is close to predictability. Functions induce simple state-transitions (what is does) from the states upon which they apply (in which circumstances). Due to the extended flexibility and autonomy of agents, this similarity does not appear and a new principle has to be introduced. Scope awareness is extremely important, at least when the user or pilot is in control of the cooperation modes and of task migrability: the pilot must be able to assess that the agent is performing reliably (reliability principle) and correctly (accordingly to the task) in its domain (scope awareness) of competence. Designers must consequently provide the user or pilot with this knowledge through the interface, documentation, and/or training courses.

Feedback—How Is It Working and What Are the Effects? This is another essential principle for agents. Due to the new problems introduced with predictability of the agent and the correlated needs to model its behavior, the visibility of agent status increases in importance. As already reported, mode awareness (in mechanical models) is a condition for real cooperative work between the user or pilot and the agent.

> Pilots reported that they are surprised by uncommanded mode transitions that occur on reaching a target state or for protection purposes. Most often, the reports referred to the automatic reversion from vertical speed mode to LVL CHG mode, which occurs if the airspeed deviates from the target range due to an excessive rate of climb or descent.
> Pilots' reports seem to indicate that such uncommanded changes are difficult to track given current cockpit displays and indications. (Sarter & Woods, 1992, p. 311)

Visibility of the agent's effects are of equal importance. Because agents display autonomy, any change introduced on the dynamic world or objects by the agent should be reported or be at least visible to the user, especially in cooperative modes where both the agent and the user are in charge of interacting tasks. It is also due to this visibility that the adequacy of the decision to activate the agent as well as its reliability can be assessed. See also Billings (1991) and the concept of fail-passive control automation situations that describe hazardous conditions where visibility of the agent effects is lowered.

Controllability—How the Hell Do I Stop This Thing Grounding My Airplane? Because of the autonomy of agents, and their ability to induce disastrous effects on the controlled world or objects, controllability should remain high in all circumstances. Several authors described how "clumsy automation" can contribute to lower controllability in circumstances where it is especially needed.

> Significantly, deficits like this can create opportunities for new kinds of human error and new paths to system breakdown that did not exist in simpler systems. (Woods, Cook, & Sarter, 1993, cited in Woods, 1993, p. 2)

It is clear evidence that users or pilots should have the capability to disengage automation (agents) or at least change the current cooperative mode to some mode where they are more involved whenever they think it is needed. Billings (1991) stated this very precisely; as shown in Fig. 25. 4.

Premise
The pilot bears the ultimate responsiblity for the safety of
any flight operation

Axiom

The human operator must be in command

FIG 25.4. Principles of human-centered automation
(from Billings, 1991, p. 12).

Billings (1991) also expressed serious concerns about recent examples of
violation of such principles. The flight control system of the A320 and its
envelope limitation operate at all times, and they cannot be disengaged by the
pilot. In the MD–11, major aircraft systems operate autonomously to a large
extent.

(Civil aircraft) do on occasion have to take violent evasive action,
and they may on extremely rare occasions need control or power
authority up to (or even beyond) structural and engine limits to cope
with very serious failures. The issue is whether the pilot, who is
ultimately responsible for safe mission completion, should be
permitted to operate to or even beyond airplane limits. (p. 29)

Error Management—What if I Err? As pointed out by Billings (1991),
system operation errors are responsible for roughly two thirds of air carrier
accidents. It thus mandatory, as for functions, to design error-resistant and error-
tolerant agent interfaces that attempt to minimize the effects of human error.
Monitoring capabilities in the automation, system envelope limitations and
procedural control are among the currently investigated techniques to enhance
safety.

Task Conformance—Does It Really Cover All My Needs? Here again,
there is no significant difference with the functional counterpart.

Flexibility—Can I Do It the Other Way? The multiplicity of ways a given
resource can be used is usually a rather desirable feature, especially because it
provides the user or pilot with a choice within several different strategies to
achieve the same goal.

For example, an automated cockpit system such as the Flight
Management System (FMS) is flexible in the sense that it provides

pilots with a large number of functions and options for carrying out a given flight task under different circumstances. There are at least five different methods that the pilot could invoke to change altitude. (Sarter & Woods, 1995, p. 6).

However, with complex cooperative agents, flexibility can strongly contribute to the clumsiness of automation and lead to very serious problems, as is witnessed by the Sarter and Woods (1995):

> This flexibility is usually portrayed as a benefit that allows the pilot to select the mode best suited to a particular flight situation. But this *flexibility has a price*: the pilots must know about the functions of the different modes, how to coordinate which mode to use when, how to "blumplessly" switch from one mode to another, how each mode is set up to fly the aircraft, and how to keep track of which mode is active. These new cognitive demands can easily congruate at high tempo and high criticality periods of device use thereby adding new workload at precisely those time periods where practitioners are most in need of effective support systems.

Clumsy use of technological possibilities such as the proliferation of modes, creates the potential for new forms of human–machine system failure and new paths towards critical incidents, e.g., the air crashes at Bangalore (e.g., Lenorovitz, 1990) and Strasbourg (Monnier, 1992). (p. 2)

Reliability—Is This Stuff Really Working as It Pretends? The reliability principle is extremely important with agents, especially because of their capability for autonomy and of the corresponding tendency of users to rely blindly on them.

As an example of overconfidence in automation, we would like to mention the accident of a Scandinavian Airlines DC–10–30 that occurred at Kennedy Airport on February 2, 1984. In this accident, the airplane touched down 4,700 feet beyond the limit of an 8,400 feet runway, was then steered to the right, and landed in water 600 feet beyond the runway. The accident was due mainly to a failure of the throttles to respond to the autothrottle speed control system commands and to the excessive confidence of the captain in the reliability of that autothrottle system, in spite of a one-month history of malfunctions. As noted by the NTSB among other causes, the "performance was either aberrant or represents a tendency for the crew to be complacent and over-rely on automated systems" (cited in Billings, 1991, p. 99).

As pointed out by Billings (1991), the ability to assess reliability is related to visibility of the world or objects and to predictability of the agent behavior (either based on mechanical or intentional models):

> It is thus necessary that the pilot be aware both of the function (or dysfunction) of the automated system, and of the results of its labors, on

an ongoing basis, if the pilot is to understand why complex automated systems are doing what they are doing. (p. 83).

However such a strategy might fail simply because of the stable condition the controlled process is in. Automatic monitoring of the agent reliability and visibility on its status is badly needed in such situations.

> "Fail-passive" control automation represents a particular potential hazard, in that its failure may not change aircraft performance at the time if the airplane is in stable condition. Such failures must be announced unambiguously to insure that the pilots immediately resume active control of the machine. (Billings, 1991, p. 85)

Learnability—Can I Avoid Headaches? As with functional interfaces, comments must be made about the strong relation between the learnability of agent interfaces and their success measured in terms of acceptance by users as means to access the full capabilities of the resources, in a safe and error-free fashion, and without the side effects (clumsy automation, shift or loss of expertise, etc.) usually observed. Given the amount of knowledge that must be learned to interact cooperatively with an agent (this point is developed later), learnability of agent interface and behavior must be (very) high. A few possible solutions are described in the section about generic interface design.

GENERIC FORMS OF INTERFACES

To begin with interfaces, some special points must be made about the amazing amount of knowledge required to interact fruitfully with agents. Users or pilots must be educated about the availability of agents (when they can be used), their accessibility (how they can be used), their scope or competence (what they can and cannot do), their predictability (how they will behave or act on the objects under control), the related mental models of their functioning, and finally their controllability (how they can be controlled). Moreover, they must develop skills or mental processes dealing with how to communicate with them, how to evaluate their reliability through predictability and visibility of the controlled world or objects, how to use or require feedback to enhance predictability itself, how to manage errors when they occur, and so on. To gain this knowledge or develop the means to access it is an extremely important task that user or pilot faces (hence, the clumsy automation problems and shift of workload toward more cognitive tasks reported by many authors). Moreover, to add to the task, the knowledge is required for each agent the user or pilot is interfaced with.

Our claim is that many problems described in modern glass cockpits could be avoided if these simple—but overwhelming—considerations were taken into account.

A possible and promising solution, as already demonstrated with function interfacing through Direct Manipulation and metaphoric interfaces, is to provide the user with a lot of the necessary knowledge embedded in the interface itself and with the means to extract it whenever needed.

A second and complementary approach is to reduce the amount of knowledge required to interact with agents. This is especially true at the level of the cockpit considered as a global work environment (or *macro interface* with functions, functional patterns, or agents provided as resources to interact with the airplane, the airspace, and the ATC). Introducing intra-and interagent coherence into cockpits seriously contributes to limiting the necessary knowledge to use them: Agents can be classed according to the kind of cooperative modes they entertain with the crew, and coherent communication protocols, feedback techniques, and support to mental modeling can be established. The current situation with cockpits might be similar to the situation of interactive computers prior to the introduction of coherent GUIs, when every application had its own way to interact with the user.

Another important issue already considered by designers as decreasing the amount of knowledge that is not intuitive is familiarity. Thanks to the introduction into cockpits of more "natural" cooperative and communication modes e.g., multimodal and multimedia), the everyday life experience of interaction situations could be made more usable.

CONCLUSION

The influences of the introduction of new and sophisticated automation technologies in the last generations of commercial aircraft regarding the pilots–systems interactions has been extensively described by numerous experts in aeronautics and human factors engineering.

Technology allows a proliferation of interaction possibilities with an increasing level of automation autonomy and poor feedback capabilities. These changes create new cognitive demands for the pilots, demands that are highest precisely during the most critical flight phases, when one would have expected the automation to be of the highest utility (Sarter & Woods, 1995; see also Moll van Charante, Cook, Woods, Yue, Howie, 1992, for similar results in the medical domain).

In summary, the complexity and lack of transparency of current automation challenge the pilot's ability to cooperate with the sophisticated systems he or she is provided with. At least three sets of measures can be explored to tackle the difficulties showed between current technologies and designs. The first set of measures would aim at improving the crew–automation interface as previously suggested. A second approach to improve the quality of the cooperation is to decrease the cognitive demand on the pilot. More natural cooperative and communication modes are considered by cognitive psychologists as rather effortless processes, thanks to the many years spent learning and automating them to interact with other humans. Improving mutual models of each other (it reduces the need to communicate), increasing reliability and the means to assess it, giving agents the possibility to provide an awareness of their own scope or competence, or providing dynamic feedback for important modes or intentional changes (e.g., Billings, 1991; Onken, 1992b; Sarter & Woods, 1995) are among the several paths designers follow. The third set of measures is to conceive the interactions between the pilots, the various automated resources, and even the ATC and other airplanes as a distributed cooperative multiagent architecture in which each partner is engaged, in collaboration with all other agents, in the pursuit of a common system goal.

To sketch the current problems encountered with the technology-centered automation, Wiener (1989) reported that the most common questions asked by pilots in glass cockpits are the following: What is it doing?, Why did it do that? and What will it do next? Sarter and Woods (1995) added: "how in the world did I ever get into that mode ?"

We believe that all those interrogations could be reinterpreted in the light of the concepts and methodology developed in this chapter. According to the analysis made on the effects of current automation in cockpits, we suggest extending that list by adding: *How can I coax agents into performing what I want them to?*

But as we have tried to highlight, this is possibly not the right way to envisage operator–automation interactions. We here suggest that a shift in view could be fruitful, which would envisage both human and artificial agents as collaborative partners. New technologies should facilitate that shift.

The question to be asked should rather be: How can we together perform the missions *I am* in charge of?

Facing that new complexity, we suggest that the certification of future equipment and designs could benefit from a systematic methodology aimed at identifying the most critical problems in pilot–automation interactions. This chapter constitutes one attempt to come up with such a methodology.

REFERENCES

Amalberti, R. (1992). Safety in process-control: An operator-centred point of view. *Reliability Engineering and System Safety, 38* 99–108.

Baddeley, A. D. (1986). *Working memory.* Oxford, UK: Oxford University Press.

Baddeley, A. D. , & Hitch ,G. (1974). Working memory. In G. H. Bower (Ed.), *Advances in Learning and Motivation,* (Vol. 8., pp. 47-90) New York: Academic Press.

Bainbridge, L. (1987). Ironies of automation. Increasing levels of automation can increase, rather than decrease, the problems of supporting the human operator. In J. Rasmussen, K. Duncan & J. Leplat (Eds.). (pp. 276–283). *New technology and human error.* Chichester, UK: Wiley..

Billings, C.E. (1991). *Human-centered aircraft automation.* (NASA Tech. Memo No. 103885). Moffett Field, CA: NASA–Ames Research Center.

Booth P. (1989). *An introduction to human–computer interaction.* Hillsdale, NJ Lawrence Erlbaum Associates.

Card, S. K., Moran, T. P., & Newell, A. (1983*). The psychology of human–computer interaction.* Hillsdale, NJ: Lawrence Erlbaum Associates.

Degani, A. (1996). *Modeling human-machine systems: On modes, error, and patterns of interaction.*Unpublished doctoral thesis, Georgia Institute of Technology, Atlanta.

Degani, A., Shafto, M., & Kirlik, A. Modes in human-machine systems: Review, classification, and application. *The International Journal of Aviation Psychology 9(2)* 125–138.

De Keyser, V. & Javaux, D. (2000). Mental workload and cognitive complexity. In N. B. Sarter & R. Amalberti (Eds.), *Cognitive Engineering in the Aviation Domain.* Mahwah, NJ: Lawrence Erlbaum Associates. To appear.

Dix, A., Finlay J., Abowd, G., & Beale, R. (1993). *Human–computer interaction.* Cambridge, UK: Prentice Hall.

Erceau, J., & Ferbe,r J. (1991, June). L'intelligence artificelle distribuée (Distributed artifical intelligence). *La Recherche. 22.* 750–758.

Flores, F., Graves, M., Hartfield, B., & Winograd, T. (1988) Computer systems and the design of organizational interaction. *ACM Transactions on Office Information Systems, 6*(2), 153–172.

Hutchins E. L., Hollan J. D., & Norman, D. A. (1986). Direct manipulation interfaces. In D. A. Norman & S. W. Draper (Ed.), *User centered system design,* (pp. 87–124). Hillsdale, NJ: Lawrence Erlbaum Associates. 87-124.

Javaux, D. (1997). Measuring cognitive complexity in glass-cockpits: A generic framework and its application to autopilots and their modes. In R. Jensen & L. Rakovan (Eds.), *Proceedings of the Ninth Symposium on Aviation Psychology* (pp. 397–402) Columbus, OH: Ohio State University.

Javaux, D., & De Keyser, V. (1997). Complexity and its certification in aeronautics. In *Proceedings of the 1997 IEEE International Conference on Systems, Man and Cybernetics,*Orlando, FL: Institute of Electrical and Electronic Engineers (IEEE).

Javaux, D. (1998). Explaining Sarter & Woods' classical results. The cognitive complexity of pilot–autopilot interaction on the Boeing 737–EFIS. In N. Leveson & C. Johnson (Eds.), *Proceedings of the 2nd Workshop on Human Error, Safety and Systems Development,* Seattle, WA: University of Washington.

Kay A. (1990). User interface: A personal view. In B. Laurel (Ed*.) The art of human–computer interface design.* (pp. 191–207). Reading, MA: Addison-Wesley.

Kieras, D. E., & Polson, P. G. (1985). An approach to formal analysis of user complexity. *International Journal of Man–Machine Studies, 22,* 365–394.

Laurel, B. (1990). Interface agents: Metaphors with character. In B. Laurel (Ed.) *The art of human–computer interface design* . Reading, MA: Addison-Wesley.

Lenorovitz J.M. (1990, June 25). Indian A320 crash probe data show crew improperly configured the aircraft. *Aviation Week & Space Technology, 132,* 84–85.

Maes P. (1989). *How to do the right thing.* (A.I. Memo N° 1180). Massachusetts Institute of Technology, Artificial Intelligence Laboratory. December 1989.

Malone T.W. & Crowston K. (1990). What is Coordination Theory and How Can It Help Design Cooperative Work Systems? In F. Halasz (Ed.) *Proceedings of CSCW 90: Conference on computer–supported cooperative work,* Association for Computing Machinery, 357–379.

Masson, M. & De Keyser, V. (1993). Preventing human error in skilled activities trough a computerized support system. In *Proceedings of HCI International '93, 5th International Conference on Human Computer Interaction* (pp. 802–807). Orlando, FL: Elsevier Science Publishers.

Miller, G. A. (1956).The magical number seven, plus or minus two: Some limits on our capacity for processing information. *Psychological Review, 63,* 81–93.

Moll van Charante, E., Cook, R. I., Woods, D. D., Yue, L., & Howie, M.B. (1992). Human-computer interaction in context: Physician interaction with automated intravenous controllers in the heart room. In H. G. Stassen (Ed.) *Analysis, design, and evaluation of man–machine systems* (pp. 263–274).Pergamon Press.

Monnier A. (1992). Rapport préliminaire de la commission d'enquête administrative sur l'accident du Mont Sainte Odile du 20 janvier 1992. (preliminary report of the administrative enquiry commission on the Mont-Saint Odile accident January 20th, 1992) Ministere de l'équipement, des transports et du tourisme, ISSN No. (pp. 1148–4292) Paris, France.

Newell, A., & Simon, H. A. (1972). *Human problem solving.* Englewood Cliffs, N.J: Prentice Hall.

Norman D. A. (1988). *The design of everyday things.* New York: Basic Books Norman D .A. (1990). The 'problem' with automation: Inappropriate feedback and interaction, not "over-automation". In D. E. Broadbent , J. Reason, & A. Baddeley (Eds.) *Human factors in hazardous situations* (pp. 137–145). Oxford, UK: Claredon Press.

Onken, R. (1992a, September). *New developments in aerospace guidance and control: Knowledge-based pilot assistance.* Paper presented at the IFAC Symposium on Automatic Control in Aerospace, Munich, Germany.

Onken ,R. (1992b). *Pilot intent and error recognition as part of a knowledge based cockpit assistant.* Paper presented at the AGARD GCP / FMP Symposium, Edinburgh, Scotland.

Payne, S. J., & Green, T. R. G. , (1986) Task-action grammars: A model representation of task languages. *Human–Computer Interaction, 2*(2), 93–133.

Reason, J.T. (1987). Generic error-modelling system (GEMS): A Cognitive framework for locating common human error forms. In J. Rasmussen, K. Duncan & J. Leplat (Eds.). *New Technology and Human Error.* Wiley: Chichester, UK.

Reason, J. T. (1990). *Human Error.* Cambridge, UK : Cambridge University Press.

Sarter, N. B., & Woods, D. D. (1991). Situation awareness: A critical but ill-defined problem. *The International Journal of Aviation Psychology, 1,* 45–57.

Sarter, N. B. & Woods, D. D. (1992). Pilot interaction with cockpit automation: Operational experiences with the flight management system. *The International Journal of Aviation Psychology, 2*(4), 303-321.

Sarter, N. B. & Woods, D. D. (1994). Pilot interaction with cockpit automation II: An experimental study of Pilots' Model and Awareness of the Flight Management System (FMS). *The International Journal of Aviation Psychology, 4,* 1–28.

Sarter, N .B. & Woods, D. D. (1995). "How did I ever get into that mode ?" Mode Error and Awareness in Supervisory Control, *Human Factors, 37,* 5–19.

Sheridan, T.B. (1988). Task allocation and supervisory control. In M. Helander (Ed.). *Handbook of Human Computer Interaction.* (pp. 159–173). Amsterdam, North Holland: Elsiever Science Publishers.

Simon, H. A. (1957). *Models of man.* New York: Wiley.

Vakil, S. S., Hansman, J., Midkiff, A. & Vanek, T. (1995). Feedback mechanism to improve mode awareness in advanced autoflight systems. In R. S. Jensen & L. A. Rakovan (Eds.), *Proceedings of the 8th International Symposium on Aviation Psychology,* (pp. 243–248). Columbus, OH:Ohio State University.

Van Daele, A. (1992). *La réduction de la complexité par les opérateurs dans le contrôle des processus continus. Contribution à l'étude du contrôle par anticipation et de ses conditions de mise en oeuvre.* (The reduction of complexity by operators in continuous process control) Unpublished doctoral Thesis, University of Liège: Liege, Belgium.

Wiener, E. (1989). *Human factors of advanced technology. ("glass cockpit") transport aircraft* (NASA Contractor Rep. No. 177528). Moffett Field, CA: NASA–Ames Research Center.

Wiener, E. , & Curry, D. (1980). *Flight deck automation: Promises and problems.* (Tech. Memo No. 81206). Moffett Field, CA: NASA.

Woods, D.D. (1993). The price of flexibility in intelligent interfaces. *Knowledge-Based Systems, 6* (pp. 159–173) Elsevier Science Publishers.

Woods, D. D., Cook, R. I. & Sarter, N. (1993). *Clumsy automation, practitioner tailoring and system failures.* (Cognitive Systems Engineering Laboratory Rep.), Columbus, OH: Ohio State University.

VIII

CONCLUSION

Human Factors Certification of Advanced Aviation Technologies: Overview

V. David Hopkin
Embry-Riddle Aeronautical University
Human Factors Consultant UK

The origins of this volume lie in a workshop held in 1993 on human factors aspects of certification, the proceedings of which were published with limited circulation and are now out of print (Wise, Hopkin, & Garland, 1994). This workshop in its turn had several origins. Three deserve particular mention. One origin was a meeting and text on verification and validation (Wise, Hopkin, & Stager, 1993), which was unable to deal fully with the topics of verification and validation without raising certification issues. As second origin of the workshop was the inclusion in the orginal comprehensive first draft of the United States National Plan for Aviation Human Factors in 1989 of a recommended detailed program of human factors work on certification issues. This program was not pursued at its initial detailed level, and, together with the other parts of the initial document, it was condensed greatly before the official version of the National Plan was finally issued (Federal Aviation Administration[FAA], 1995). The third origin of the workshop was the recognition and treatment of certification as a human job, with the consequence that the application of human factors principles to it should offer prospects of benefits, because human factors is the main discipline applied to people at work to improve what they achieve, how they achieve it, and how the work affects them personally.

This present text represents the continuing interest in the application of human factors certification. Specific papers in it report recent advances and progress. A recent handbook of aviation human factors make reference to certification, mainly in the context of system evaluation (Garland, Wise, and Hopkin, 1999). In addition, Radio Technical Commision for Aeronautics

(RTCA) in the United States has been tasked with examining certification and activities related to it, with particular emphasis on safety and including the human factors aspects of certification. The RTCA Report (RTCA, 1999) is in a late draft stage at the time of writing. Activity applying human factors to certification, and pressure to link the disciplines more closely and productively are therefore continuing. Many of those at the original workshop brought broad and relevant knowledge to it, but only a few of them at that time had direct previous experience actually applying human factors to certification because of the paucity of such work. The position has improved since then. However, the fact that the chapters in this volume differ in addressing the main theme either directly or indirectly is still a fair reflection of the current state of the art regarding human factors and certification.

The roles of certification are themselves under review, partly in response to criticisms of it. Among the main criticisms are the following. Certification procedures have lagged behind growth in aviation. They have not dealt well with whole functioning complex systems. They can cause delays in system procurement. It can be difficult to pinpoint what certification procedures actually achieve that is not covered by other procedures. They are not well integrated with other related global processes. Their cost-effectiveness is unproven. The application of human factors to certification must acknowledge that certification itself is being pressed to review and change its roles and procedures. On the other hand, the fact that certification is in a state of flux may provide a window of opportunity to introduce human factors procedures and expertise into it.

RELATIONS BETWEEN HUMAN FACTORS AND CERTIFICATION

Human factors as a discipline can be related to certification in two distinct ways. In the first, existing certification processes provide the starting point, and the objective is to apply human factors data and principles to improve the products of certification processes without significantly changing those processes themselves. In this application, the human factors specialist would work mainly as a member of an interdisciplinary team. In the second relation, certification is considered primarily as human work, and human factors considers the methods, measures, and processes of certification in terms of their effectiveness as tasks, the training for them, their success in utilizing known human capabilities well and avoiding known human limitations, and their facilitation of the formation of valid human judgments on certification issues. In this application, human factors is applied directly to certification, and the human factors specialist therefore would tend to be working independently of an interdisciplinary team. The human factors perspective on certification has to include the need in principle to certify everything, the questionable feasibility of doing so, and the consequences of not doing so.

If we could view the current state of the application of human factors to certification retrospectively from a future state in which human factors was being applied routinely and successfully to certification processes, most of our current efforts and achievements would probably seem primitive. Despite authoritative statements from time to time to the effect that the application of human factors to certification would be a good thing, progress toward such a goal and willingness to fund it has been piecemeal and fragmentary at best, and in many contexts is still nonexistent. The pioneering work that has been done has been generally encouraging, sufficiently so to suggest that persistence would be rewarded. But perhaps it is still too early to suggest that the work that has been done so far coalesces sufficiently to permit standard specifications and rubrics for the application of human factors to certification to be formulated and applied.

HUMAN FACTORS CONTRIBUTIONS

As a discipline, human factors has not always thrived as much as it should have because of the lack of established techniques to show the cost-effectiveness of the benefits of its achievements. Even where human factors contributions have been conspicuously successful, their value, and especially their financial value, has been difficult to ascertain, partly because of uncertainties about what the corresponding outcomes would have been without human factors interventions and contributions. The world of certification does not owe human factors a living. The burden of proof of the value of human factors rests squarely on those within the discipline of human factors itself to prove its worth, its applicability, its independence, and its cost-effectiveness. It is reasonable for others to demand tangible evidence that human factors can be beneficial when applied to certification.

Human factors typically has made its most effective contributions during the planning, specification, and design stages of systems, when sources of human factors problems are identified and solutions introduced, and when the system can still be changed relatively easily. Human factors contributions to certification would normally be made at a much later stage of system evolution. However, this is not a human factors problem unique to certification. For example, the established human factors contributions to incident and accident investigation are made at a later stage of system evolution than its contributions to certification would be.

Often the best evidence that a human factors contribution has been successful is that no more is subsequently heard about it, because the successful solution is incorporated into the operational system, is fully accepted, and leads to no further difficulties. People seldom remark on the comfort of a comfortable chair, but accept and use it without comment. If any source of discomfort remains, this is cause for comment. If the designers got it right and applied correctly the ergonomic data on seating, the comments are all about other matters. . However, silence is not usually construed as evidence of success or of cost-effectiveness. Human factors in certification does not escape the problems of how to prove and cost its benefits and of how to ascertain what would have happened without it. Nevertheless, in certification as elsewhere, successful human factors contributions to it are liable to mean that no more is heard of the aspects of certification to which human factors has been applied, just because the applications have been successful.

It is rare for the full range of human factors contributions to be made during the evolution of a system from initial plans and concept formation through the procurement cycles to its introduction into operational use. Certification procedures may therefore be applied most commonly to systems where the full human factors contribution has not already been made. The most crucial human factors decisions during system evolution have usually been how to deploy limited and insufficient human factors resources to maximum advantage, and how to make the most cost-effective contribution to the solution of each identified problem without omitting anything important. The human factors contributions seek to effect improvements, but sufficient resources to find the optimum solution to every human factors problem are seldom available. Therefore there is the likelihood that many human factors applications to systems through their certification procedures will be to systems that are not already optimum in human factors terms. Past precedents suggest that there will be insufficient human factors resources to make the optimum human factors contributions to certification, and that the familiar human factors problem of having to try and deploy limited human factors resources to best advantage will arise again in regard to certification. A difference is that there will be less relevant human factors experience than usual that could serve as guidelines on how to deploy the resources most effectively.

Questions that arise concern the availability now of suitable human factors resources for certification requirements, the cost and cost effectiveness of human factors contributions to certification, and the nature and amount of training required for human factors specialists before they can apply their professional knowledge productively to certification. An important question is what the human factors specialist needs to know to become effective in applying that knowledge to certification. Should the human factors certifier become a full-time occupation, or would it be prudent for people doing human factors certification to keep their professional knowledge of broader aspects of systems up to date by continuing to have practical experience of other jobs? Is the training in human factors of people whose profession is system certification a viable alternative, and should it be in addition to or instead of the employment of human factors specialists in certification work? Whatever form of training evolves for human factors in certification, the training itself would have to be certified, so that those who completed it successfully could be recognized as qualified for the job.

IS THERE A NEED FOR HUMAN FACTORS CONTRIBUTIONS?

Potentially there is plenty of evidence about what happens without human factors inputs into certification, because much of the certification done now lacks any human factors contributions. Certification is not a theory or a philosophy, but is a current practical activity, usually with some form of legal backing. Most current certification is not a guarantee that human factors requirements have been met, or even a guarantee that some human factors contributions have been made to the certification process. If no human factors requirements have been included in the certification procedures, then the certification is irrelevant to human factors requirements. Under these circumstances, there can be no direct evidence from the certification itself about whether the human factors aspects of the system have been improved or could be improved. Perhaps of more importance is that the certification will not reveal any human factors weaknesses.

There is also the issue of whether human factors would convey genuine benefits if applied to certification or would merely be a sop, a form of social reassurance (Hancock, Chap. 4). If the certification took the form of a superficial human factors audit that did not lead to anything except to confirm that certain anodyne human factors procedures had been followed and some general standards met, then the human factors contributions could indeed degenerate into little more than an authentication procedure applied to the work of others with no real human factors input and no real attempt to seek a human factors optimum. Many current certification procedures do not include any requirement for a human factors evaluation or a thorough human factors audit. If they ignore human factors requirements, they may indeed be a sop in relation to human factors, ostensibly suggesting that human factors requirements must have been met when in fact they have not. A sop may be beneficial if the reassurance that it represents has a genuine basis, because the reassurance itself is desirable. But if the reassurance is spurious it may not only mislead, but also discourage the proactive introduction of human factors requirements into certification, because they could be perceived as unnecessary or already taken care of.

The existence of certification processes suggests a perceived need, shared by those concerned with what is certified and by many other interested parties who desire to have some checking of advanced aviation technologies prior to their operational use. Human factors is one of many disciplines contributing to such checking processes, but it differs from all other disciplines insofar as part of its expertise covers the reasons why people may need the reassurance of certification, and also the forms of certification that are likely to be most efficacious in providing reassurance.

It could be argued that it would be unfair to introduce such human factors requirements for the first time at the certification stage unless they had been specifically mentioned at earlier stages of the system evolution, including its specification. The phrasing in some human factors standards and in other definitive sources of human factors data can be too general to be very useful in certification procedures applied to a particular system. These general human factors data would often need to be converted into the specific forms applicable within the system being certified before they could serve directly as human factors requirements within that particular system.

CERTIFICATION AS BIAS

Applications of human factors to certification may share a bias that is intrinsic to certification itself, unless positive steps are taken to redress this bias. Given that an overriding objective must be to achieve certification, any aspect of the system that is subject to certification is liable to receive more attention than any aspect of the system that is not subject to certification. Separate and independent certification procedures have evolved for some subsystems within the system, such as selection procedures for controllers, or software in support of particular functions. However, separate certification procedures have not evolved for every subsystem. Procedures intended to certify the whole system as a functioning entity are inevitably vulnerable if the system includes subsystems that have not been subject to previous certification procedures, as the only certification that these subsystems will receive is that applied to the system as a whole.

Certification procedures applicable to the whole system are also vulnerable if there are interactions between independently certified subsystems, because these previous more limited certification procedures are unlikely to have been broad enough to take account of these interactions. Indeed there may be no suitable techniques for taking these interactions into account, because the certification of interactions within the whole system can seldom be included within the certification procedures for one of its subsystems. It is therefore apparently easy for many aspects of the system, particularly those dealing with interactions between subsystems, to slip through any certification net intended for application to the system as a whole.

CERTIFICATION AS WISHFUL THINKING

The expectation that the application of human factors to certification would improve certification could be construed by others as wishful thinking on the part of human factors professionals. Nevertheless, experience in other contexts may lend some support to this expectation. Usually improvements made on the basis of human factors evidence are in addition to, but not instead of, improvements originating in other kinds of evidence. For example, there was much initial opposition to the introduction of confidential incident reporting, on the grounds that procedures already in place were adequate to gather all the relevant evidence about any incidents that occurred. However, in the event, confidential incident reporting did succeed in tapping additional sources of relevant and applicable data about incidents that were not being covered by existing procedures and sources. It did this as an additional source of evidence without invalidating any of the other sources of evidence that were already in use. It is reasonable to expect that the application of human factors to certification would have a similar effect. Any additional source of evidence that could make certification better should at least be tried, and the application of human factors principles and evidence seems likely once again to tap new evidence not otherwise gathered.

Procedures such as confidential incident reporting, applied to incidents that have occurred or could have occurred in aircraft cockpits or in air traffic control systems, form part of the raw data about human factors problems that can still arise within systems that have already been certified. Together with many other sources of information, including those from other kinds of incident reports, these can reveal problems that have not been resolved either during the system procurement cycle or during its certification. A few of these problems are so potentially hazardous that they lead to urgent action to remove the source of them, even though the retrospective modifications entailed can often be costly. They should routinely be related to the certification processes, to see if appropriate modification of these processes could have detected them. Although this is not always done, it could forestall the need for many retrospective modifications by introducing the appropriate changes during procurement, and it would thus be highly cost-effective.

CERTIFICATION AS OPINION

Confirmation of human factors improvements effected through certification has to rely considerably on the professional opinions of human factors specialists. According to human factors principles, this is not the ideal approach. More objective and impartial measures and evaluation techniques should be developed and applied. However, human factors is not out of line with other disciplines in this regard. Most current certification processes rely on professional expertise for the correct conduct of the certification, for the validity of the judgments made in the course of certification, and sometimes for the selection of the aspects of the system to be certified. This professional expertise is primarily the expertise of those whose job it is to certify, but other disciplines also make contributions either directly or by advising the certifiers. Those from other disciplines who contribute to certification procedures customarily also rely mainly on their expert professional opinions. In aviation, there is strong reliance on the judgments of the test pilot, and many judgments may be made using a mock up and not a real aircraft (Paries, Chap. 23). Perhaps there should also be more reliance on the professional expertise of a test air traffic controller (Westrum, Chap. 16).

Human factors evidence and recommendations are often perceived, misleadingly, to rely too much on 'soft' data in the forms of opinions and guidelines rather than on 'hard' data in the form of facts. As a result, derogation and relaxation of human factors requirements by others can become common in the context of certification (McClumpha & Rudisill, chap. 22). Part of the unique contribution that human factors can make to certification therefore consists in the fact that the making of judgments itself falls within the professional remit of human factors as a discipline. Human factors recommendations can therefore include advice on ensuring the impartiality and fairness of judgments, the avoidance of known sources of bias in making judgments, the range of evidence applicable to judgments, and the weighting of different kinds of evidence.

AVIATION CONTEXTS

Many applications of human factors have been pioneered in civil aviation. Examples include human factors work on aircraft cockpits, head-up displays, maps, conventional and electronic cockpit instruments, air traffic control, maintenance procedures, and incident and accident investigation. The examples do not include certification because much of the pioneering human factors work on certification has been done on other large human–machine systems such as military systems and nuclear power plants. This does not mean that there has been no work on applying human factors to certification in aviation systems, and the six chapters in the section on aviation and several other chapters in this volume testify otherwise, but the aviation work in this field has not led the way as much as it often does.

The Nuclear Regulatory Commission in the United States has had for some years a human factors program involving certification, using both in-house human factors specialists and external consultants (Hanes, Chap. 5). The pioneering of human factors applications involves more than the application of human factors data, although it does include that. Pioneering usually entails the development and proving of new human factors methods, measures, and techniques suitable for the new application. The human factors professionals involved have to gain a detailed understanding of the discipline to which human factors is applied, so that the human factors contributions to it do not seem naive to professionals in that discipline but are acceptable because they are well informed, helpful, and practical.

Large and complex human–machine systems, in aviation, the military, nuclear power, electricity generation, chemical processing, maritime, space, and other environments, tend to be marketed in several nations, and even worldwide. Recent human factors concerns with cultural ergonomics apply to them, and have often arisen from them. Because certification procedures rely heavily on human knowledge, understanding, judgment, evaluation, and the legal or moral authority underpinning them, cultural differences would be expected in the ways in which the human activities that constitute certification procedures are applied, conducted, administered, and supervised in different nations. If these cultural differences result in differences in the certification procedures, they tend to be countered by pressures for agreed international standardization of certification wherever what is certified is intended for international use.

In aviation, international bodies such as the International Civil Aviation Authority therefore take an interest in certification and in any cross-cultural differences in it (Maurino & Galotti, Chap. 20). Their interest is in addition to the different standards in certification commonly implemented at the national level, where each nation may insist on forming its own judgments on the matter. Cross-cultural differences in certification practices include their aims, methods, procedures, and the allocation of responsibility. There are often cultural differences in the legal status of certification. Any effective human factors contributions to certification must take account of its cultural context, and must clearly indicate whether the human factors contributions are intended to apply at international, national, or a more regional level, even to the extent of being plant or system specific.

New technologies pose new challenges for human factors certification. Many of these are exemplified by the introduction of glass cockpits. Commonly, new technologies bring new kinds of potential human errors. Design decisions primarily predetermine what these can be, although the kinds of human error that can and will occur as a consequence of design decisions are not usually acknowledged sufficiently to have much influence over the design decisions themselves. Common sense is not of much help here either. The relation between human error and accident risk is not a linear one (Amalberti &Wilbaux, Chap. 24).

CERTIFICATION AS A MINIMUM

Implicit in the notion of certification seems to be the prescription of an achievable minimum. It has been claimed that certification is a guarantee that a minimum or standard has been achieved, that minimum requirements have been met or that a predetermined procedure for testing or verification has been followed, and that the essential purpose of certification relates to this defined minimum but does not go beyond it. There may be no credit for exceeding the minimum. This applies to the system as a whole, to the subsystems within it, to the various kinds of equipment and facilities, and to the people who have jobs within the system. Human factors applications to certification may therefore have to examine most closely the performance not of the best operator or of the average operator but of the worst operator who is accepted within the system, because the system must continue to function effectively and safely not merely under average conditions but under the worst conditions to which it may be subjected.

Treating certification as a minimum seems to imply that the granting of certification absolves everyone from seeking further improvements. Given this, there seems considerable merit in the principle of granting interim certification for a fixed period such as a year, during which comments by users and others with direct experience of the system are not merely invited but actively sought. This practice is probably most widespread in military contexts. Many current systems are complex, and many unexpected interactions can arise within them when they are brought into operational use. It is unrealistic to expect that any certification process, no matter how thorough, can detect and remedy, before a system comes into operational service, every aspect of it that could still be improved. It is more realistic to presume that human certification activities share a degree of fallibility with all other human activities and to make allowance for this. Further possible improvements to the certification are then actively sought rather than discouraged as a source of embarrassment and a sign that certification is fallible or has failed. This can be particularly important when applied to the human factors aspects of the system. If there is a tendency to construe certification only as an end in itself, an unacceptable corollary may be that the achievement of certification seems to remove any subsequent need for critical scrutiny of what has been certified. Certification therefore could, in some circumstances, become an excuse for subsequent inactivity and even complacency.

ITERATIVE CERTIFICATION

Linked to this is the growing need for some impermanence in regard to certification. The days have almost gone when systems were built and not expected to evolve during their lifetime. Under such circumstances, certification could be treated as a permanent once-and-for-all activity. The norm now is for systems to evolve during their lifetime in their objectives, their equipment, their facilities, the demands on them, their tasks, their technology, their information sources, and even their legal status. Some of the ways in which they are likely to evolve can be foreseen when they are first introduced, because certain changes, particularly in capacity demands, technologies, and information sources, can be recognized some years ahead. Other evolutionary changes cannot be anticipated because they depend on options that did not exist when the system was originally specified, but have become available and reliable through advances in technology. Minor system changes may be accommodated, but major changes, whether foreseen or not in the initial system design, will probably require some form of recertification of the whole system or of the aspects of it that have changed. System certification, even of a system that is operational, will thus tend to be a more iterative process than it has been in the past.

Certification procedures are also expected to become more iterative during system procurement. They could be applied at many stages of system evolution, from concept formation and initial planning, through specification and design, to modeling, simulation, the building of prototypes, and field testing. Certification might help to integrate these various stages in system evolution better by providing a focus for them, which they often lack at present. There is a tendency to conduct each of these stages rather independently, and particularly to take insufficient account at each later phase of what has been done in earlier phases. Ideally products of modeling should guide the detailed specifications for simulations, especially for real-time simulations (Small & Bass, Chap. 11), and the simulations should then guide the specifications of prototypes, but these phases are often not well coordinated, to the extent that they may not even be conducted in the optimum order. Certification procedures will have to come to terms with the trend to introduce field testing not merely to verify that the planners were right, but as an iterative process during system development and as an aid to define system specifications and requirements before too many resources are committed to further system development (Harwood & Sanford, Chap. 19).

Whether certification is best treated as an iterative activity during system design and procurement or as an end product has been extensively discussed. If certification occurs during system evolution, this raises issues about how it should relate to other concurrent activities of testing, assessment, checking, auditing, verification, and validation that are also taking place during system procurement. If certification is an end product, it may be viewed by those concerned with system development and procurement as a hurdle or barrier that has to be overcome. If certification is construed as a means of excluding people, practices, or equipment, the addition of human factors as a further discipline included in certification processes would tend to increase the role of certification as a barrier. Certification might then exclude on human factors grounds additional people or practices formerly acceptable on all other grounds, without encouraging the inclusion of people or practices already legitimately excluded on other grounds.

If certification activities do become more iterative, their level of detail would not, of course, be the same throughout the procurement cycle. In general, the later in the cycle the certification occurs, the more detailed it can be, and the more thorough the verification and validation procedures that can be applied to it become. A live issue at the present time is when and how often certification could and should be applied routinely during system design and evolution. If system validation and self-checking procedures become a facet of design, it seems logical to certify these procedures also within the system design phase.

The general consensus from the chapters in this volume seems to favor a top-down approach to human factors in certification in which the system is treated as a whole, rather than a bottom-up approach that focuses on separate parts of the system or on subsystems and tends to equate the whole system with the sum of its parts. Wise and Wise (Chap. 2) discussed these alternative approaches in some detail. Unfortunately, almost all previous human factors certification within systems has been of aspects of systems, such as selection or software, rather than of whole systems. It has therefore favored a bottom-up approach. There seems no quarrel with the idea that the top-down approach is preferable in principle. Rather it is its practicality that is in doubt, and the sheer complexity involved.

Envisaged forms of progress are seldom entirely and exclusively beneficial, without bringing any new problems, particularly human factors problems, with them. Certification is no exception to this. If it does become a more iterative activity, on balance this would be beneficial, but the more iterative certification becomes, the more difficult it will be for certification to continue to be perceived as wholly independent of other activities during system evolution, and to remain genuinely independent. Yet this is one of its great strengths, although increasing system complexity makes it ever more difficult to achieve. An urgent practical problem, which will have to be addressed more comprehensively, is how to retain the actual and perceived independence of certification as it becomes incorporated more closely within system evolution activities.

CERTIFICATION AS CHANGE

An associated issue also needs urgent consideration. This is the role of certification in identifying and promoting feasible beneficial system changes. In principle, certification can be passive or interventionist. In passive form, it can simply check whether the system meets its specification in all respects. If it does not, it can state where and how it does not, and perhaps suggest what changes would be necessary to meet the specification, but in passive form it would go no further. It would not normally criticize aspects of the system that met the specification, for instance. Active interventionist certification would also point to changes required to meet certification requirements, but additionally would identify other changes that would be desirable because they would result in improvements, even sometimes where the original specification has been met.

This distinction between active and passive certification is an oversimplification, because there are not just two options but many, and numerous possible combinations of passive and active involvement. Some of the benefits of certification that can be realized only if it retains its complete independence seem to imply that certification must have a proactive role to some extent, because any form of independent auditing, such as that which certification represents, is bound to reveal from time to time unresolved problems or issues that other techniques have missed.

In considering the application of human factors to certification, it seems natural to start with certification as it is now. A legitimate human factors question is to ask what further human factors roles certification could be adapted to fulfill, now or in the future. For example, could the certification process with human factors contributions identify planned roles for humans that are in fact unsuitable for them? Could certification be used to alert human factors specialists to circumstances when the human operator would have to rely too much on monitoring but would be essentially inactive? Could certification identify inherent sources of human error in the planned human functions? The general point is that if human factors is introduced more widely into certification, the opportunity should be seized to appraise all the kinds of interactions there might be between human factors and certification, to their mutual benefit.

With any test or assessment procedure that has to be passed as a condition for acceptance, certification shares the characteristic that its existence inevitably changes the nature of that to which it is applied. In other words, the act of certification tends to change what is being certified. Whatever is included in the certification process thereby acquires importance; whatever is excluded is diminished, often to the point of insignificance. In some contexts, certification can represent the culmination and ultimate judgment of all other processes in that if certification is not granted the other processes are nullified. This may discourage the expenditure of effort on any processes not included in certification, no matter how vital or significant they may be. It raises the issue of how to make certification sufficiently comprehensive to avoid this trap of concentrating resources exclusively on those aspects of the system subject to it. Considerable human factors experience and evidence can be applied to predict and explain how certification as a process may change whatever it is applied to, to advise on safeguards against this, and to try and ensure that certification is fully comprehensive and does not omit anything of vital human factors significance.

COGNITIVE FUNCTIONS

The crucial role of human cognitive functions in all large systems that include human operators has become progressively more apparent in recent years, so that it is now probably the most dominant human factors issue in regard to certification. Several of the chapters in this volume single out cognitive processes as presenting particular difficulties in certification (e.g., Stager, Chap. 8). Everyone who has actually tried to certify human cognitive functions has found problems in doing so, many of which are recounted in this volume. The cognitive functions themselves are complex, and they are subject to so many influences, many of which are in their turn the subject of certification. Human cognitive functions are influenced by the workspace, the facilities, the data entry devices, the information presented or available on calldown, the procedures, the tasks, the instructions, the communications, and especially the training. The functions depend on professional knowledge, skills, experience, understanding, memory, situational awareness, judgment, and attention. Knowledge has to be comprehensive and cover how the system works, what has to be done, the timing and scheduling of all activities, what is correct and what is an error, and the expected norms and standards of human achievements. Some human functions in systems are entirely cognitive (MacLeod & Taylor, Chap. 14). Therefore if certification cannot deal with them, they will be omitted altogether from the certification processes.

Human cognitive functions give meaning to everything in the workspace. Despite efforts through training and other means to minimize individual differences among operators in their cognitive functioning, some differences will remain. Human cognitive functions are not themselves constants, but evolve, generally fairly slowly, in response to changes in experience, workload, skills, tactics and strategies, and the selective automaticity of human functions whereby the performance of some of them gradually involves less thinking and attention but more overlearning and habit. It is not surprising that human cognitive functions should be difficult to certify. More surprising perhaps is the willingness of human factors specialists to make the attempt.

Even the preceding explanation oversimplifies human cognitive functioning in relation to certification. All system changes, and especially technical innovations, affect human cognition. Some effects of system changes are direct, for example by changing what an operator needs to know about the functioning of the system to do a task, or by changing the way in which a task has to be done. Other effects of system changes on human cognition are more subtle but can be just as powerful. A few examples can illustrate this. A certification requirement may be that the system and the operator, acting in conjunction as an integrated entity, are capable of functioning in innovative and flexible ways. Another certification requirement may be that the actual human cognitive processes are appropriate, and that it is possible to teach them during training. Further certification requirements may be to confirm that attempts to incorporate human cognitive functioning into the safe and efficient functioning of the system as a whole have been successful, and that attempts to assist essential human cognitive functioning within the system have also been successful.

There has been extensive discussion of the possible role of certification in the detection of resident pathogens in complex systems wherever certification is applied. Human factors researchers have borrowed the concept of pathogens to describe certain kinds of latent or dormant human error, which are not obviously apparent but are errors waiting to happen because their origins are inherent in aspects of the specification, design, or intended functioning of the system. The concept of resident pathogens has proved useful in explaining particular incidents where the causes have been latent and partly predictable in terms of retrospectively recognized deficiencies in system designs, procurements, and procedures. Specific incidents in specific circumstances then trigger these resident pathogens, often in circumstances in which the human operator cannot understand their nature, their consequences, or how their occurrence or their effects could be prevented.

The concept of pathogens has provided a rationale for ensuring that incidents of a particular type do not recur; however, as a concept it may be much more useful for providing explanations of incidents that have occurred than for predicting incidents that will occur, because resident pathogens are potentially so numerous. When systems become very complex and include many advanced aviation technologies and their interactions, they must contain thousands of resident pathogens, many of which may never appear at all, and most of which appear very rarely and perhaps only after a long time. The potential number of pathogens may be too large to be dealt with comprehensively by any certification process that treats them individually instead of in categories. Therefore an issue that arises is whether the concept of resident pathogens could ever be as helpful when applied to certification as it has sometimes been when applied to incidents and their investigation.

New technologies, and particularly new forms of automation, can introduce completely new kinds of human error. An example is mode errors. When these have occurred in the past, the people involved often seem to have been completely unprepared for them, to the extent of being unable to recognize what they are, especially when automated agents have changed the mode without human awareness of this (Jauvaux et al., Chap. 25). Certification, as an independent process, should have a role in identifying such new forms of human error, and this role is both a human factors one and a cognitive one. The notion of mode itself refers to computers and the mode of operation they are currently in, as with a landing mode in an aircraft. The human error is to mistake the mode that the aircraft is in and to try to take an action, such as abandoning a landing, that the mode will try to counter or overrule. The human and the machine are then trying to do contrary things, which is dangerous. Certification techniques are needed to identify potential mode errors, and this is a proactive role for certification.

The application of human factors to certification must also deal with human cognitive functioning at another level. Certification is human cognitive functioning. It remains a very human-centered activity. The professional judgments of the certifier, the selection and weighting of information employed in the human judgments made during certification, redressing the well-known biases associated with human judgments, the methods for training certifiers, and the content of what they are taught are all matters for human factors comment and recommendations. Human factors studies on the formation of attitudes, on human difficulties in weighting and using correctly more than three or four independent sources of data at the same time, and on the typical overemphasis of the most recent information in human judgments are all pertinent to certification processes. They can all illustrate sources of fallibility in the human judgments on which certification depends.

CERTIFICATION AND THE LAW

Certification must ensure that each operator has available everything that is needed for that operator to exercise his or her legal responsibilities. Many system changes introduced for other reasons, such as technical advances, increased system capacity, and the whole or partial automation of human functions, have the incidental consequence of changing human functionality in ways that make it more difficult for the human operator to fulfill his or her legal obligations, and in some instances the changes can make it impossible to meet legal obligations. Somehow this problem must be addressed and resolved. Certification may be the only means to tackle it that is independent and external to the system.

There seems to be general agreement that certification must not only be independent, but be seen to be independent, and that certification processes should be open and available for inspection and part of the public record so that they are seen to be fair and it is possible to confirm by verification that they have all been followed correctly. This seems sensible and desirable, and a matter of common sense. However, human factors evidence suggests that although it should be implemented as the main policy, it is unlikely to prove sufficient. In many contexts it has been found necessary to provide some form of confidential incident reporting system whereby potentially hazardous actions or omissions that occurred, or could have occurred, or were observed, or the system did not prevent, can be reported anonymously in the interests of safety and efficiency, without attributions of blame and possible loss of employment. Much of this evidence would not be forthcoming if it all had to be part of the public record. This issue is likely to arise in certification whether there is human factors input or not. Precedents from other contexts suggest that certification is not likely to be an exception to this recognized need. Therefore some additional channels for gathering information outside formal public records may also be required.

CERTIFICATION AND SAFETY

An important perspective in dealing with advanced aviation technologies concerns the generally high levels of safety of existing aviation systems. Although they are not perfect and may never become so because they are designed and operated by humans, existing aviation systems are nevertheless very safe, and they compare favorably with many other kinds of system. Serious incidents are rare enough to be newsworthy. To what extent this safety record is wholly or partly attributable to certification processes is not easy to determine. In many instances the importance of certification and its effectiveness as an ultimate safeguard could depend on its legal status, rather than on the effectiveness of the certification processes themselves.

In extreme cases, the certification of an aircraft or a person may be withdrawn. If it is, that aircraft does not fly and that person does not practice. A significant aspect of certification is therefore as a threat with legal backing. Many ascribe overriding importance to it for this reason. Certification can then become a process to be feared. In this context, if certification is not granted or is withdrawn, there will be strong pressures to find a culprit to blame for the failure. This would be counterproductive and engender negative attitudes toward certification that would be difficult to overcome. In very complex systems, the notion that the blame for failure can be attributed wholly and exclusively to a single individual or group is highly unlikely. Malevolence may be ascribed to certification as a process if it is also perceived to have lost some of its independence.

Any human factors approach to certification must question how much existing certification processes actually add to safety, check that no certification process impairs safety, and recommend possible modifications to the certification process that could enhance safety still further. If the existence of independent checks in the form of certification is what is important, rather than the detailed nature of the certification procedures themselves, satisfactory checks could in principle take many forms because their forms would not be critical. They could be as comprehensive as certification is intended to be; they could aim to be complete, which is at best a major undertaking and perhaps ultimately impossible to achieve; or they could be arbitrary in the sense that a series of checks would be made but the particular checks chosen would be selected at random or according to other criteria not known beforehand. Because certification must be seen to be independent, a possible human factors contribution could be to prescribe the conditions that would have to be met to achieve perceived independence for certification, by explaining how judgments of independence are made, taking account of both rational and irrational influences.

Alternative explanations for safety may be in the professionalism of those employed in aviation coupled with their training, which together are primarily responsible for the high standards achieved and the consequent high levels of safety, rather than the certification that may not necessarily add much. A possible corollary is that changes in certification might not lead to major system changes, but that large changes in the professional attitudes of those who work within systems toward the systems within which they work might have major consequences for safety. The acceptability of certification processes by those who apply them and by those to whom they are applied would seem to be essential preconditions for their practicality and success. It is difficult to impose processes that people do not agree with and are not prepared to follow. This implies that in some contexts the notion of certification may have to be promulgated and positively advocated in terms of its benefits.

APPLYING HUMAN FACTORS EVIDENCE

Human factors evidence can help to explain how certification has evolved and why current certification processes take their present form. The application of human factors to certification can be extended to consider the adaptability of system processes to novel situations, to estimate how recoverable they are if the system fails, and to judge the extent to which the need for certification originated in human variability, with the consequence that reductions in human variability should affect the nature of and need for certification processes. It is reasonable to expect that many existing human factors data, practices, methods, and procedures should be applicable to certification processes, although this expectation requires verification.

A problem encountered elsewhere when human factors is applied is likely to recur with certification. This concerns the general utilization of what is already known, and the practical application of existing evidence. One of the means by which human factors as a discipline can prove its worth is to show that existing evidence can be applied to good effect, but an educative process will be needed so that others know that this evidence exists and of what it consists. In some instances, significant progress could apparently be made by examining the possible relevance to certification of human factors recommendations already applied in other contexts.

Human factors as a discipline has the problem of great variability in the strength of the evidence on which its recommendations depend. For many traditional ergonomic applications, existing data of known high validity and generalizability could be applied to certification processes. However, for other aspects of certification, particularly those concerned with higher cognitive functions, a much lower degree of certainty and confidence often applies to the human factors recommendations that can be made. Nevertheless, even the poorest of these recommendations constitutes a substantial improvement over the ill-informed options, guesses, or random choices that are the main alternatives.

Human factors specialists tend to view this problem as peculiar to their own discipline, whereas it is often widespread in other disciplines, but it does imply that human factors recommendations may include advice on the appropriate level of confidence that should be placed in them. Most of this advice has to come from human factors specialists themselves because they are the people most likely to possess this further essential information about the strength of the evidence underpinning their recommendations. With human factors data of limited validity, the question arises of how to treat human factors recommendations in a certification context when they rest on somewhat tentative evidence.

A related point is that because certification is a real activity, it is essential to reach practical compromises whenever the requirements of different disciplines related to certification conflict. The origins of such conflicts have to be ascertained in the interests of reaching working compromises. For this reason also, the human factors specialists involved need not only a knowledge of the human factors evidence but also a knowledge of its strengths, its generalizability, the conditions attached to it, and the extent to which it can be compromised without being invalidated. This point raises the issue of how well the human factors profession is equipped to furnish this sort of advice, and of whether those giving it would need some special training in certification processes and in the applicability of human factors evidence to them. This is not a new issue applicable only to certification, but an inevitable aspect of normal human factors practice in all its applications because the strength of the evidence supporting different human factors recommendations is not uniform. This does not mean that human factors evidence is shaky. Far from it: The evidence underlying some human factors recommendations is very strong indeed. Nor is this a problem specific to human factors. The recommendations from most other disciplines also vary greatly in the amount and strength of the evidence on which they rest.

The role of the human factors specialist can be considered in terms of the status of the human factors evidence and whether it should be advisory or mandatory. The ultimate responsibility for the certification must rest with the professional employed as a certification expert; yet human factors has a role equivalent to that of several other disciplines in which occasionally a human factors requirement may take precedence over all others. Normally and whenever possible, human factors requirements would be met without overriding the requirements of others. Certainly, those making human factors contributions must be perceived by others as maintaining their independence, but they must also be perceived as being reasonable to work with.

A fundamental problem arises because someone has to pay for the human factors time, effort, and resources devoted to certification. Naturally, whichever organization is paying claims some say in how the money is spent. This is a recurrent problem not confined to certification but arising in many other human factors and safety contexts, whenever any findings potentially clash with the policies, practices, or wishes of those providing the funding. This problem has to be solved elsewhere, and can be solved in certification, too, but it can be a permanent source of dissension.

There is also a problem in deciding how much effort and money should be devoted to the application of human factors to certification. One kind of solution would be to allocate a fixed percentage of the total certification budget for human factors work. This is better than no policy or no human factors effort at all, but it is not the best means of ensuring high-quality human factors contributions. Some certification procedures require far more human factors resources than others, so that any chosen percentage will be too high for some applications and insufficient for others. Additionally, funding alone does not guarantee quality of effort or of its deployment. As human factors contributions become more established and acceptable in certification, the increasing body of evidence about them should lead to some reduction in the costs of providing them and to more reliable and standardized procedures for specifying and costing the human factors contributions in advance. Such objectives entail positive management of the human factors work programs in certification (Baldwin, Chap. 18).

TRAINING

Certification procedures have not only to be devised and validated, but they have to be taught. Whatever else the certification depends on, it should not be on the whim of the particular individual or group tasked with the conduct of the certification procedures. This implies a definition of the knowledge required for effective certification, appropriate training, verification that the knowledge is present and can be applied by each individual, and some means of establishing the reliability with which the certification procedures are followed. The training of certifiers should culminate in a formal qualification earned by independent assessment. The issue of the training of those whose profession is certification therefore has to be addressed, and this should rest on human factors learning principles. The degree of knowledge of human factors that they need to possess will depend on the extent to which human factors specialists participate alongside them in certification procedures, but a human factors syllabus in some form will be required in certification (Hunt, 1997). Training for certification will encounter such familiar human factors problems as the degree of fidelity necessary for valid training and the establishment of criteria to measure training effectiveness (Gibson, chap. 12). Solutions to these problems should therefore be among the human factors contributions to certification.

A related point concerns the certification of human factors specialists for work in certification and the appropriate training for them. The efficacy of the human factors contributions to certification must not rely heavily on the characteristics of the individual specialist who makes those contributions. There are problems in the training of human factors specialists for certification work in defining required knowledge, testing that all possess it and can apply it, and ensuring uniformity of training and professional standards. Perhaps there is a requirement for an official compendium or textbook of human factors data and procedures applicable to certification. An additional contribution of the human factors specialist in relation to certification should be to ensure that what has been defined as comprising the certification processes can be taught to those who will carry them out, so that those who certify all reach an agreed required standard and have a comparable understanding of what they are doing. This is the familiar issue of teachability.

RESEARCH

Although it was expected before the original workshop was held that the introduction of human factors into certification would be a contentious issue, it did not prove to be, perhaps because those who would accept an invitation to attend a workshop on that theme must already be sympathetic to the general idea. What was unexpected was the lack of discussion during the workshop about human factors research on certification issues, and particularly about any supporting laboratory studies on the application of human factors to certification. Perhaps this is simply a reflection of contemporary funding priorities, or it may even be thought that laboratory studies do not offer a particularly profitable research approach to human factors and certification. However, the groups of human factors specialists that collectively formulated the initial version of the U.S. National Plan for Aviation Human Factors more than 10 years ago did consider that supporting research was important, and they drafted a research program on human factors in certification accordingly.

Without a research program, the applications of human factors to certification are bound to be somewhat uncoordinated and arbitrary. A research program can be related to existing human factors knowledge and can help to pinpoint where existing knowledge is insufficient to support its application to certification without some form of confirmation or verification. A program can be used to prioritize the certification problems that are most urgent in human factors terms. It can serve as a guide for the deployment of limited human factors resources on certification issues to maximum advantage and effectiveness. Each human factors study on certification can be set within the framework provided by the other studies in the program, which can share common methods, measures, techniques, and scenarios, so that each study not only answers its particular questions but contributes effectively to the sum of knowledge accumulated within the framework. A framework can provide a rationale for the generalization of research findings and for suggesting the boundaries beyond which findings should not be generalized without further verification. Parallel findings in separate studies within the same research framework can help to support the validation of the findings of all of them.

The formulation of a human factors research program on certification can provide a focus for the funding of research. A research framework can be used to demonstrate that requests for funding do not entail an open-ended commitment because the research program has boundaries. In the longer term, a human factors research program on certification could provide the framework that would allow the human factors contributions to certification to be standardized in their forms, contents, and applications. This would guarantee that all the appropriate human factors contributions to certification had been made, that no irrelevant or superfluous human factors contributions had been included, and that a comprehensive human factors audit of certification procedures and processes had been completed successfully, showing that each particular certification did indeed comply with all human factors requirements.

REFERENCES

Federal Aviation Administration (1995). *National plan for civil aviation human factors: an initiative for research and application.* Washington, DC: U.S. Department of Transportation.

Garland, D. J., Wise, J. A. , & Hopkin, V. D. (Eds.). (1999). *Handbook of aviation human factors.* Mahwah, NJ: Lawrence Erlbaum Associates.

Hunt, G. J. F. (Ed.). (1997). *Designing instruction for human factors training in aviation.* Aldershot, UK: Avebury Aviation.

Radio Technical Commision for Aeronautics (RTCA). (1999). *Final Report of RTCA Task Force 4: Certification.* (Working Draft Version 2.0). Washington, DC: Author.

Wise, J. A., Hopkin, V. D., & Garland, D. J. (Eds.). (1994). *Human factors certification of advanced aviation technologies.* Daytona Beach, FL: Embry-Riddle Aeronautical University Press.

Wise, J. A., Hopkin, V. D., & Stager, P. (Eds.). (1993). *Verification and Validation of complex systems: Human Factors Issues.* Berlin: Springer-Verlag.

SUBJECT INDEX